Case Studies in Mathematical Modeling -- Ecology, Physiology, and Cell Biology

Hans G. Othmer

Fred R. Adler

Mark A. Lewis

John C. Dallon

(each affiliated with The University of Utah)

PRENTICE HALL, Upper Saddle River, New Jersey 07458

Library of Congress Cataloging-in-Publication Data

Case studies in mathematical modeling in ecology, physiology, and cell
  biology / Hans G. Othmer...[et al.].
      p. cm.
   Includes bibliographical references and index.
   ISBN: 0-13-574039-8
     1. Ecology--Mathematical models--Case studies. 2. Physiology--
   Mathematical models--Case studies. 3. Cytology--Mathematical
   models--Case studies. I. Othmer, H. G. (Hans G), 1943-
   QH541.15.M3C37   1997
   577'01'5118--dc21   96-47909
                                                          CIP

Acquisitions Editor: **GEORGE LOBELL**
Editorial Assistant: **GALE EPPS**
Editorial Director: **TIM BOZIK**
Editor-in-Chief: **JEROME GRANT**
Assistant Vice-President of Production
    and Manufacturing: **DAVID W. RICCARDI**
Editorial/Production Supervision: **RICHARD DeLORENZO**
Managing Editor: **LINDA MIHATOV BEHRENS**
Executive Managing Editor: **KATHLEEN SCHIAPARELLI**
Manufacturing Buyer: **ALAN FISCHER**
Manufacturing Manager: **TRUDY PISCIOTTI**
Marketing Manager: **JOHN TWEEDDALE**
Marketing Assistant: **DIANA PENHA**
Creative Director: **PAULA MAYLAHN**
Art Director: **JAYNE CONTE**
Cover Designer: **BRUCE KENSELAAR**

 ©1997 by Prentice-Hall, Inc.
Simon & Schuster / A Viacom Company
Upper Saddle River, NJ 07458

Printed in the United States of America

10 9 8 7 6 5 4 3 2 1

ISBN 0-13-574039-8

Prentice-Hall International (UK) Limited, London
Prentice-Hall of Australia Pty. Limited, Sydney
Prentice-Hall Canada Inc., Toronto
Prentice-Hall Hispanoamericana, S.A., Mexico
Prentice-Hall of India Private Limited, New Delhi
Prentice-Hall of Japan, Inc., Tokyo
Simon & Schuster Asia Pte. Ltd., Singapore
Editora Prentice-Hall do Brasil, Ltda., Rio de Janeiro

Contents

PREFACE

As its name indicates, the field of mathematical biology is inherently interdisciplinary. Students and researchers seeking to enter this field, or to broaden their knowledge, face special challenges. How does one strike an appropriate balance between learning the details of the underlying biology and the intricacies of the mathematics? Are explorers in this area beset, like Odysseus, by the twin dangers of Scylla and Charybdis? Or are they at risk of being seduced by the sirens of biological complexity and mathematical elegance?

Books that seek to introduce this field must guide readers along a safe path between the same dangers. This volume finds the path not by following some pedagogical theory or by believing rumors about past shipwrecks, but by tracing the wake of successful researchers who have survived to tell the tale. Each scientist who has addressed biological questions with mathematical methods has found a different way, and this book presents this diversity. The case studies presented herein invite readers to join a researcher as he or she attacks a significant biological problem from start to finish. Each chapter combines the focus on cutting-edge research characteristic of the professional literature with the emphasis on teaching characteristic of a textbook. The authors provide a synthetic view of the biological problem, illustrating the multiple approaches attempted, and the strengths and weaknesses of each. The goal is to motivate and explain biological problems and their mathematical solution, while simultaneously exemplifying the process of developing a successful research program.

These case studies guide advanced undergraduates, beginning graduate students, and researchers along several such paths. Students who have worked to build mathematical skills will able to set sail in quest of important problems. The goal is to initiate them into both the diversity of approaches to mathematical biology and the breadth of the field. This book thus has two unique features, summarized as **case studies** in **mathematical biology**.

As a guide to both student and teacher, we suggest that the chapters can be read with the following general framework for modeling in mind.

1. The first step is to identify a biologically interesting problem which has a significant aspect that requires mathematical modeling, to identify critical observations on which to base a model, and to distinguish between dependable and undependable experimental results.

2. The second step is to formulate the model conceptually, making reasoned judgments regarding which processes to include and which to exclude.

3. The next step is to convert the conceptual model into a mathematical model and estimate parameters in the model, keeping in mind differing levels of certainty regarding their values.

4. The final step is to analyze the model, both qualitatively and perhaps numerically, if analytical solution is impossible, and use the results to interpret the original critical observations, make new predictions and propose new experiments.

This complex process cannot be distilled into an algorithm, because it is as much art as science. As in the experimental sciences, the techniques of modeling are learned by example and by hands-on experience. This book aims to provide both.

Coverage

Powerful new techniques in molecular biology, physiology, genetics and ecology are making biology more quantitative and more unified. Mathematical methods are needed both to analyze increasing volumes of data, and to forge connections between data shedding light on common problems from different angles.

Due to the expanding role of mathematics as a unifying force in biology, broad coverage of the field of biology is essential. The topics chosen in this volume fall into three overlapping categories: ecology, cell biology, and physiology. Although the areas have their characteristic concerns, each area uses mathematics to link levels of biological organization, depends on related mathematical techniques from dynamical systems, and, more broadly, emphasizes that mechanisms which act at the cellular or molecular level manifest themselves in the functioning of the whole organism in its environment. Each of the three sections of the book begins with an introduction by the editors that elucidates specific themes and concerns characteristic of that section.

How to use this book

We envision three primary uses for this book: as a supplementary text to accompany a mathematical biology course, as a primary text for an intensive mathematical modeling course, and as a reference volume for researchers and students.

The mathematical level of the book is graded, becoming more advanced in the later chapters. Every chapter requires that students be familiar and comfortable with differential equations and linear algebra (short appendices outlines relevant aspects of these techniques). Partial differential equations and functional differential equations are used more and more in the later chapters, and previous background or concurrent study is necessary for full comprehension. Numerous other techniques, ranging from stochastic processes to statistics, are used by the authors as needed. Like researchers, students must realize that research is not an idealized romance, with pre-defined problems and techniques that are "made for each other." Rather, techniques must be developed and modified constantly, and we hope that students learn best when forced to learn the way researchers do.

These case studies are written with more concision and demands than an ordinary textbook. The chapters are thus ideally suited to serve as starting points for group or individual projects, providing sufficient background and explanation to stand on their own, but constantly motivating students to read alternative literature and experiment with novel mathematical methods. In conjunction with an ordinary textbook, students will see how standard topics are the foundation for working mathematical biologists, but that building on that foundation requires constant imagination and ingenuity to choose and modify the appropriate method.

As a stand-alone text, the book will be most appropriate for a more advanced course, where students have a strong background in ordinary and partial differential equations that they are yearning to put to use. Such students should find the book to be an amiable but insistent companion, a source of new ideas, and, at times, a source of creative irritation.

Acknowledgments

During the academic year 1995–96, the Department of Mathematics, in cooperation with the Departments of Biology, Bioengineering, Human Genetics and the Nora Eccles Harrison Cardiovascular Research and Training Institute at the University of Utah, ran a special educational program entitled "A Special Year in Mathematical Biology". This volume is the outcome of the lectures given during the special year. We are grateful to the lecturers, who contributed their time and energy to what proved to be a very worthwhile educational experiment, and then further agreed to record their lectures in the chapters herein.

We also gratefully acknowledge the financial assistance provided by the National Science Foundation, the Department of Mathematics, the College of Science, the Office of the Vice President for Research, the Departments of Biology, Bioengineering, Human Genetics and the Nora Eccles Harrison Cardiovascular Research and Training Institute at the University of Utah for making the Special Year, and thus this book, possible.

A large measure of the success of the program is due to the participation of visiting graduate students and the local graduate students and post-doctoral fellows. In addition to attending the course and seminars, the local graduate students and post-doctoral fellows created a pleasant ambiance for the visiting students and faculty, and acted as informal reviewers of the chapters. We thank Pat Corneli, Barry Eagan, Daniel Grunbaum Andrew Kuharsky, "Colonel" Tim Lewis, Eric Marland, Steve Parrish, Bradford Peercy, Steve Proulx, Pejman Rohani, Peter Spiro, Min Xie, Toshio Yoshikawa, and Haoyu Yu for their participation and work.

We thank Eleen Collins and Jill Heersink for their patience, skill and diligence in getting this manuscript into final form as promptly as they did. Without their persistent efforts throughout the numerous revisions and corrections the volume would have appeared much later. Nelson Beebe performed extraordinary service in dealing with all the technical aspects of LaTeX, EMACS and POSTSCRIPT that arose throughout the course of preparing the book. Numerous new LaTeX macros and awk scripts resulted from his work and will make the preparation of books via LaTeX much easier in the future. Without their work, this book would still be a sheaf of loose paper and assorted computer files. Thanks also to George Lobell and Rick DeLorenzo for nursing this book through at Prentice-Hall, and dealing with numerous and constant annoyances.

Finally, the editors would like to thank their wives and families for patience throughout the highs and lows of the Special Year and its editing aftermath. We dedicate this volume to them and our students.

H. G. Othmer
F. R. Adler
M. A. Lewis
J. C. Dallon

Salt Lake City, Utah
November, 1996

PART I: ECOLOGY AND EVOLUTION

Frederick R. Adler

Mathematical modeling has played a fundamental role in the development of the sciences of ecology and evolutionary biology. The predator-prey equations of Lotka and Volterra in the 1920s helped to establish the ecological study of population dynamics. The population genetics models developed by J. B. S. Haldane, R. A. Fisher, and S. Wright at around the same time were critical in demonstrating the efficacy of Mendelian genetics. Since then, mathematical models have guided both measurement and intuition describing numerous phenomena: life histories, kin selection, sex-ratio evolution, foraging optimization, and many others.

It might seem surprising that the precision of mathematics has found so comfortable a home amid the complexity that characterizes ecological and evolutionary interactions. The model of successful mathematical science has traditionally been physics, where elegant equations are extraordinarily powerful in describing and predicting the behavior of relatively simple systems. The case studies in this section demonstrate a different strength of mathematical models: uncovering the essential structures governing complex interactions.

The authors in this section address a range of important current problems in ecology and evolutionary biology, united in their efforts to understand observed patterns based on partially understood mechanisms. S. P. Ellner begins with an analysis of the strategic responses of copepods to attacks by fish, using optimization methods to determine the "best" strategy and game-theoretic methods to show that multiple strategies will coexist when types compete. R. D. Holt and R. Gomulkiewicz ask what ecological factors determine niche breadth, showing that many evolutionary forces favor conservatism and niche constriction. O. Diekmann compares deceptively similar models of disease spread, showing that they have radically different effects on populations. R. M. Nisbet et al. use simple models of individual feeding and growth to show that some, but not all, facets of the population can be predicted with quantitative accuracy. Finally, S. Tavaré shows how to make accurate deductions of the time since human beings had a common ancestor using only current genetic data.

Several themes should be looked for throughout the chapters in this section. The authors pay particular attention to identifying what *can* be measured, often different from what ideally *would* be measured. The logic which makes these real measurements usable is necessarily quantitative. For example, we might wish to know the underlying genetics but have access only to the trait (Ellner), wish to know the structure of dispersal and mutation but have access only to the surviving types (Holt and Gomulkiewicz), wish to understand individual behavior but have access only to the whole population (Diekmann), or wish to know the history of a population but have access only to the current genetic structure (Tavaré).

Second, each author links different levels of biological organization. Again, mathematical models are essential in establishing and quantifying these links. For example, models establish the connections between individual physiology and population dynamics (Nisbet et al.), individual life histories with population dynamics and evolution (Ellner, Holt and Gomulkiewicz, Diekmann), and population dynamics with genetics (Tavaré). These links illustrate the central art of the modeler: choosing the level of realism in each component of the model appropriate to the question. Recognizing the levels of biological mechanism included and excluded constitutes the art of reading and interpreting models.

These general themes are common to mathematical modeling throughout biology. The following specific concerns in ecology and evolution are stressed throughout this section.

- **Population structure.** Not all individual are alike, differing in age, location, and response to the environment. How much of this detail is required to understand the dynamics and evolution of a population? Which factors are included, and which are ignored?

- **Genetic structure.** The genes underlying traits of individuals and the distribution of traits in a population can be very complicated and difficult to determine. How much detail is needed for accurate modeling?

- **Local interactions and global patterns.** Many ecological interactions occur locally (with neighbors) while the resulting patterns arise over broader scales. What is the scale of the mechanism and what is the scale of the pattern?

- **Stochastic mechanisms and deterministic models.** Many ecological and evolutionary processes are governed by chance events, yet we wish to model them with deterministic models. Under what conditions is this a reasonable approach? Which stochastic factors are included, which are approximated, and which are ignored?

- **Simulation versus analysis.** Many models are too complex to solve mathematically but can be simulated on the computer. How does even limited mathematical analysis help us in interpreting computer output?

- **Understanding models assumptions.** Many models have been applied outside the realm where they are appropriate. What errors of biological interpretation have been incurred by failing to fully appreciate the assumptions that underlie models?

1. YOU BET YOUR LIFE: LIFE-HISTORY STRATEGIES IN FLUCTUATING ENVIRONMENTS

Stephen P. Ellner

1.1 INTRODUCTION

Consider a zygote that is about to begin its life and imagine all the possibilities open to it. At what age and size should it start to reproduce? How many times in its life should it attempt reproduction; once, more than once, continuously, seasonally? When it does reproduce, how much energy and time should it allocate to reproduction as opposed to growth and maintenance? Given a certain allocation, how should it divide up those resources among the offspring? Should they be few in number but high in quality and large in size, or should they be small and numerous but less likely to survive? (Stearns 1992)

Life-history theory was launched when Cole (1954) proposed that these decisions, and the resulting age-specific survival and reproductive rates, should be viewed as the solution to a constrained nonlinear optimization problem. The constraints are imposed by limits on time, energy, nutrients, etc., which create tradeoffs: for example, between offspring size and number, and between foraging for food and guarding young against predators. The nonlinearity occurs because the quantity defining the "best" life history is a nonlinear, implicitly defined function of survival and reproduction rates. Cole's theory was based on a life-table or Leslie-matrix type model in which vital rates (survival and reproduction) vary with age $x = 0, 1, 2, \ldots$. The "best" among the feasible life histories is the one that maximizes the solution r of the Euler-Lotka equation,

$$\int_0^\infty e^{-rx} l_x m_x \, dx = 1,$$

(Appendix A) where l_x is the survivorship to age x, and m_x is the fecundity (daughters per female) at age x (Charlesworth 1980). r is the population's asymptotic growth rate, i.e., the total number of individuals of all ages, $n(t)$, grows as $n(t)e^{rt}$. Cole's original formulation, and the parallel theory (Caswell 1989) in which vital rates depend only on *stage* (some trait such as size that

3

changes over an organism's lifetime), are still the basis for most of life-history theory (Stearns 1992; Roff 1992; Charnov 1993).

This optimization approach assigns payoffs (fitness) to each strategy without regard to the mix of strategies in the population, so there is a tacit assumption that each individual is "playing against nature." If individuals "play against each other" so that payoffs depend on what others are doing, then evolution is not expected to maximize the population growth rate r. An artificial but instructive example is a one-locus, two-allele haploid organism in which the fecundity of the wild-type allele A_1 is $W_1 = W_0 - p$, and that of a mutant A_2 is $W_2 = W_0 - p^2$, where p is the frequency of the mutant in the population. Then, once the mutant arises ($p > 0$), the mutant allele has higher reproductive success and increases in frequency up to $p = 1$. This leaves the population with the minimum possible reproductive success. The optimization approach fails in this case because vital rates are *frequency-dependent*. Instead of seeking an optimal life history, we seek one that is an *evolutionarily stable strategy* (ESS), meaning a strategy that is better than any other when the entire population adopts that strategy. The optimality and ESS versions of life history-theory are now central to much of evolutionary and behavioral ecology, with an enormous literature. Fortunately, some good introductory surveys are available (Stearns 1992; Roff 1992; Bulmer 1995; Krebs and Davies 1987; Krebs and Davies 1993).

My goal here is to explain and illustrate how the optimality and ESS versions of life-history theory can be extended to deal with environmental "noise": unpredictable fluctuations over time in environmental conditions, and hence in reproduction or survival rates. As modelers we often choose to ignore small random fluctuations, but temporal variation in vital rates can be enormous. In a review of published studies on variation in recruitment, Hairston, Jr. et al. (1996) found that reproductive success of long-lived adults varied from year to year by factors up to 333 in forest perennial plants, 4 in desert perennial plants, 591 in marine invertebrates, 706 in freshwater fish, 38 in terrestrial vertebrates, and 2200 in birds. These figures represent the variation among years when some reproduction occurred; many of the studies also report years in which reproduction failed completely. Similarly, the recruitment success of diapausing seeds or eggs varied by factors of up to 1150 in chalk grassland annual and biennial plants, 614 in chapparal perennials, 1150 in freshwater zooplankton, and 31,600 in insects. In such cases, a theory based on average conditions seems hard to justify. To introduce "noisy" life-history theory in a simple setting, I will first review the classical theory of emergence-from-dormancy strategies for seeds or eggs in temporally fluctuating environments (Sections 1.2 and 1.3), emphasizing methods over biology. This provides the necessary background for the main case study, the evolution of going-into-dormancy strategies in a freshwater copepod, *Diaptomus sanguineus*. I asked the students in the course what they hoped to learn from my lectures, and the consensus was (in the words of one student) "what you do and how you do it." In that spirit, the case study is presented roughly as it happened, with explanations of the techniques insofar as space here permits. The related topic of spatial variation will not be addressed; see Ludwig and Levin (1991) for a study of ESS dispersal that parallels the dormancy theory presented here.

1.2   HOW TO GAMBLE IF YOU'RE A SEED

Seeds of desert annual plants were one of the main motivations for Cohen's (1966) model, sketched in Figure 1.1. At the start of growing season

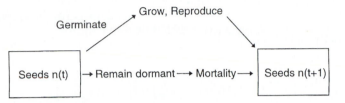

Figure 1.1. Cohen's (1966) model for optimal dormancy strategies in seeds of annual plants. At census times 0, 1, 2, ... the population consists entirely of dormant seeds. The state variable for the model is $n(t)$, the total number of dormant seeds at census time t. Seeds that remain in dormancy have a constant risk of death, while seeds that germinate face a random and unpredictable set of conditions affecting their chances of completing the adult life cycle and producing a crop of new seeds.

t, there are $n(t)$ buried seeds (or eggs) that have the choice of "hatching" or not. The seeds that hatch each produce Y_t new seeds that reenter the seed bank and survive until the start of growing season $(t + 1)$. For simplicity I will follow Cohen's original model and assume that the Y_t values in successive years are independent and identically distributed (but really nothing is changed under the weaker assumption that Y_t is stationary and ergodic). Of those that don't hatch, a fraction s survive to time $(t + 1)$. Then if H is the fraction that hatch, the population dynamics are

$$n(t + 1) = [HY_t + (1 - H)s]\,n(t). \qquad (1.1)$$

We assume for now that H is constant: seeds are not able to predict Y_t and adjust their germination accordingly. Under what conditions is it then best for some seeds to "sit it out" until next year ($H < 1$)?

This may seem like an odd way to specify an evolutionary model: where are the genes, Mendel's laws, and the fitness differentials that drive evolution? But (1.1) has all the necessary information, if we adopt the simplifying assumption that "like begets like": all offspring have the same H as their "mother". This can be given some respectability by calling it a haploid asexual model. Each genotype's abundance then obeys (1.1) with its own value of H, and the changes in abundance translate into changes in genotype frequencies. If we're lucky, we can get away with this, and the outcome of selection will be the same under more reasonable assumptions about inheritance (see below).

Under "like begets like," the winner is the type with the highest long-run growth rate, which we can calculate as follows. Denote $\lambda(t) = [HY_t + (1 - H)s]$, so that

$$n(t + 1) = \lambda(t)n(t). \qquad (1.2)$$

Then to compute the long-term growth rate, we take logarithms in (1.2) (which gives $\log(n(t + 1)) = \log(n(t)) + \log(\lambda(t))$), iterate forward from $t = 0$, and divide by t. This gives us

$$\frac{1}{t}\log(n(t)) = \frac{1}{t}\log(n(0)) + \frac{1}{t}\sum_{j=0}^{t-1}\log(\lambda(j)). \tag{1.3}$$

As $t \to \infty$, the first term on the right-hand side in (1.3) goes to zero. The second term converges, by the strong law of large numbers (or the ergodic theorem, if you prefer), to $E\log(\lambda(t))$, which is therefore the long-term growth rate analogous to r defined by the Euler-Lotka equation (above). Thus the winning H is the one that maximizes

$$\rho \stackrel{\text{def}}{=} E\log(\lambda(t)) = E\log[HY_t + (1 - H)s]. \tag{1.4}$$

One immediate consequence of (1.4) is that the per capita reproductive success Y_t must be greater than 1 on average. Applying Jensen's inequality to the logarithmic function in (1.4), we have

$$\rho = E[\log(\lambda(t))] < \log[E(\lambda(t))] = \log[HE[Y_t] + (1 - H)s] \tag{1.5}$$

so long as the variance of $\lambda(t)$ is nonzero. So when $E[Y_t] \leq 1$, it follows that $\rho < \log(1) = 0$, a negative long-term "growth rate" implying that the population is (in the long run) decreasing to zero. Thus any viable population must have $E[Y_t] > 1$, and from here on out I assume that to be true.

   In addition, (1.4) can be used to qualitatively characterize and approximate the winning H by curve-sketching ρ as a function of H. We can find the derivatives of $\rho(H)$ by differentiating inside the expectation in (1.4) — the fastidious will observe that this does require some mild assumptions about Y_t — to get

$$\rho'(H) = E\left[\frac{Y_t - s}{HY_t + (1 - H)s}\right] \tag{1.6}$$

$$\rho''(H) = E\left[\frac{-(Y_t - s)^2}{(HY_t + (1 - H)s)^2}\right] < 0 \tag{1.7}$$

Because the second derivative is negative, there are only three possibilities (Figure 1.2). The optimum is at $H = 0$ if $\rho'(0) < 0$, i.e., if $E(Y_t) < s$; this would mean that a seed should follow Peter Pan and avoid the trials of adulthood by refusing to grow up. The optimum population growth rate is then $\rho(0) = \log(s) < 0$, implying that the population decreases to zero abundance for any value of H. So for any real population this possibility is actually impossible. The optimum is at $H = 1$ if $\rho'(1) \geq 1$, i.e., if $1 - sE[1/Y_t] \geq 0$, and otherwise at some $0 < H < 1$. Thus we have found the condition under which some seeds should "sit it out" each year:

$$\text{Optimal } H < 1 \text{ if } sE\left[\frac{1}{Y_t}\right] > 1. \tag{1.8}$$

   Having derived (1.8), we need to make sense of it. It is often helpful to look at two special cases: *small variance*, and *good years/bad years*. The

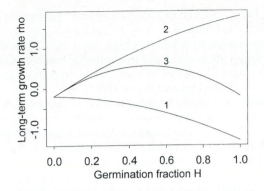

Figure 1.2. Curve-sketching $\rho(H)$ for Cohen's (1966) model. Because $\rho(0) = \log(s) < 0$ and $\rho''(H) < 0$, there are only three possibilities, determined by the sign of ρ' at 0 and 1. If $\rho'(0) < 0$, then $\rho(H)$ is everywhere negative and the population cannot survive for any H (curve 1). If $\rho'(0) > 0$ and $\rho'(1) > 0$ (curve 2), then ρ is maximized at $H = 1$. Only for $\rho'(0) > 0$ and $\rho'(1) < 0$ is the optimal H between 0 and 1 (curve 3).

small-variance case proceeds by writing random quantities as their mean plus random deviations, e.g., $Y_t = \bar{Y} + \sigma\varepsilon_t$, where $E(\varepsilon_t) = 0$, $\text{Var}(\varepsilon_t) = E(\varepsilon_t^2) = 1$, and σ^2 is therefore the variance of Y_t. We then pretend that the variance is "small" and do a truncated Taylor expansion about $\sigma = 0$:

$$E\left[1/Y_t\right] = E\left[\frac{1}{\bar{Y} + \sigma\varepsilon_t}\right] = \frac{1}{\bar{Y}}E\left[\frac{1}{1 + (\sigma/\bar{Y})\varepsilon_t}\right] \tag{1.9}$$

$$= \frac{1}{\bar{Y}}E[1 - (\sigma/\bar{Y})\varepsilon_t + (\sigma/\bar{Y})^2\varepsilon_t^2 - \cdots] \cong \frac{1}{\bar{Y}}[1 + (\sigma/\bar{Y})^2].$$

Substituting into (1.8), the "small-fluctuations" condition for $H < 1$ to be optimal is that

$$(s/\bar{Y})\left[1 + (\sigma/\bar{Y})^2\right] > 1. \tag{1.10}$$

Because we are assuming $\bar{Y} > 1$ (as argued above), we have $s/\bar{Y} < 1$. The interpretation of (1.10) is therefore that all seeds should hatch each year if the variance in reproductive success (σ^2) is small, while if the variance is large some seeds should sit it out.

The *good years/bad years* case is to suppose that $Y_t = M$ or m with probabilities p and $1 - p$ respectively, where $m \ll 1 \ll M$. Then from (1.6) we calculate directly

$$\rho'(H) = p\left[\frac{M - s}{HM + (1 - H)s}\right] + (1 - p)\left[\frac{m - s}{Hm + (1 - H)s}\right]$$

$$\cong \frac{p}{H} - \frac{1 - p}{1 - H} \tag{1.11}$$

using the assumption $m \ll s \ll M$ to simplify the fractions. Solving for $\rho'(H) = 0$, the optimum is approximately at $H = p$: the fraction of seeds "sitting it out" should be approximately equal to the fraction of bad years, $1 - p$.

Because $\bar{Y} > 1$ but $s < 1$, the *expected* number of seeds next year (new plus survivors) is always maximized at $H = 1$. Thus a conclusion from both cases is that when environmental variation is high, organisms should *bet-hedge*: reduce their average yield in order to have a hedge against occasional misfortune. The same principle applies to many traits other than dormancy (Seger and Brockmann 1987; Phillipi and Seger 1989).

   One unpleasant aspect of this analysis is that population density increases without limit, for many "losers" as well as the "winner" phenotype. This behavior results from the simplifying assumption that the per capita fecundity Y is unaffected by crowding. The simplest way of imposing limits to growth is to assume that per capita reproductive success is affected by the overall density of competitors, say $Y_t = K_t F(N_t)$, where K_t is random and $N_t = \sum_i H_i n_i(t)$ is the total number of seeds emerging from dormancy (i running over all types in the population). The key simplifying assumption in this model is that the population is well mixed; i.e., it assumes that spatial variation in the density of competitors is small enough to ignore. Because seeds are now pitted against each other, instead of an "optimal" strategy we seek a "winning" ESS: a type H such that any other type is at a disadvantage when type H dominates the population. Formally, let $Y_t^*(H)$ denote the value of Y_t in a population consisting entirely of type-H individuals. Then the logic behind (1.2) and (1.4) still applies to give the growth rate of a rare type h invading an otherwise all-H population:

$$\rho(h|H) = E \log \left[h Y_t^*(H) + (1-h)s \right] \tag{1.12}$$

$$\frac{\partial \rho}{\partial h} = E \left[\frac{Y_t^*(H) - s}{h Y_t^*(H) + (1-h)s} \right] \tag{1.13}$$

(the formal derivation considers a mixed population with type h rare, and linearizes the dynamics of type h around $h = 0$ (Chesson and Warner 1981; Ellner 1985; Chesson and Ellner 1989)). As in (1.7) we then have $\partial^2 \rho / \partial h^2 < 0$, so for any H there is a single "best" invader, and an ESS H^* is defined by the property of being the best invader of itself:

$$\frac{\partial \rho}{\partial h} = 0 \quad \text{at} \quad h = H = H^*. \tag{1.14}$$

The conditions favoring dormancy can be found as in the density-independent model, by asking whether $H = 1$ can be invaded by $h < 1$ (i.e., whether $\partial \rho / \partial h < 0$ at $h = H = 1$). The formal calculations are the same as before, and the result is (1.8) with Y_t replaced by the Y_t^* for $H = 1$.

   The analysis of (1.8) is more difficult in the density-dependent case, because Y_t^* is only defined implicitly through the nonlinear dynamics of a single-type population. So before we can evaluate the expectation in the ESS criterion and thereby determine the success of the invader, we first have to find the long-term (stationary) distribution of the resident type. So it's time for one more gambit: we will choose an especially convenient form of density dependence, $F(N) = 1/N$. This is convenient because it implies that the total number of new seeds in year t is exactly K_t. Then when the population consists of a single type with $H = 1$, we have $N_t = K_{t-1}$ — all seeds on hand "now" were produced last year, because those produced earlier have all germinated already. The per capita reproductive success in year t is therefore

$Y_t^*(H) = K_t/N_t = K_t/K_{t-1}$. Substituting these into (1.8) and simplifying, the condition for dormancy to be advantageous is $sE[K_t]E[1/K_t] > 1$.

As in the density-independent case, one can get a good feel for what this formal condition says by looking at special cases. I leave that for you to do by yourself as an **EXERCISE** to see how well you have followed the story so far:

> Derive and interpret the small-fluctuations and good years/bad years conditions for $H^* < 1$. You should discover that the good years/bad years result is now very different: $H^* = 1$ if good years are either very frequent or very infrequent. Why is this? — give an *intuitive, nonmathematical* explanation. Check your results by writing a program to simulate competition among multiple (10–25) types with different H-values; run it first at parameter values where $H = 1$ should win, and then at parameters where $H = 1$ should lose.

There's a useful way of rewriting the ESS condition $\partial \rho / \partial h = 0$. From (1.13) we see that the ESS condition is equivalent to

$$E\left[\frac{Y_t^*(H)}{hY_t^*(H) + (1-h)s}\right] = E\left[\frac{s}{hY_t^*(H) + (1-h)s}\right] \qquad (1.15)$$

and therefore each of the expectations must in fact equal 1 (because $h \times$ (right hand side) $+ (1 - h) \times$ (left hand side) $= 1$). It's hard to wring any deep biological meaning from this way of characterizing the ESS, but it often turns out to be the most useful form of the ESS condition for mathematical purposes. It was used by Sasaki and Ellner (1995) and Haccou and Iwasa (1995) to analyze models for ESS bet-hedging with a continuum of options, and McNamara (1995) has used this characterization to derive numerical methods for obtaining ESS's in a general class of models that cannot be solved analytically (McNamara et al. 1995).

1.3   PROCEED WITH CAUTION

We know that evolution involves

1. Dynamics governed by Mendelian genetics.

2. Competition among a suite of genotypes for different trait values.

To do an ESS analysis we *pretend* that it involves

1. "Like begets like": uniparental reproduction, effectively clonal.

2. Pairwise competition between an "established" and a rare "invader" type.

These gambits don't always succeed. (1) "Like begets like" is especially dicey if the trait being modeled concerns differences among offspring or between parent and offspring, such as allocation of resources within the family. It is then essential to figure out who's the boss — which individuals' genotypes determine the outcome — and to model the power structure correctly. Germination can have this complication if competition is localized in space (Ellner 1986). (2) An ESS can't be dislodged once it is established, but that doesn't guarantee that evolution will move the population to the ESS, even under "like begets like" (Eshel and Motro 1981; Takada and Kigami 1991). A strategy x^* with the latter property is called a CSS (*continuously stable strategy*). For a scalar strategy parameter, a CSS is defined by the property that if x is near x^* and y is between x and x^* and near to x, then y can invade x.

Therefore the essential last step in an ESS analysis is to *check ourselves*. The available checks are general theory, special cases, and simulation. "General theory" is a set of results giving conditions under which an ESS analysis agrees with the outcome of evolution in a proper genetic model (Charlesworth 1980; Taylor 1989). Almost all of these rely on weak-selection approximations, so they offer comfort but not certainty. Convenient special cases of the genetics can often be used to check the "like begets like" gambit by raising the likely complications in the simplest possible setting (e.g., few locus, few allele diploid with strong selection). Simulation is often the only way to check the pairwise competition gambit, because multitype competition models are high dimensional.

1.4   HOW TO GAMBLE IF YOU'RE A COPEPOD

A conflict between existing theory and experimental data usually means that there's useful work to be done. This conflict walked into my office in the form of Nelson G. Hairston, Jr., explaining a mismatch between population genetics theory and the diapause behavior of the copepod *Diaptomus sanguineus* in Bullhead Pond, Rhode Island.

Hairston studied the *switch-date* of an adult female: the date each spring when she switches from laying clutches of *immediate-hatching* eggs (that hatch in a few days) to laying *diapausing* eggs that remain dormant at least until the next fall. Copepods don't carry calendars, but they can use photoperiod and temperature to tell the time of year. Diapausing eggs are safe from predation by fish, which intensifies when the fish become more active as the pond warms up in spring. The increase in fish activity is rapid enough that as a first cut we can imagine all fish suddenly becoming active on a single *catastrophe date* each year (Figure 1.3a).

The evolutionary "problem" confronting a female is how to time her switch date relative to the catastrophe date, in order to maximize her fitness. If each year's catastrophe date were known in advance, the solution to this problem would be easy. Roughly speaking, a female should lay immediate-hatching eggs so long as those eggs will be able to complete their life cycle before the catastrophe date, laying many eggs themselves; thereafter she should lay diapausing eggs. Unfortunately, this strategy cannot be implemented: the catastrophe date varies widely from year to year and cannot be predicted far enough in advance. Any given switch date might be good, bad, or indifferent, depending on what the catastrophe date turns out to be in that year. The selection against those with bad switch dates is strong enough to cause year-to-year changes in mean switch date: the mean switch date shifts earlier in the season

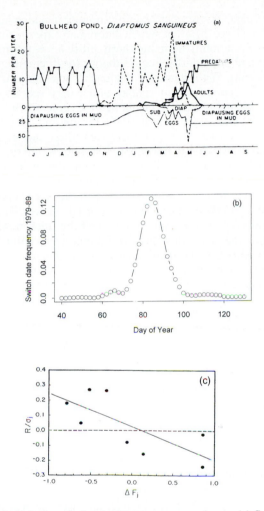

Figure 1.3. Switch-date evolution in *Diaptomus sanguineus*. (a) Summary of the life cycle. From late spring through fall, the population consists entirely of diapausing eggs. In early winter, fish predation declines and eggs begin to hatch out, producing immatures and then adults in the water column. Adults initially lay immediate-hatching ("subitaneous") eggs and then switch to producing diapausing eggs. (b) Average frequency distribution of switch date for 1979–1989. (c) Switch date evolves in response to year-to-year fluctuations in predation intensity. The *x*-axis is fish density relative to its mean and the *y*-axis is the between-year change in population mean switch date, scaled by the standard deviation of switch date in the initial year. The regression (solid line) is significant (from Hairston, Jr., and Dillon (1990)).

following a year of high fish density and shifts later after a year of low fish density (Figure 1.3c).

The fact in conflict with theory was the high level of variation in switch date among females in the population (Figure 1.3b). In principle the variation could be either heritable or nonheritable, i.e., either a genetic polymorphism for switch date maintained by selection favoring different switch dates in different years, or else a bet-hedging strategy in which individuals with a common genotype have different switch dates. Lab experiments showed that a large part of the variation is heritable (heritability $h^2 \sim 0.5$ for photoperiod response). The extant population genetics theory for fluctuating selection was interpreted as showing that temporally fluctuating selection on a quantitative trait only maintains genetic polymorphism under very special circumstances (primarily, if the fluctuations act to create heterozygote advantage) (Hedrick 1986; Bull 1987; Karlin 1988; Turelli 1988; Barton and Turelli 1989; Frank and Slatkin 1990; Gillespie 1991). So there were three (nonexclusive) possibilities: the variation was due to nonadaptive mutation; the theory was incorrect; or the theory was mathematically correct but had been misinterpreted, and the conclusions were in fact not applicable to *Diaptomus*.

We suspected that the theory was correct but not applicable to *Diaptomus*. Like most population genetics theory it used the conventional assumption of discrete nonoverlapping generations, and the interpretation of the results tacitly ignored the possibility that overlapping generations could change the results. *Diaptomus* has overlapping generations due to eggs remaining in diapause for several years (Stasio 1989; Hairston, Jr. et al. 1995); the annual survival rate of diapausing eggs is over 98%, and eggs can remain viable in the sediment for up to 300 years (Hairston, Jr. et al. 1995), although eggs older than a decade or two probably have slim chances of hatching because they are too deeply buried in the sediment. In ecological models for interspecific competition, overlapping generations greatly increase the range of conditions under which environmental fluctuations can lead to coexistence (Chesson and Warner 1981; Chesson 1994; Shmida and Ellner 1984).

We therefore set out to show that the effect of generation overlap is similar in genetic models, beginning with a "cartoon" genetic model for switch date derived by modifying (1.1) as little as possible. We assumed that the competing types differ in switch date but all have the same hatching fraction (H). Relative values of reproductive success Y_t for each genotype were assumed to be a function $w(d - \theta_t)$, where d is a female's switch date, θ_t is the optimal switch date in year t, and w is a selection function that penalizes wrong guesses, such as the Gaussian $w(z) = \exp(-z^2/2v^2)$. The essential properties of w are that it is maximized when the female uses the best switch date for this year's catastrophe date ($d = \theta_t$) and decreases with worse guesses (larger $|d - \theta_t|$). It is convenient to shift the time scale so that $E(\theta_t) = 0$. Absolute reproductive success was defined in the model by assuming (for convenience) that the total egg production in each year is constant (K).

For ESS analysis of this model, as usual we consider a rare type ($z = d$) invading a common type ($z = D$). Let N_t denote the abundance of the common type and n_t the abundance of the invader — as in the seed-bank model, these are the numbers of diapausing eggs just before hatching time in year t. In competition with the invader, the common type's share of the total egg production K is

$$\frac{KHN_t w(D - \theta_t)}{Hn_t w(d - \theta_t) + HN_t w(D - \theta_t)}. \tag{1.16}$$

Because the invading type's abundance is taken to be negligible, this reduces to simply K. Therefore $N_{t+1} = K + s(1 - H)N_t$, and in the long run N_t converges to the constant value $K/(1 - \gamma)$, where $\gamma = s(1 - H) < 1$. We can now compute the growth rate ρ for the invader. The invading type's share of the total egg production is

$$\frac{Kn_t w(d - \theta_t)}{n_t w(d - \theta_t) + N_t w(D - \theta_t)} \tag{1.17}$$

so the per capita egg production by invading-type eggs is

$$\frac{Kw(d - \theta_t)}{Hn_t w(d - \theta_t) + HN_t w(D - \theta_t)}. \tag{1.18}$$

In the denominator n_t is again negligibly small compared to N_t, and $N_t = K/(1 - \gamma)$, so the per capita egg production of the invader reduces to

$$Y_t = \frac{(1 - \gamma)w(d - \theta_t)}{Hw(D - \theta_t)}. \tag{1.19}$$

Thus the growth rate for the invader is

$$\begin{aligned}
\rho(d|D) &= E \log[s(1 - H) + HY_t] \\
&= E \log \left[\gamma + (1 - \gamma)\frac{w(d - \theta_t)}{w(D - \theta_t)} \right]. \tag{1.20}
\end{aligned}$$

where the expectation is over the distribution of the random environment θ_t. Note that $\rho(x|x) = 0$ for all x, as it should be.

These minor changes to the optimal diapause models turned out to have the consequence we were hoping for: coexistence of multiple switch dates when $\mathrm{Var}(\theta_t)$ is large, rather than a single ESS switch date (Ellner and Hairston, Jr. 1994). To hold the algebra down I will consider the Gaussian selection function, choosing time units so that $v^2 = 1$ and therefore $w(z) = \exp(-z^2/2)$; essentially the same qualitative behavior holds for any smooth symmetric selection function. As usual we can characterize the ESS via the derivatives of ρ: to be an ESS, a trait value D^* must satisfy $\partial \rho / \partial d = 0$ and $\partial^2 \rho / \partial d^2 < 0$ at $d = D = D^*$.

For this model we get

$$\frac{\partial \rho}{\partial d} = E \left[\frac{(1 - \gamma)w'(d - \theta)/w(D - \theta)}{\gamma + (1 - \gamma)w(d - \theta)/w(D - \theta)} \right], \tag{1.21}$$

which at $d = D = D^*$ simplifies to

$$\frac{\partial \rho}{\partial d} = -(1 - \gamma)D^*. \tag{1.22}$$

Thus the only possible ESS is $D^* = 0$, the optimal switch date under average conditions (recall that $E(\theta_t) = 0$). Differentiating within the expectation in (1.21) and evaluating for Gaussian w at $d = D = D^*$ gives

$$\frac{\partial^2 \rho}{\partial d^2}(D^*, D^*) = (1 - \gamma)\left[\sigma_\theta^2 \gamma - 1\right] \tag{1.23}$$

where σ_θ^2 is the variance of θ_t. When $\sigma_\theta^2 \gamma < 1$, this quantity is negative, and D^* is an ESS. Similar calculations show that $\partial^2 \rho / \partial D^2$ is always positive at

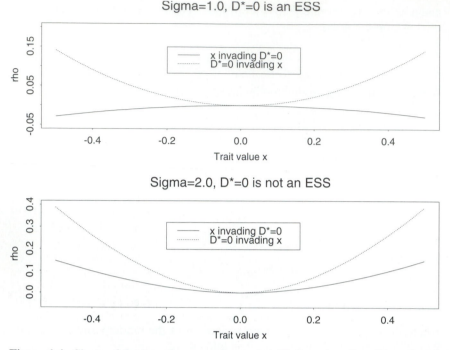

Figure 1.4. Shape of the boundary growth rates $\rho(x|0)$ for x invading $D^* = 0$, and $\rho(0|x)$ for $D^* = 0$ invading x. The top panel shows the situation below the threshold for maintaining genetic variance. D^* is an ESS because it cannot be invaded by nearby types ($\rho(x|0) < 0$). In our model D^* can also invade nearby types ($\rho(0|x) > 0$), though in general an ESS need not have this property. Above the threshold (bottom panel) D^* is no longer an ESS because it can be invaded by nearby types, but it retains the ability to invade nearby types ($\rho(0|x)$ and $\rho(x|0)$ are both positive).

$d = D = D^*$ in this case, which implies that $\rho(D^*|d) > 0$ for d near D^*. Because $\rho(D^*|d) > 0$ is the growth rate for the ESS invading a non-ESS type, we can conclude that the ESS can always invade nearby types. On the other hand, when $\sigma_\theta^2 \gamma > 1$, (1.23) is positive, so D^* is not an ESS.

The corresponding picture is in Figure 1.4. For $\sigma^2 \gamma < 1$, $D^* = 0$ resists invasion by nearby types (i.e., $\rho(x|D^*) < 0$) so it is an ESS; moreover it can itself invade nearby types (i.e., $\rho(D^*|x) > 0$). For $\sigma^2 \gamma > 1$, D^* is still able to invade nearby types, but it is vulnerable to a reinvasion by those same types ($\rho(x|D^*) > 0$) so it is not an ESS. Consequently there aren't any possible types that cannot be invaded by some other nearby type, so we expect that genetic variability will be maintained indefinitely. Simulations of multigenotype competition with a wide range of parameter values matched the results of this ESS analysis (Ellner and Hairston, Jr. 1994).

When σ_θ^2 crosses above the threshold value $\sigma_{\mathrm{crit}}^2 = 1/\gamma$, the single ESS is replaced by a set of coexisting types. Simulations suggested an initial bifurcation to two alleles of opposite effect $\{x, -x\}$, so we began by finding conditions under which there existed a value of x such that $\rho(z|\{-x, x\}) < 0$ for all z near x or $-x$ (Ellner and Sasaki 1996). Assuming "like begets like" as usual leads to

$$\rho(z|-x, x)$$

(1.24)

$$= E \log \left[\gamma + (1 - \gamma) \frac{w(z - \theta_t)}{p_t w(x - \theta_t) + (1 - p_t)w(-x - \theta_t)} \right]$$

where p_t is the frequency of type x in year t, and the expectation is with respect to the joint distribution of (θ_t, p_t), which is a mess. It's a bit simpler to assume that the values of θ_t in different years are independent, implying that θ_t and p_t are independent, but we still need the distribution of p_t to evaluate (1.25). We failed to find a general solution, and resorted to treating $\delta = (1 - \gamma)$ as a small parameter to allow small-fluctuations approximations to the dynamics of p_t and to the integrand in (1.25). For σ_θ^2 just above the threshold the necessary derivatives can be gotten from Taylor expanding in x and z (which are near 0). Far past the threshold a different approximation is needed. If x is very large, then in most years one of the pair $\{x, -x\}$ will fare much better than the other, and it is a good approximation to pretend that the other has no offspring at all. I'll spare you the technical details — in fact we spared ourselves the details, doing most calculations by programming them in MAPLE — but in the end it works, giving approximations that allowed us to derive the first few bifurcations in the ESS when σ_θ^2 exceeds the threshold (Figure 1.5).

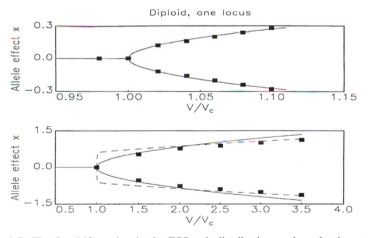

Figure 1.5. The first bifurcation in the ESS trait distribution as the selection variance increases past the threshold for maintaining genetic variance. These plots compare simulation results on competition among a large set of competing types (solid squares) with the asymptotic approximations that are the basis for studying subsequent bifurcations. Solid line: approximation for near the threshold. Dashed line: approximation for far above the threshold. See the text and Ellner and Sasaki (1996) for further details.

Above the threshold value σ_{crit}^2, the "like-begets-like" gambit fails because the underlying genetics affects the set of competing phenotypes. In a diploid model, for example, if the invader z is at the same locus as the established types $\{-x, x\}$ then there are six competing phenotypes: (x, x), $(-x, x)$, $(-x, -x)$, (x, z), $(-x, z)$, (z, z). But if z is at a different locus, then there are nine competing types: all possible combinations of (x, x), $(-x, x)$, $(-x, -x)$

at the first locus with $(0, 0)$, $(z, 0)$, (z, z) at the second (0 denotes the unmutated wild-type allele at the second locus). These kinds of complications mean that the precise form of variation is sensitive to hard-to-estimate quantities, and to questionable simplifying assumptions (symmetric selection, additive gene effects, etc.).

However, all the results for specific models that we examined have the same qualitative pattern: the variation consists of a few alleles of large effect at each locus affecting the trait (Ellner and Sasaki 1996; Ellner 1996). Another general result (Sasaki and Ellner 1996) is that the covariance components of the genetic variance (i.e., the correlations between loci affecting the trait, and between maternal and paternal contributions at each locus) are all positive for all polymorphic loci determining the trait.

These qualitative predictions are important because they allow experimental tests of the fluctuating-selection hypothesis. For example, covariance components can be estimated by conceptually simple laboratory breeding experiments. Under random mating in the absence of selection, the changes in trait variance from generation to generation are a function of the underlying covariance components (Falconer 1981). This makes it possible to estimate the genetic covariance components simply by estimating the distribution of trait values in the population for several successive generations. These estimates need not be terribly precise, because it is their sign rather than their numerical value that distinguishes between the fluctuating-selection hypothesis and the alternative of mutation-selection balance.

1.5   OLD AND IN THE WAY

But before rushing out to test the predictions from a deliberately simplified model, it's a good idea to check that they aren't artifacts of the simplifying assumptions. We therefore examined models with nonheritable variation (Sasaki and Ellner 1995), models with density-dependent fecundity and fluctuating population size instead of "constant K" (Ellner and Sasaki 1996; Babai and Ellner, in prep), and models with age or stage structure (Ellner 1996).

In this section I will briefly consider models with population structure, because they involve some important new concepts and techniques, and because the assumption that "an egg is an egg is an egg" — whether one year old or twenty — is probably the most egregious error in the cartoon copepod and seed-bank models. Older eggs or seeds may deteriorate or become too deeply buried to encounter the conditions required for successful emergence from dormancy.

As a first step, consider a *deep-shallow model* (Easterling 1995) in which any eggs that don't hatch the year after they were produced become covered by sediment and can only hatch if some disturbance mixes them back up to the top (Figure 1.6). These older eggs then have a far lower hatching rate than newly produced eggs. Equation (1.1) for a single type x is replaced by:

$$n_0(x, t+1) = [H_0 n_0(x, t) + H_1 n_1(x, t)] Y_t \qquad (1.25)$$
$$n_1(x, t+1) = (1 - H_0)s n_0(x, t) + (1 - H_1)s n_1(x, t)$$

where $n_0(x, t)$ is the number of newly produced type-x eggs (age < 1 year) and $n_1(x, t)$ is the number of older eggs. Reverting to "constant K," the per capita

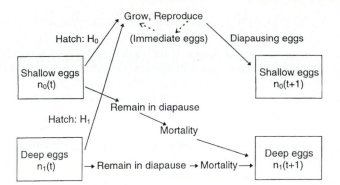

Figure 1.6. The deep-shallow egg-bank model motivated by *Diaptomus sanguineus*. This model distinguishes between newly produced "shallow" eggs, and older "deep" eggs that have become covered by a layer of sediment at the pond bottom.

yield for type x is

$$Y_t = \frac{K w(x - \theta_t)}{\sum_y [H_0 n_0(y, t) + H_1 n_1(y, t)] w(y - \theta_t)}. \tag{1.26}$$

For an ESS analysis we consider as usual an infinitesimally rare invader d competing with established type D; the per capita yield for the invader (1.26) then becomes

$$Y_t = \frac{K w(d - \theta_t)}{[H_0 \bar{n}_0 + H_1 \bar{n}_1] w(D - \theta_t)} \tag{1.27}$$

where $\bar{n}_i$ is the equilibrium value of $n_i(D, t)$ in the absence of competitors ($\bar{n}_0 = K, \bar{n}_1 = (1 - H_0)sK/(1 - (1 - H_1)s)$). With (1.27) substituted into (1.26) the equation for the invader is linear (as always) and can be written in matrix notation as

$$\begin{bmatrix} n_0(d, t+1) \\ n_1(d, t+1) \end{bmatrix} = \begin{bmatrix} H_0 Y_t & H_1 Y_t \\ (1 - H_0)s & (1 - H_1)s \end{bmatrix} \begin{bmatrix} n_0(d, t) \\ n_1(d, t) \end{bmatrix}. \tag{1.28}$$

The good news about (1.28) is that the invader's growth rate $\rho(d|D)$ still exists, and is a nonrandom quantity under fairly mild technical conditions; the bad news is that there is no longer a general closed formula for ρ analogous to (1.4) (Furstenberg and Kesten 1960; Tuljapurkar 1990). Only in very special cases is there an explicit formula for ρ; for example, if all eggs hatch either in their first year or in their second (Tuljapurkar and Istock 1993).

The only practical way available for getting around this problem is to use the small-fluctuations approximation to $\rho(d|D)$. This approximation is derived by writing $A(t) = \bar{A} + \varepsilon B(t)$ where $\bar{A} = E[A(t)]$, followed by a long string of clever calculations to get the first two terms in a Taylor expansion for ρ about $\epsilon = 0$ (Tuljapurkar 1990; Derrida et al. 1987). If the matrices $A(t)$ are independent and identically distributed (as in the models here), the result is:

$$\rho = \log \lambda_1 - \frac{\varepsilon^2}{2\lambda_1^2} \sum_{i,j,k,l} v_i w_j v_k w_l \text{ Cov } (ij, kl) \tag{1.29}$$

where λ_1 is the dominant eigenvalue of $\bar{A}$, $\vec{v}$ $\vec{w}$, are the corresponding left and right eigenvectors normalized to $\langle \vec{w}, \vec{v} \rangle = 1$ and $\mathrm{Cov}(ij, kl)$ is the covariance between $B_{ij}(t)$ and $B_{kl}(t)$. If the matrices $A(t)$ are not independent, then there are additional terms resulting from covariances between $B_{ij}(t)$ and $B_{kl}(t - m)$, $m = 1, 2, 3, \ldots$. Tuljapurkar (1990) presents a general derivation; Derrida et al. (1987) use a different approach that leads to a simpler derivation for independent matrices.

The most important thing to realize about (1.29) is that it's not nearly as bad as it looks. After a few practice problems to get yourself oriented it really is a usable tool. Moreover, for many ESS analyses, (1.29) can actually be used to derive exact (rather than approximate) local stability results, by getting the leading terms in a Taylor expansion of $\rho(D + \varepsilon | D)$ in ε, and inferring from those the partials of ρ that figure in the ESS analysis. Ellner (1996) uses this approach to carry out the ESS analysis for a more general model of a population with age, stage, or spatial structure.

The conclusion (Ellner 1996) is that structure doesn't matter very much at all for the maintenance of genetic variance. As in the unstructured model, the threshold for maintaining genetic variance is $\sigma_\theta^2 \gamma > 1$ for the deep-shallow model, and also for models with far more general age or stage structure. The only difference is that the generation overlap γ must be computed in units of Fisher's reproductive value (Appendix A) — i.e., instead of counting all eggs equally, weight them by their reproductive value and let γ be the fraction of the current total that survives to next year. Reproductive values show up because they are given by the dominant left eigenvector of A_0 (Caswell 1989), which appears in both the small-fluctuations approximation and the eigenvalue sensitivity formula. That's another reason why (1.29) isn't so bad: you can probably give a meaningful interpretation of the results even if you can't calculate $\vec{v}$, $\vec{w}$ or λ_1 explicitly, because you know what they represent about the population.

1.6   THEORY MEETS DATA

Back to the copepods: does the theory actually apply to *Diaptomus* in Bullhead Pond? We have made two checks of the fluctuating-selection hypothesis against the available data.

The first check was to ask whether our hypothesis about the selective forces generates accurate predictions of the short-term responses to selection. We did this by constructing an a priori selection model based on field and laboratory studies of the underlying processes (Ellner et al., in prep). There are generally two cohorts each year. In the model we treat the cohorts as completely discrete (all first-cohort adults die before any second-cohort individuals mature), but in fact there is usually a brief period of overlap each year. Females lay immediate-hatching eggs until their switch date, and thereafter lay diapausing eggs. Immediate-hatching eggs from first-cohort females grow up to be the second-cohort females. Immediate-hatching eggs from second-cohort females emerge so late in the spring that they have no chance of surviving to reproduce. Fitness is measured by the number of diapausing eggs produced by the end of the active season (R_0), which is a function of a female's hatching date, her switch date, and the catastrophe date (onset date of fish predation). The computation of R_0 is based on a stage-structured matrix population model, with parameters estimated directly from field or lab studies: stage durations from bottle experiments in the field; survivorships of preadult stages from population sample data; reproductive schedule from field data on clutch size and lab observations of the interclutch interval. The onset date and intensity of predation each year were taken from Hairston, Jr. (1988), adding subsequent years' data. The average R_0 for each switch date (averaged over the observed distribution of hatching dates) is then the measure of fitness. This was applied to each year's switch-date distribution to predict the total diapausing egg production, the per capita diapausing egg production, and the mean switch date in the following year.

The results for egg production are quite good (Figure 1.7). The model

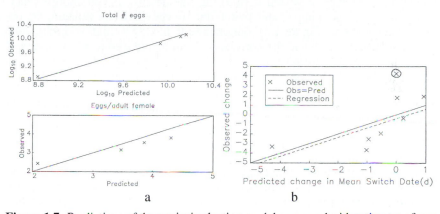

Figure 1.7. Predictions of the a priori selection model compared with estimates of egg production within a year, and year-to-year changes in mean switch date. The solid line in each panel is observed = predicted. The dashed line in panel (b) is the unconstrained linear regression of observed on predicted, omitting the "outlier" (circled).

also predicts the trend in year-to-year changes in switch date, but there is considerably more scatter. This can partially be explained by a factor omitted from the model: eggs hatching out from deep in the sediment which did not experience the most recent round of selection. Indeed, the extreme high outlier was from a year when pond drying resulted in an exceptionally high hatch-out of more deeply buried eggs.

Our second check was a laboratory test of the prediction that the egg bank acts as a "reservoir" of genetic variance in switch date. Eggs were collected from the sediment and the water column, and the distribution of switch date was determined in the lab for each of these subpopulations (after several generations of within-subpopulation random mating in the lab, to eliminate effects of the environment experienced by the founding eggs). The prediction was that switch-date variance should be higher in the sediment-derived subpopulation. Instead, there were no detectable differences in variance of switch date, but rather large differences in the mean switch date (Hairston, Jr. et al. 1997).

Where did we go wrong? The data revealed an unexpected correlation between diapause traits in eggs and adults. After the fact, the results make sense in terms of the theory developed in Sections 1.2 and 1.4 of this chapter. Genotypes with an early switch date never encounter fish predation, so these genotypes perceive a relatively constant environment and should have a high value of H. Similarly, late-switching genotypes perceive a highly variable environment and should have a low value of H. This correlation is what we appear to see in the experimental data (Hairston, Jr. et al. 1997). Eggs with a higher propensity to hatch (either earlier hatching in the lab, or being found in the water column rather than the egg bank) also had an earlier mean switch date in the lab. These observations do not show unequivocally that the correlations are due to genetic differences rather than maternal effects or other forms of nonheritable variation. Regardless of the mechanism, our basic assumption that all switch-date genotypes have the same H turned out to be incorrect.

With this added layer of complexity, quantitative predictions of the model become even more sensitive to the fine details. If you choose everything just right, allowing H and switch date to coevolve can generate the observed correlations (Figure 1.8); but change things a bit (high recombination rate, constant K, ...) and the correlations can be reduced to nil or reversed. So for now we can only say that the fluctuating-selection hypothesis is consistent with all available experimental data and generates good predictions of the short-term responses to selection.

1.7   CODA

Even if you don't care especially about copepods or genetic variance, this case study illustrates some benefits of combining theoretical and experimental approaches in biology. The most important benefit is the opportunity for creative tension when theory and experiments don't line up as perfectly as hoped. This can be simply a matter of theoreticians using preexisting experimental data, or experimental ecologists browsing the latest issues of theoretical journals for inspiration. But much of the "added value" accrues from real feedback between theory and experiments, theory provoking experiments that in turn force changes in the theoretical models, generating new predictions and calling for further experiments. Although we still cannot present compelling evidence that our (improved) theories are a complete and correct explanation for our (more extensive) data on *Diaptomus*, the new theory and the new data are both valuable in themselves.

The new theory was developed in order to address specific questions about diapause in *Diaptomus*, but the results are relevant to the general problem of understanding the maintenance of genetic variability. Genetic variance among individuals, and variance over time in environmental conditions that affect reproductive success and survival, are both ubiquitous in nature. Yet there has been little empirical study of temporally fluctuating selection and its consequences in natural populations. One likely reason for this is that temporally fluctuating selection was generally considered to be relatively unimportant for maintaining genetic variance in quantitative traits. The new theory asserts that temporally fluctuating selection can be important if generations overlap, and it has yielded some explicit testable predictions that distinguish cleanly between the fluctuating-selection and mutation-selection balance hypotheses. Our unexpected prediction that ESS trait and genotype distributions

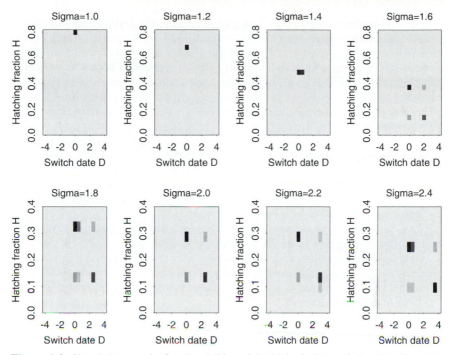

Figure 1.8. Simulation results for a haploid model with heritable variation in both hatch fraction (*H*) and the switch date (*D*). Each of the traits was assumed to be controlled by a single locus. The squares show how the long-term average genotype frequencies change as the selection variance increases. Initially there is a single type, with *H* decreasing as the variance increases. The initial bifurcation is to two values each for the *H* and *D* genotypes, with a negative correlation between *H* and *D*. The model is essentially the same as the basic model in Section 1.4, except that the total diapausing egg production (*K*) is assumed to be higher in years when the optimal switch date is later (because the predation rate is lower in those years). This assumption is essential for generating the negative correlation between *H* and *D*.

should be discrete applies to any system with temporal or spatially fluctuating selection (Sasaki and Ellner 1996); molecular genetic evidence consistent with this prediction has already been found in a number of terrestrial plant and animal species (Mitchell-Olds 1995).

Because the theory identified long-term diapause as a potentially key component in the evolutionary dynamics, experiments followed to describe and quantify the evolutionary and population dynamics in the egg bank. From those experiments, we first found the correlation between juvenile and adult traits and discovered what a drastic effect it had on our standard-type model for diapause strategies (Figure 1.8). We also learned for the first time that zooplankton eggs can remain viable in diapause for decades or even centuries — far longer than anyone had suspected (Hairston, Jr. et al. 1995) — a finding that has since been confirmed in several other species (Hairston personal communication). This is not merely a curiosity, but rather a key piece of information for anyone trying to predict the response of aquatic ecosystems to environmental change, or trying to understand how those systems are now changing in response to pollution, acidification, and nutrient enrichment.

This chapter has focused on a single case study, but the literature holds a lot of required reading for anybody intending to work on life history evolution in fluctuating environments. Here are a few starting points that will point you toward the rest of what's out there. For models with heritable variation, Charlesworth (1980) is a self-contained compendium of the rigorous (and as-rigorous-as-possible) theory for discrete and overlapping generations, though it is mostly about constant selection. Gillespie's (1991) book similarly summarizes the theory for random selection with discrete, nonoverlapping generations. The overwhelming majority of the literature is about bet-hedging strategies where the variation is nonheritable, as in Section 1.2 of this paper. The reviews by Seger and Brockmann (1987) and Phillipi and Seger (1989) are excellent background, and the book edited by Yoshimura and Clark (1993) is a good introduction to the scope of recent work. Recently, McNamara (1995; 1996) has written a series of papers that unify and extend many prior results on optimal bet-hedging strategies. McNamara also uses his results to derive dynamic programming methods to calculate optimal strategies in models too complex for analytic solution, and I expect that these methods will inspire a burst of applications in the next few years.

ACKNOWLEDGMENTS I am grateful to all the organizers of the special year course, with special thanks to Fred for serving as the hospitality committee; to the students for keeping me on track; and to Paul, Eric, and Pej for getting me camping in the Uintas by somehow neglecting to mention the snow. Eric Charnov and Jon Seger attended the lectures and kept me on my toes, and several of their comments and questions have been incorporated into the chapter.

REFERENCES

Barton, N. H. and Turelli, M. 1989. Evolutionary quantitative genetics: How little do we know? *Annual Review of Genetics*, **23**, 337–370. {*12, 22*}

Bull, J. J. 1987. Evolution of phenotypic variance. *Evolution*, **41**, 313–315. {*12, 22*}

Bulmer, Michael. 1995. *Theoretical Evolutionary Ecology*. Sunderland, MA: Sinauer Associates, Inc. ISBN 0-87893-078-7 (paper), 0-87893-079-5 (cloth). Pages xi + 352. {*4, 22*}

Caswell, Hal. 1989. *Matrix Population Models: Construction, Analysis, and Interpretation*. Sunderland, MA: Sinauer Associates, Inc. ISBN 0-87893-094-9 (hardcover), 0-87893-093-0 (paperback). Pages xiv + 328. {*3, 18, 22, 353, 356*}

Charlesworth, Brian. 1980. *Evolution in Age-Structured Populations*. Cambridge, UK: Cambridge University Press. ISBN 0-521-29786-9 (paperback), 0-521-23045-4. Pages xiii + 300. {*3, 10, 22, 353, 356*}

Charnov, Eric L. 1993. *Life History Invariants: Some Explorations of Symmetry in Evolutionary Ecology*. Oxford: Oxford University Press. ISBN 0-19-854072-8 (hardback), 0-19-854071-X (paperback). Pages xv + 167. {*4, 22*}

Chesson, P. L. 1994. Multispecies competition in varying environments. *Theoretical Population Biology*, **45**, 227–276. {*12, 22*}

Chesson, P. L. and Ellner, S. 1989. Invasibility and stochastic boundedness in monotonic competition models. *Journal of Mathematical Biology*, **27**, 117–138. {*8, 22*}

Chesson, P. L. and Warner, R. R. 1981. Environmental variability promotes coexistence in lottery competitive systems. *American Naturalist*, **117**, 923–943. {*8, 12, 22*}

Cohen, D. 1966. Optimizing reproduction in a randomly varying environment. *Journal of Theoretical Biology*, **12**, 119–129. {*5, 22*}

Cole, L. C. 1954. The population consequences of life history phenomena. *Quarterly Review of Biology*, **29**, 103–137. {*3, 23*}

Derrida, B., Mecheri, K., and Pichard, J. L. 1987. Lyapounov exponents of products of random matrices: weak disorder expansion. *Journal de Physique*, **48**, 733–740. {*17, 18, 23*}

Easterling, M. E. 1995. *Optimal Dormancy Strategies in Structured Population*. M.Phil. thesis, North Carolina State University, Raleigh, NC. {*16, 23*}

Eisen, E. J., Goodman, M. M., Namkoong, G., and Weir, B. S. (eds). 1988. *The Second International Conference on Quantitative Genetics*. Sunderland, MA: Sinauer Associates, Inc. ISBN 0-87893-900-8, 0-87893-901-6 (paperback). Pages xii + 724. {*23, 24*}

Ellner, S. 1985. ESS germination strategies in randomly varying environments. *Theoretical Population Biology*, **28**, 50–116. {*8, 23*}

Ellner, S. 1986. Germination dimorphisms and parent-offspring conflict in seed germination. *Journal of Theoretical Biology*, **123**, 173–185. {*10, 23*}

Ellner, S. 1996. Environmental fluctuations and the maintenance of genetic diversity in age or stage-structured populations. *Bulletin of Mathematical Biology*, **58**, 103–128. {*16, 18, 23*}

Ellner, S. and Hairston, Jr., N. G. 1994. Role of overlapping generations in maintaining genetic variation in a fluctuating environment. *American Naturalist*, **143**, 403–417. {*13, 14, 23*}

Ellner, S. and Sasaki, A. 1996. Patterns of genetic polymorphism maintained by fluctuating selection with overlapping generations. *Theoretical Population Biology*, **50**, 31–65. {*14, 16, 23*}

Eshel, I. and Motro, U. 1981. Kin selection and strong evolutionary stability of mutual help. *Theoretical Population Biology*, **19**, 420–433. {*10, 23*}

Falconer, D. S. 1981. *Introduction to Quantitative Genetics*. Second edn. Harlow, Essex, UK: Longman Scientific and Technical. ISBN 0-470-20474-5. Pages viii + 340. {*16, 23*}

Frank, S. A. and Slatkin, M. 1990. Evolution in a variable environment. *American Naturalist*, **136**, 244–260. {*12, 23*}

Furstenberg, H. and Kesten, H. 1960. Products of random matrices. *Annals of Mathematical Statistics*, **31**, 457–469. {*17, 23*}

Gillespie, J. H. 1991. *The Causes of Molecular Evolution*. Oxford: Oxford University Press. ISBN 0-19-506883-1. Pages xiv + 336. {*12, 23*}

Haccou, P. and Iwasa, Y. 1995. Optimal mixed strategies in stochastic environments. *Theoretical Population Biology*, **47**, 212–243. {*9, 23*}

Hairston, Jr., N. G. 1988. Interannual variation in seasonal predation: its origin and ecological importance. *Limnology and Oceanography*, **33**, 1245–1253. {*19, 23*}

Hairston, Jr., N. G. and Dillon, T. A. 1990. Fluctuating selection and response in a population of freshwater copepods. *Evolution*, **44**, 1796–1805. {*11, 23*}

Hairston, Jr., N. G., van Brunt, R. A., Kearns, C. M., and Engstrom, D. R. 1995. Age and survivorship of diapausing eggs in a sediment egg bank. *Ecology*, **76**, 1706–1711. {*12, 21, 23*}

Hairston, Jr., N. G., Ellner, S., and Kearns, C. M. 1996. Overlapping generations: the storage effect and the maintenance of biotic diversity. *Pages 109–145 of:* Rhodes, Jr., Olin E., Chesser, Ronald K., and Smith, Michael H. (eds), *Population Dynamics in Ecological Space and Time*. Chicago: University of Chicago Press. {*4, 23*}

Hairston, Jr., N. G., Kearns, C. M., and Ellner, S. 1997. Phenotypic variation in a zooplankton egg bank. *Ecology*. In press. {*19, 20, 23*}

Hedrick, P. W. 1986. Genetic polymorphism in heterogeneous environments: a decade later. *Annual Review of Ecology and Systematics*, **17**, 535–566. {*12, 23*}

Karlin, S. 1988. Non-Gaussian phenotypic models of quantitative traits. *In:* (Eisen et al. 1988). {*12, 23*}

Krebs, J. R. and Davies, N. B. 1987. *An Introduction to Behavioral Ecology*. Second edn. Sunderland, MA: Sinauer Associates, Inc. ISBN 0-87893-431-6, 0-87893-428-6 (paperback). Pages ix + 389. {*4, 23*}

Krebs, J. R. and Davies, N. B. 1993. *Behavioral Ecology: An Evolutionary Approach.* Third edn. Oxford, UK: Blackwell Scientific Publications. ISBN 0-632-02701-0 (hardback), 0-632-02702-9 (paperback). Pages xi + 482. {*4, 24*}

Ludwig, D. and Levin, S. A. 1991. Evolutionary stability of plant communities and the maintenance of multiple dispersal types. *Theoretical Population Biology*, **40**, 285–307. {*4, 24*}

McNamara, J. M. 1995. Implicit frequency dependence and kin selection in fluctuating environments. *Evolutionary Ecology*, **9**, 185–203. {*9, 22, 24*}

McNamara, J. M. 1996. Optimal life histories for structured populations in fluctuating environments. *Theoretical Population Biology.* In press. {*22, 24*}

McNamara, J. M., Webb, J. N., and Collins, E. J. 1995. Dynamic optimization in fluctuating environments. *Proceedings of the Royal Society of London. Series B. Biological Sciences*, **261**, 279–284. {*9, 24*}

Mitchell-Olds, T. 1995. The molecular basis of quantitative genetic variation in natural populations. *Trends in Ecology and Evolution*, **10**, 324–328. {*21, 24*}

Phillipi, T. and Seger, J. 1989. Hedging one's evolutionary bets, revisited. *Trends in Ecology and Evolution*, **4**, 41–44. {*8, 22, 24*}

Roff, Derek A. 1992. *The Evolution of Life Histories: Theory and Analysis.* London, UK: Chapman and Hall, Ltd. ISBN 0-412-02391-1 (paper), 0-412-02381-4 (cloth). Pages xii + 535. {*4, 24*}

Sasaki, A. and Ellner, S. 1995. The evolutionarily stable phenotype distribution in a random environment. *Evolution*, **49**, 336–350. {*9, 16, 24*}

Sasaki, A. and Ellner, S. 1996. Quantitative genetic variance and covariance maintained by fluctuating selection with overlapping generations. *Evolution.* Submitted. {*16, 21, 24*}

Seger, J. and Brockmann, H. J. 1987. What is beg-hedging? *Oxford Surveys in Evolutionary Biology*, **4**, 182–211. {*8, 22, 24*}

Shmida, A. and Ellner, S. 1984. Coexistence of plant species with similar niches. *Vegetatio*, **58**, 29–55. {*12, 24*}

Stasio, B. T. De. 1989. The seed bank of a freshwater crustacean: copepodology for the plant ecologist. *Ecology*, **70**, 1377–1389. {*12, 24*}

Stearns, Stephen C. 1992. *The Evolution of Life Histories.* Oxford: Oxford University Press. ISBN 0-19-857741-9. Pages xii + 249. {*3, 4, 24*}

Takada, T. and Kigami, J. 1991. The dynamical attainability of ESS in evolutionary games. *Journal of Mathematical Biology*, **29**, 513–529. {*10, 24*}

Taylor, P. B. 1989. Evolutionary stability in one-parameter models under weak selection. *Theoretical Population Biology*, **36**, 125–143. {*10, 24*}

Tuljapurkar, S. and Istock, C. 1993. Environmental uncertainty and variable diapause. *Theoretical Population Biology*, **43**, 251–280. {*17, 24*}

Tuljapurkar, Shripad. 1990. *Population Dynamics in Variable Environments.* Lecture Notes in Biomathematics, vol. 85. Berlin: Springer Verlag. ISBN 0-387-52482-7. Page 154. {*17, 18, 24*}

Turelli, M. 1988. Population genetic models for polygenic variation and evolution. *In:* (Eisen et al. 1988). {*12, 24*}

Yoshimura, J. and Clark, C. W. (eds). 1993. *Adaptation in Stochastic Environments.* Lecture Notes in Biomathematics, vol. 98. Berlin: Springer Verlag. ISBN 3-540-56681-3 (Berlin), 0-387-56681-3 (New York). Pages vi + 193. {*22, 24*}

2. THE EVOLUTION OF SPECIES' NICHES: A POPULATION DYNAMIC PERSPECTIVE

Robert D. Holt and Richard Gomulkiewicz

2.1 INTRODUCTION

A full understanding of evolution requires one to consider the absence of evolutionary change as well as its presence. A surprising feature of the history of life is that populations exposed to novel environments often seem to fail to adapt to, or even persist in, those environments. For instance, Bradsaw (1991) has noted that although many plant species have evolved resistance to herbicides, many others have failed to evolve resistance despite repeated exposure. Such evolutionary "failures" span short and long time scales (Holt and Gaines 1992). There are many possible explanations for evolutionary conservatism. In this chapter we present one class of explanations, emphasizing how population dynamics can constrain species' evolutionary responses to novel environments.

We will examine the evolution of a species' "fundamental niche," which is intimately tied to population persistence and extinction. If N_t denotes population size at time t, a population goes extinct if $N_t \to 0$ with increasing t. The basic model for population growth in a closed, discrete-generation population is $N_{t+1} = \lambda N_t$, where λ is the finite rate of increase per generation. A population deterministically goes toward extinction if, at low densities, $\lambda < 1$. A crisp definition of a species' niche is thus: all sets of conditions, resources, etc. for which $\lambda > 1$ (Figure 2.1). A given habitat is within a species' niche if $\lambda > 1$. Conversely, if $\lambda < 1$, the habitat is outside the species' niche, and any population found there deterministically reaches densities where, in the real world, it would face inevitable extinction due to "demographic stochasticity" (i.e., chance demographic events in small populations; see, e.g., Renshaw (1991).) We refer to populations inside their species' fundamental niche (i.e., $\lambda > 1$) as "source populations" and those outside the niche ($\lambda < 1$) as "sink populations." In this terminology, niche evolution occurs when a population evolves such that a sink environment becomes a source environment. Our primary interest is in determining the circumstances in which populations will evolve sufficiently to permit persistence in an initially unfavorable environment.

A species' fundamental niche can evolve either as a correlated evolutionary response of populations that occupy source environments ("indirect

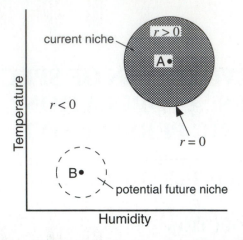

Figure 2.1. A hypothetical species' niche. The shaded region indicates combinations of temperature and humidity conditions within which population size can increase ($\lambda > 1$). Environmental conditions in habitat A permit a population to persist deterministically. By contrast, in habitat B, the same population becomes extinct — unless the niche itself evolves.

niche evolution") or through the evolution of populations directly exposed to sink environments ("direct niche evolution"). In this chapter we analyze four circumstances with the potential for direct niche evolution: (1) an isolated sink population, (2) a sink population maintained by recurrent immigration from a source population, (3) coupled sink and source populations, and (4) a network of source and sink populations ("metapopulation").

The material presented here focuses on our own past and present research. Important contributions to understanding niche evolution are also being made by other investigators (Pease et al. 1989; Brown and Pavlovic 1992; Lynch and Lande 1993; Burger and Lynch 1995; Kawecki 1995).

2.2   NATURAL SELECTION AND EXTINCTION IN A CLOSED ENVIRONMENT: DETERMINISTIC APPROACHES

We begin with a population that has suddenly encountered a novel environment outside its fundamental niche. This could characterize a colonizing group that encounters an inhospitable habitat, or a closed population experiencing abrupt environmental change. In either situation, the initial population occupies a sink habitat. When will such a population evolve into a source, thereby expanding its species' niche?

Consider an isolated sink population with discrete generations. We assume the population's finite rate of increase at time t is identical to the mean fitness of its members, $\bar{W}_t$. Population size thus changes according to

$$N_{t+1} = \bar{W}_t N_t. \tag{2.1}$$

Since the population is initially a sink, $\bar{W}_0 < 1$. If the finite rate of increase does not change, then

$$N_t = \bar{W}_0^t N_0 \tag{2.2}$$

where N_0 is initial population size. Although the population deterministically tends toward extinction, $N_t \neq 0$ for any finite t. A more complete treatment that included demographic stochasticity (Renshaw 1991) would show that extinction becomes increasingly certain as the population declines to small sizes. For now, we will bypass the complexities of demographic stochasticity entirely by using a crude deterministic approximation for its effects. Specifically, we assume there is some "critical density" N_c below which a population is highly vulnerable to rapid extinction. This approximation can be justified on the grounds that the probability of rapid extinction often rises dramatically as population size declines (MacArthur and Wilson 1967). (The results in the next section offer a more direct justification.) One can thus approximate the time until extinction, t_E, by solving $N_c = N_0 \bar{W}_0^t$ for t: $t_E = (\ln N_c - \ln N_0)/\ln \bar{W}_0$ generations until extinction.

Without evolution, extinction is inevitable. However, a population may be able to avoid extinction if it can adapt sufficiently rapidly in the sink environment. This follows from Fisher's fundamental theorem of natural selection which suggests, roughly, that given genetic variation in fitness, mean fitness, $\bar{W}_t$, should increase through evolution by natural selection (Fisher 1958; Burt 1995). Consider the following very simple model of this adaptive process: assume mean fitness increases by a fixed amount δ each generation, i.e., $\bar{W}_{t+1} = \bar{W}_t + \delta$. Then t generations after a population first encounters the sink environment, $\bar{W}_t = \bar{W}_0 + t\delta$. Population density will increase whenever mean fitness exceeds one. Thus, the first time at which a population grows, t_R, can be found by solving $1 = \bar{W}_t = \bar{W}_0 + t\delta$ for t: $t_R = (1 - \bar{W}_0)/\delta$. Adaptation will "rescue" a population from extinction if $t_R < t_E$, whereas extinction is likely, even with adaptive evolution, if $t_R > t_E$.

This simple model gives us a sense of the important time scales involved in the race between evolution and extinction, but it is much too simplistic because mean fitness does not generally increase at a constant rate per generation. We now consider a more realistic evolutionary model that more accurately describes changes in a population's mean fitness under selection, viz., a quantitative genetics model.

This model makes the following basic assumptions. First, we assume individual fitness depends on a trait, z, with polygenic autosomal inheritance (e.g., body size). We assume fitness in the novel environment has the form $W(z) = W_{max} \exp\left[-z^2/2\omega\right]$, where ω is a parameter inversely related to the strength of selection, so that (without loss of generality) the optimum phenotype lies at $z = 0$ with fitness W_{max}. This "Gaussian" form can represent a variety of biological situations, including directional and stabilizing selection (it is also mathematically convenient). The distribution of phenotypes in generation t, $p_t(z)$, is assumed normal with mean d_t (the distance of the mean phenotype from the local optimum, $z = 0$) and variance P:

$$p_t(z) = (2\pi P)^{-1/2} \exp\left[-(z - d_t)^2/2P\right].$$

Quantitative traits are often normally distributed when measured on an appropriate scale (Falconer 1989). By completing the square in the exponent and simplifying, it can be shown that the mean fitness in generation t is

$$\bar{W}_t = \int W(z)p_t(z)\,dz = \hat{W}\exp\left[\frac{-d_t^2}{2(P+\omega)}\right] \tag{2.3}$$

where $\hat{W} = W_{max}\sqrt{\omega/(P+\omega)}$ is the population growth rate when the mean phenotype is at the local optimum ($d = 0$). Finally, we assume the effects of other evolutionary forces (e.g., drift) are negligible.

A standard result from quantitative genetics (Falconer 1989; Lande 1976) predicts that the mean phenotype, d_t, of a quantitative trait z changes between generations (i.e., evolves) according to the equation

$$\Delta d_t \overset{\text{def}}{=} d_{t+1} - d_t = h^2 s, \tag{2.4}$$

where h^2 is the "heritability" of z and s is the "selection differential." Roughly, h^2 measures the degree to which offspring phenotypes resemble their parents' phenotypes, in the absence of common environmental influences. The selection differential is the difference between the mean phenotype of individuals selected to be parents and the mean phenotype before selection. For our model,

$$s \overset{\text{def}}{=} \int z\left[W(z)/\bar{W}_t\right]p_t(z)\,dz - d_t = \frac{-d_t P}{P+\omega} \tag{2.5}$$

which implies that (2.4) has the form

$$\Delta d_t = -h^2\frac{d_t P}{P+\omega}. \tag{2.6}$$

Therefore,

$$d_{t+1} = d_t + \Delta d_t = \frac{\omega + (1-h^2)P}{P+\omega}d_t = kd_t \tag{2.7}$$

where $k = [\omega + (1-h^2)P]/(P+\omega)$ is the "evolutionary inertia" of the mean; $0 \le k \le 1$. Note that k will be near 1 if heritability is low ($h^2 \approx 0$) or selection is weak ($\omega \gg P$). If P and h^2 are constant, then

$$d_t = k^t d_0. \tag{2.8}$$

We now turn to the dynamics of population density, which are described once again by (2.1). Because $\bar{W}$ changes through time, $N_t = N_0\prod_{i=0}^{t-1}\bar{W}_i$. Recall that our goal is to determine the times t_R and t_E. To do this, we try to write $N_0\prod_{i=0}^{t-1}\bar{W}_i$ as an explicit function of t. Substituting (2.3) for $\bar{W}_t$ into (2.8) results in an expression with the geometric series $\sum_{i=0}^{t-1}k^{2i}$ in the exponential term. This series can be rewritten in the closed form $(1-k^{2t})/(1-k^2)$, giving

$$N_t = N_0\hat{W}^t\exp\left[\frac{-d_0^2(1-k^{2t})}{2(P+\omega)(1-k^2)}\right]. \tag{2.9}$$

We now use (2.9) to compute t_E, the time to reach N_c, and t_R, the first time at which $\bar{W} \geq 1$. By definition, t_E must satisfy

$$N_c = N_0 \hat{W}^{t_E} \exp\left[-\frac{1}{2}\left(\frac{d_0^2}{P+\omega}\right)\left(\frac{1-k^{2t_E}}{1-k}\right)\right]. \qquad (2.10)$$

While it is not possible to solve for t_E explicitly, (2.10) provides an *implicit* definition for t_E as a function of the parameters. Two parameters can be eliminated by defining $v_0 = N_c/N_0$ and $\beta_0 = d_0^2/(P+\omega)$. These give "natural" scales for measuring, respectively, the relevant initial population density and distance from the optimum phenotype. This rescaling reduces (2.10) to

$$v_0 = \hat{W}^{t_E} \exp\left[-\frac{\beta_0}{2}\left(\frac{1-k^{2t_E}}{1-k}\right)\right]. \qquad (2.11)$$

The time t_R must be a solution of the equation $\bar{W}_t = 1$. Using (2.3) and (2.8) leads to

$$t_R = \frac{\ln\left(\ln\hat{W}\right) - \ln(\beta_0/2)}{2\ln k}. \qquad (2.12)$$

With expressions for t_E and t_R in hand, we can examine in detail the influence of adaptive evolution on the chances of population persistence.

We begin by using (2.11) and (2.12) to determine whether an adapting population is likely to persist (because $t_E > t_R$) or risk rapid extinction ($t_E < t_R$). We are especially interested in the respective influences of initial population size, N_0, and the magnitude of d_0, which indicates the initial degree to which the population is maladapted to the novel environment. We want to know the set of N_0 and d_0 at which $t_E = t_R$ separates the combinations of initial values consistent with likely persistence ($t_E > t_R$) versus extinction ($t_E < t_R$). We determined these critical values by substituting (2.12) into (2.11). This eliminates t, resulting in an equation that defines the relationship between v_0 (the scaled form of N_0) and β_0 (the scaled version of d_0) at which $t_E = t_R$. Graphs of this relationship, for various values of heritability, h^2, are shown in Figure 2.2. Populations whose initial density and degree of maladaptation lie below the curve of critical values for a particular h^2 are likely to persist, whereas populations with combinations falling above the critical curve have a high risk of rapid extinction.

While evolution will not rescue all initially maladapted populations from extinction, evolution might slow the decline of populations destined to reach critically low densities. To assess this potential effect, we compared the times t_E to reach N_c with evolution (2.11) to the times to reach N_c without evolution:

$$t_E = \frac{\ln N_c - \ln N_0}{\ln \bar{W}_0} = \frac{\ln v_0}{\ln \bar{W}_0} \qquad (2.13)$$

Some of the results are shown in Figure 2.3. Evolving populations with initial degree of maladaptation less than a critical level (β^* in Figure 2.3) never drop below N_c. This figure also shows that, in contrast to our expectations, evolution does little to slow the decline of populations destined to reach critically

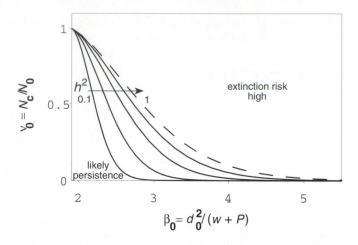

Figure 2.2. Combinations of scaled initial population densities (v_0) and degrees of initial maladaptation (β_0) leading to likely persistence or high extinction-risk heritability, $h^2 = 0.1$ (solid curve) and $h^2 = 1$ (dashed curve). For a given level of heritability, populations with v_0 and β_0 below the curve persist deterministically because they remain above N_c; those with v_0 and β_0 above the curve decline below N_c and become highly vulnerable to rapid extinction by stochasticity. (Adapted from Gomulkiewicz and Holt (1995).)

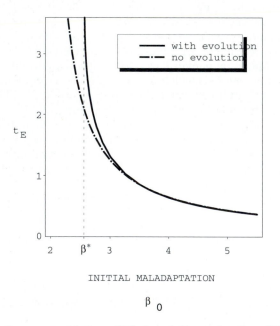

Figure 2.3. The time t_E a population will first reach N_c as a function of its initial degree of maladaptation (2.11). The dot-dashed curve indicates analogous times for the case of no evolution (2.13).

low densities. Apparently, the main effect of evolution is to prevent extinction altogether, rather than slow the approach of populations that are destined to reach critically low densities and face rapid extinction risks.

While a population whose size drops below N_c faces a high risk of extinction, it may nevertheless avoid this fate. Such a population would continue to adapt and, at time t_R, begin to grow. Provided the population persists, at some later time, t_P, its size will climb above N_c. Thereafter, the population will be relatively invulnerable to extinction by demographic stochasticity. The time t_p needed for a population to grow above N_c is, like t_E, a solution of (2.11), because $N_{t_P} = N_c$. In fact, (2.11) has either zero or two roots for (essentially) every pair of initial conditions. When there are two roots, the lower one is t_E while the upper root is t_P. The difference, $t_P - t_E$, defines the maximum period that a population's density will be below N_c. It can be shown (Gomulkiewicz and Holt 1995) that this "period of extinction risk" is longer for populations that are initially more highly maladapted. Equations (2.11) and (2.12) can also be used to explore the dependence of t_E, t_R, and t_P on initial density, N_0. Initially small populations are similarly likely to face a period of extinction risk.

The above results lead to four main biological conclusions. First, only initially large and mildly maladapted populations are expected to evolve sufficiently to persist in novel environments, while small or severely maladapted populations are likely to face a high risk of rapid extinction. The range of initial population sizes consistent with likely persistence shrinks rapidly for intermediate degrees of initial maladaptation (Figure 2.2). Second, populations that are more severely maladapted (or initially small) will face a high extinction risk sooner and, if they manage to avoid chance extinction, remain at high risk for a longer period of time. Third, the main effect of local adaptation in a sink environment is to allow some populations to avoid facing high extinction risk altogether. Evolution by natural selection does little to slow the approach of populations destined to reach critically low densities (Figure 2.3). Finally, if niche evolution occurs primarily through isolated colonizations of novel environments outside a species' current fundamental niche (or through populations exposed to rapid environmental deterioration), then niche expansion will occur only rarely, unless colonizing groups tend to be very large.

The above models permit extensive analysis. An important task for future work will be to examine the robustness of our conclusions to relaxation of genetic and ecological assumptions. (We have elsewhere considered a standard one-locus diallelic model of a continuously breeding population and reached thematically similar conclusions (Gomulkiewicz and Holt 1995)). In the following sections, we first examine stochasticity, then permit recurrent immigration into the sink, and then emigration back to the source.

2.3 NATURAL SELECTION AND EXTINCTION IN A CLOSED ENVIRONMENT: STOCHASTIC APPROACHES

In the last section we considered the dynamics of a closed population that is suddenly exposed to an environment outside its species' niche. We found that the main issue is whether the population can evolve sufficiently rapidly to avoid reaching low densities where it is highly vulnerable to chance extinction. Our deterministic treatment of evolution and extinction in a closed environment rested on a crude (but mathematically convenient) device, the notion of a critical density N_c, to avoid dealing with the complicated probabilistic details of the actual extinction process.

Besides validating our deterministic approach, a fully stochastic analysis can address issues that a deterministic analysis cannot. For example, deterministic analyses do not provide a probability distribution of actual times to extinction which may be important in applications. However, stochastic models are typically difficult to analyze, except for certain simple cases, and usually require approximations or computer simulations. In this section we present a stochastic model of evolution and extinction that is sufficiently simple to allow mathematical analysis.

The simplest version of our basic ecological scenario involves a population of clonal organisms facing a novel environment. As it turns out, a substantial body of empirical work involves such systems. The experiments performed by Lenski, Bennett and their colleagues (e.g., Lenski and Bennett (1993)) on the evolutionary responses of *E. coli* to novel thermal and nutritional environments is one prominent example. Our mathematical model assumes that reproduction is asexual and generations do not overlap. Suppose that the population contains two genotypes, A_1 and A_2. In the novel environment, individuals with genotype A_1 have an expected absolute fitness $W_1 > 1$ while those with genotype A_2 have expected absolute fitness $W_2 < 1$. The *actual* number of offspring left by an individual is a random variable. Finally, resources and space are sufficient to allow individuals to survive and reproduce independently of one another.

Our analysis has two goals. First, we will determine how the probability of extinction depends on the size and composition of the initial population, particularly when the "adapted" clone A_1 is initially rare. Second, we want to examine how initial population size and composition affect the probability distribution of times to extinction. Besides providing useful quantitative information, these analyses will allow us to compare the results of our previous deterministic analyses with a full-blown (albeit simple) stochastic model and thereby begin to assess the adequacy of our deterministic approximations.

Our asexual model is an example of a "two-type branching process" (Karlin and Taylor 1975). The standard way to analyze branching processes is through the use of a *probability generating function*. Probability generating functions are useful for studying stochastic processes because they conveniently "package" critical information about the process.[1] In particular, it is straightforward to compute extinction probabilities and times once an appropriate probability generating function is available. We thus begin by defining and computing a generating function for our asexual model.

It is easiest to construct the probability generating function for our two-type branching process by combining two separate generating functions, one for each of the two genotypes. The generating function for genotype A_1,

---

[1]To quote Lin and Segel (1988), pg. 76: "It is not difficult to follow manipulations using the generating function, but it is amazing that anyone would have thought of this device. Amazement is lessened upon learning that the first person to make use of the generating function was the genius Euler ... . In discussing the motivation for using a generating function, Polya ... states that 'a generating function is a device somewhat similar to a bag. Instead of carrying many little objects detachedly (the individual coefficients), which could be embarrassing, we put them all in a bag (a generating function), and then we have only one object to carry, the bag.'"

$f(\cdot)$, is defined as

$$f(u) = \sum_{i=0}^{\infty} p_i u^i \tag{2.14}$$

where p_i is the probability that an individual with genotype A_1 produces i offspring ($i = 0, 1, \dots$). Similarly, the generating function for genotype A_2 is defined as

$$g(v) = \sum_{i=0}^{\infty} q_i v^i \tag{2.15}$$

where q_i is the probability that an individual with genotype A_2 produces i offspring ($i = 0, \dots$). Variables u and v are "dummy variables"; they serve as nothing more than placeholders during computations.

To make further progress, we must make specific assumptions about the offspring-number distributions, $\{p_i\}$ and $\{q_i\}$. We assume that the probability distribution of an individual's offspring number follows a Poisson distribution whose expected value depends on genotype. Assume that the offspring number of an A_1 individual is a Poisson random variable with expectation $W_1 > 1$, and the offspring number of an A_2 individual is a Poisson random variable with expectation $W_2 < 1$: $p_i = e^{-W_1} W_1^i / i!$ and $q_i = e^{-W_2} W_2^i / i!$. Not only are these assumptions mathematically convenient, they conform to the offspring distribution assumed in a frequently used model of genetic drift (Crow and Kimura 1970). With these assumptions, (2.14) and (2.15) simplify (using the result $e^{ax} = \sum_{i=0}^{\infty} (ax)^i / i!$) to

$$f(u) = \sum_{i=0}^{\infty} \frac{e^{-W_1} W_1^i u^i}{i!} = e^{W_1(u-1)} \tag{2.16}$$

and

$$g(v) = \sum_{i=0}^{\infty} \frac{e^{-W_2} W_2^i v^i}{i!} = e^{W_2(v-1)}. \tag{2.17}$$

We now use $f(u)$ and $g(v)$ to define $h_t(u, v)$, the probability generating function in generation t for an asexual population that initially consists of x_0 clones with genotype A_1 and y_0 clones with genotype A_2:

$$h_0(u, v) = u^{x_0} v^{y_0} \tag{2.18}$$

and

$$h_t(u, v) = h_{t-1}(f(u), g(v)) \qquad \text{for } t = 1, 2, \dots. \tag{2.19}$$

For example, $h_1(u, v) = [f(u)]^{x_0}[g(v)]^{y_0}$, $h_2(u, v) = [f(f(u))]^{x_0}[g(g(v))]^{y_0}$, etc. In most cases (including the present one), it is difficult to obtain a closed-form expression for $h_t(u, v)$ for any given t. We overcame this difficulty by using a symbolic manipulation program to compute the generating functions for different generations.

The generating function $h_t(u, v)$ can now be used to determine critical information about the stochastic process associated with our model. In fact, the joint probability that there are k copies of genotype A_1 and l of genotype A_2 in generation t is exactly the coefficient $c_{kl,t}$ of the product $u^k v^l$ in the bivariate power-series expansion of $h_t(u, v)$:

$$h_t(u, v) = \sum_{i=0}^{\infty} \sum_{j=0}^{\infty} c_{ij,t} u^i v^j \tag{2.20}$$

where

$$c_{kj,t} = \left. \frac{\partial^k \partial^l h_t(u, v)}{\partial u^k \partial v^l} \right|_{u=v=0} \tag{2.21}$$

(Karlin and Taylor 1975). In particular, the probability of extinction (i.e., of $l = k = 0$) at or before generation t, denoted F_t, is

$$F_t \equiv c_{00,t} = h_t(0, 0). \tag{2.22}$$

Note that F_t is the *cumulative* probability distribution function for the time to extinction.

With our simple model, it is possible that a population will never go extinct. The exact probability that this occurs is $F_\infty = \pi_1^{x_0} \pi_2^{y_0}$, where π_1 and π_2 are the respective unique non-zero solutions less than or equal to one of $\pi = f(\pi) = e^{W_1(\pi-1)}$ and $\pi = g(\pi) = e^{W_2(\pi-1)}$. It can be shown that $W_2 < 1$ implies that $\pi_2 = 1$ (clone A_2 goes extinct with probability 1), so

$$F_\infty = \pi_1^{x_0}. \tag{2.23}$$

(See Karlin and Taylor (1975), Chapter 8, for details.) We can use the above definitions and results to address our main questions about how adaptive evolution affects the probability and timing of extinction of a population facing a novel environment, given both demographic stochasticity and genetic drift. First, consider how initial population size and the initial degree of population maladaptation affect the likelihood of extinction. If there are initially x_0 individuals with genotype A_1 and y_0 with genotype A_2, then the initial population size is $N_0 = x_0 + y_0$. We let the initial frequency of the maladaptive A_2 genotype, $q_0 \equiv y_0/N_0$, indicate the initial degree of population maladaptation. It is not hard to show that, for a given q_0, the probability of extinction at or before generation t decreases with N_0, whereas, given N_0, this probability increases with q_0. The minimum N_0 that is consistent with at least a 95% chance of extinction by generation 100 varies with q_0 in a qualitatively similar manner to the deterministic results shown in Figure 2.2. This suggests that an appropriate interpretation of "likely to face a high risk of rapid extinction" in our deterministic treatment is that there is at least a $P\%$ chance of extinction by generation T, for specified P and T.

We are also interested in how initial population size and composition affect the time a population persists. We can use our generating function methods to determine E_k, which is the smallest time t such that $F_t \geq k$. The graph of $E_{0.05}$ versus q_0 for a fixed initial population size closely resembles Figure 2.3, which indicates how the "time until high extinction risk" in our deterministic treatment depends on the initial degree of maladaptation for fixed N_0. (Similar comparisons hold when the initial degree of maladaptation is fixed, rather than N_0.)

While these results support the qualitative results of our deterministic analyses, this stochastic model can provide quantitative and further qualitative information about the extinction process that our deterministic approach cannot. For example, using the cumulative distribution function F_t, it can be shown that the most rapid increase in extinction probability occurs in the first few generations. This suggests that if a population is destined to become extinct, it will likely do so quickly. Median times to extinction can also be easily computed using F_t. In fact, $M = E_{0.5}$. It can be shown that M increases rapidly with N_0, as one might expect.

Our analysis of this simple stochastic model yields two main conclusions. First, the qualitative features of the race between extinction and adaptation suggested by our deterministic analyses are supported. Second, our analysis found that populations tend to go extinct quickly, if they go extinct at all.

Our simple, asexual model could be extended in many ways, but in most cases, analysis will be sufficiently difficult to require Monte Carlo simulation methods. We have analyzed a one-locus, two-allele version of the above scenario in this way and found that the above results hold for this more complex model. However, much more work needs to be done before we will be fully convinced that the deterministic results are robust to stochasticity.

2.4 THE INFLUENCE OF IMMIGRATION ON LOCAL ADAPTATION: FRESH PERSPECTIVES ON AN OLD PROBLEM

In Sections 2.2 and 2.3 we considered a completely isolated population that finds itself outside its species' niche. How is niche evolution affected given recurrent immigration from a source? In this section, we consider a simple case we call a "black-hole sink," a sink population that recurrently receives locally maladapted immigrants that arrive from a separate source population but returns no emigrants to the source (Holt and Gomulkiewicz 1996).

A black-hole sink closely resembles the "island-continent model" used in population genetics theory to understand how one-way recurrent gene flow from a "continent" can impact local adaptation on an "island." Before presenting our model, it is instructive to consider the intuition provided by population genetics theory. Generally speaking, analysis of an island-continent model results in a "rule of thumb": for a given selective advantage of a locally favored allele, there is some rate of gene flow below which that allele will spread when rare (e.g., Nagylaki (1977), p. 125). This implies that the greatest scope for local adaptation should occur at low to zero rates of gene flow. Now suppose that the island population is a "sink" population. Without immigration, such a population goes extinct deterministically. This presents a paradox: At zero immigration — which provides the greatest scope for local adaptation — a sink population goes extinct and, thus, local adaptation is impossible! How can this be explained? To answer this, one must examine more explicitly the demographic consequences of immigration.

Consider, as in the last section, a discrete breeding asexual population with two genotypes, A_1 and A_2, where the absolute fitness on the island of genotype A_i is W_i. Assume that A_1 individuals have the higher local fitness, or $W_1 > W_2$. The mean fitness (and finite growth rate) of the island population is $\bar{W} = pW_1 + (1-p)W_2$, where p is the frequency of A_1. Immigrants arrive on the island just after reproduction at a rate m per generation, where m is the percentage of the post immigration island population that consists of immigrants (i.e., the rate of "gene flow"). We assume all immigrants have the locally less fit genotype A_2 (Nagylaki 1992). The frequency of A_1 in the next generation is

$$p' = (1-m)\left(\frac{W_1}{\bar{W}}\right)p. \tag{2.24}$$

The first term in (2.24) indicates how the frequency of A_1 is reduced by A_2 immigrants who arrive at gene flow rate m; the second term describes how selection increases the frequency of the locally fitter genotype A_1.

The conditions under which the locally favored allele A_1 will increase in frequency when initially rare are found when $p \approx 0$ and $\bar{W} \approx W_2$ in (2.24), which shows that $p' > p$ if $(1-m)W_1/\bar{W} > 1$, that is

$$\frac{W_1}{W_2} > \frac{1}{1-m} > 1. \tag{2.25}$$

Equation (2.25) is an example of the rule of thumb mentioned above: for given fitnesses (W_1, W_2), there is a rate of migration (m) below which the locally more fit genotype will spread when initially rare. Note that the maximal scope for local adaptation (spread of A_1) occurs as the gene flow rate m approaches zero.

We now reanalyze the same scenario, but follow genotype numbers rather than frequency. For this purpose, let N_i denote the *number* of genotype A_i and let N be total population density, $N = N_1 + N_2$. Consider first the dynamics of a population fixed for the less fit genotype A_2 so that $N = N_2$. Because the island population is a sink, $W_2 < 1$. We assume that a constant number I of A_2 individuals immigrate to the island each generation. The island population size changes according to

$$N' = NW_2 + I. \tag{2.26}$$

The size of this population will equilibrate when $N' = W_2N + I = N$. Solving for N gives the equilibrium density of A_2 genotypes:

$$\hat{N} = \frac{I}{1-W_2} \equiv \hat{N}_2. \tag{2.27}$$

Now assume that a few A_1 individuals are introduced to such an equilibrium population. The frequency of A_1 is $p = N_1/(N_1 + N_2)$. Because all immigrants have genotype A_2, the dynamics of A_1 are simply

$$N_1' = N_1W_1. \tag{2.28}$$

The density of A_2 will return to its equilibrium $\hat{N}_2$. The dynamics of A_1 are determined completely by the absolute fitness W_1. On the one hand, if $W_1 < 1$, then $N_1 \to 0$ and $p \to 0$; the locally favored A_1 is lost. On the other hand, if $W_1 > 1$, then $N_1 \to \infty$ and $p \to 1$; the locally favored A_1 spreads through the population. This analysis provides a criterion for determining whether a locally favored genotype will spread: its *absolute* fitness must be greater than one — regardless of the immigration rate I or the fitness of the less fit genotype. This criterion seems strikingly different from the rule of thumb provided by (2.25), which involves both the immigration rate and fitness of A_2.

The key to resolving the apparent discrepancy between these two criteria is to recognize that the migration rate m is a *variable*, not a fixed parameter as is implicitly assumed in the standard approach. Consider the recursion for total population size:

$$N' = N\bar{W} + I. \tag{2.29}$$

By definition, $m = I/N'$, which implies that $1 - m = N\bar{W}/N'$. By substituting this expression in (2.24), the recursion for p can be rewritten as

$$p' = \left(\frac{1-m}{\bar{W}}\right) W_1 p = \left(\frac{N}{N'}\right) W_1 p \tag{2.30}$$

Now if A_1 is rare and the immigrant genotype A_2 is near its equilibrium density, then $N' \approx N$ which implies from (2.30) that $p' \approx W_1 p$. Thus we have recovered from (2.24) that the necessary and sufficient condition for A_1 to increase when rare is $W_1 > 1$.

Our analysis of this simple model leads to two unexpected conclusions. First, absolute — not relative — fitness governs the spread of a locally favored allele in a sink population. Second, provided it is not zero, the immigration rate, I, has no influence over the spread or loss of the favored allele in a sink population. To what extent do these conclusions depend on the simplicity of the model we analyzed? We next analyze the same black-hole-sink scenario for a diploid sexual population.

Consider a model in which fitness is determined by variation at a diploid locus with alleles A_1 and A_2. Assume that adults immigrate after selection but before reproduction, and that the population is censused immediately after (sexual) reproduction. (Similar conclusions hold if immigration occurs before selection.) As above, let N be total population size and p be the frequency of the locally favored A_1 allele. Assume that I immigrants, all with genotype $A_2 A_2$, arrive each generation. Denote the birth-to-immigration viability of $A_i A_j$ by v_{ij}. For simplicity, we assume that the expected fecundity is f, independent of genotype. The fitness of genotype $A_i A_j$ is thus $W_{ij} = f v_{ij}$. Finally, assume that $W_{22} < 1$, so that in the absence of A_1, the population is a black-hole sink maintained only by immigration.

Recursions for the dynamics of this population can be derived as follows. Following random mating, genotype frequencies are in Hardy-Weinberg proportions: the densities of $A_1 A_1$, $A_1 A_2$, and $A_2 A_2$ among newborns are Np^2, $N2p(1-p)$, and $N(1-p)^2$. After viability selection and immigration, the density of breeding adults is

$$N^* = v_{11}Np^2 + v_{12}N2p(1-p) + v_{22}N(1-p)^2 + I \tag{2.31}$$

Following reproduction, the density of newborns, N', is

$$N' = fN^* = N\bar{W} + fI \tag{2.32}$$

where $\bar{W} = p^2 W_{11} + 2p(1 - p)W_{12} + (1 - p)^2 W_{22}$ is mean fitness. Random mating returns genotype frequencies to Hardy-Weinberg proportions without altering allele frequencies. The frequency of A_1 of the newborns, p', is equal to the frequency of A_1 of the parents, i.e., $p' = $ (number of parental A_1 alleles)$/2N^* = (2Np^2 v_{11} + 2Np(1-p)v_{12})/2N^*$. Multiplying the numerator and denominator by f and using (2.32) shows

$$p' = \left(\frac{N}{N'}\right)\bar{W}_1 p \tag{2.33}$$

where $\bar{W}_1 = pW_{11} + (1 - p)W_{12}$ is the average fitness of individuals with an A_1 allele. This recursion closely resembles the asexual equation (2.30).

Now examine the conditions under which the fitter A_1 allele will spread when rare. As above, consider first a population in which A_2 is the only allele present. Then $p = 0$ and $\bar{W} = W_{22}$. Setting the left-hand side of (2.32) equal to N and solving for N shows that the population will equilibrate at density

$$\hat{N} = \frac{fI}{1 - W_{22}} \equiv \hat{N}_{22}. \tag{2.34}$$

Now suppose a few copies of A_1 are introduced into this population so that $p \approx 0$ and $N' \approx N$. The gene-frequency recursion (2.33) is given approximately by $p' \approx pW_{12}$. Clearly, $p' < p$ if $W_{12} < 1$ and $p' > p$ if $W_{12} > 1$. The locally favored allele will increase when rare-provided the absolute fitness of heterozygous individuals exceeds one. As in the simpler asexual model, the spread of A_1 depends on absolute — not relative — fitness, and is independent of both the rate of immigration, I, and the fitness of the immigrant genotype, W_{22}.

Together, our asexual and diploid model results suggest that an absolute fitness criterion for spread of an initially rare allele is a generic feature of local adaptation and, hence, niche evolution in a black-hole-sink population. Before taking such a generalization too seriously, it is important to consider some potentially important ecological and genetic limitations of our two models.

The most obvious ecological deficiency in our models is that population size may increase without bound. How does density regulation affect the spread of a rare locally favored allele? We have analyzed a version of the above haploid model in which fitness is density dependent (Holt and Gomulkiewicz 1996) as well as the diploid model (Gomulkiewicz, Holt, and Barfield, in preparation). Our analysis shows that, once again, the locally fitter allele will spread in a black-hole sink population if the absolute fitness of heterozygotes is greater than one. Because fitness is density dependent, this criterion does depend on population size, but only indirectly through its influence on absolute fitness. (Density-dependence plays a more pronounced and complicated role in determining the eventual size and genetic composition of a population, once an initially rare locally favored allele increases.)

There are also a number of genetic limitations in our models (e.g., no mutation or genetic drift). Given our analysis in the first section, it is reasonable to ask how populations might adapt in the face of recurrent immigration if fitness depends on characters with polygenic inheritance. One could easily (and naively) extend the quantitative genetics model we considered in the first section to include recurrent immigration as follows. The population-size dynamics is described by (2.32) except that mean fitness $\bar{W}$ is defined by (2.3). It is not hard to show that the evolution of d, the distance of the current mean phenotype from the local optimum, satisfies the recursion equation $d' = (1 - M)kd + md_I$, where k is the evolutionary inertia (see Section 2.2), $M = fI/(N\bar{W} + fI)$ is a gene flow variable, I and f are as defined above, and d_I is the (fixed) difference between the mean phenotype of immigrants and the local optimum.

The dynamic features of this model are relatively simple to explore, and they are, in fact similar to those of the one-locus models. Unfortunately, there are several reasons why it seems premature to declare with any confidence that these evolutionary features are "robust" to genetic assumptions. First, this quantitative genetic analysis completely ignores departures from normality that are caused by immigration. Such departures are known to alter dynamical behavior under some circumstances. Second, this formulation also ignores the linkage disequilibrium (nonrandom associations between alleles at different loci) that is constantly generated by immigration, which in turn can affect variances and heritability. In the future, we plan to explore whether these neglected assumptions will have a noticeable impact on our main biological conclusions by analyzing more realistic (and complex) multilocus models.

2.5   NICHE EVOLUTION IN COUPLED SOURCE-SINK ENVIRONMENTS

The previous section considered evolution in a sink population maintained by immigration without dispersal back to the source. A natural generalization is to two habitats, one a source, the other a sink, with reciprocal dispersal (Holt 1996). For simplicity we consider a species with discrete generations and haploid genetics.

Census the population following dispersal. Let $N_i(t)$ be population size in habitat i at the start of generation t. In generation $t+1$ there are $N_i(t+1)$ individuals, who either immigrated from habitat j (denoted $N_{ij}(t + 1)$) or did not (denoted $N_{ii}(t + 1)$). Necessarily, $N_i(t + 1) = N_{ii}(t + 1) + N_{ij}(t + 1)$. Individuals in habitat i at time t contribute to the population there at $t + 1$ via production of offspring which do not emigrate. Let $N_{ii}(t + 1) = a_{ii}(t)N_i(t)$. The quantity a_{ii} is the per capita contribution of habitat i to itself. Likewise, $N_{ij}(t+1) = a_{ij}(t)N_j(t)$ defines the per capita contribution of habitat j to habitat i. With this notation, a 2×2 matrix model describes the dynamics of the two coupled habitats over a single time step: $N(t + 1) = A(t)N(t)$, where the vector $N(t) = (N_1(t), N_2(t))$, and the ijth element of matrix A is $a_{ij}(t)$. In general, $a_{ij}(t)$ may vary as the external environment changes, or because dispersal or local growth rates are density dependent and population size changes. Consider first the case of constant transition rates, $A(t) \equiv A$.

As with any matrix model (Caswell 1989), as t increases the population settles into a *stable patch distribution* , defined by the right eigenvector of A, and changes in size at a constant rate λ (the dominant eigenvalue of A),

$$\lambda = \frac{1}{2}(a_{11} + a_{22} + \sqrt{(a_{11} - a_{22})^2 + 4a_{12}a_{21}}). \qquad (2.35)$$

(Because of the low dimensionality of the above model, one can explicitly solve for eigenvalues and eigenvectors as a function of arbitrary matrix elements a_{ij}; this is not generally possible). The stable patch distribution has a defined fraction of the population in each patch. A right eigenvector for A (with elements summing to unity) is

$$(w_1, w_2) = \left(\frac{\lambda - a_{22}}{\lambda - a_{22} + a_{21}}, \frac{a_{21}}{\lambda - a_{22} + a_{21}} \right). \qquad (2.36)$$

A familiar interpretation of the left eigenvector in an age-structured matrix model is that it gives "reproductive value" (Caswell 1989) — the relative contribution of an individual to future generations — as a function of age. A left eigenvector of this stage-structured patch model is

$$(v_1, v_2) = \left(\frac{\lambda - a_{22}}{\lambda - a_{22} + a_{12}}, \frac{a_{12}}{\lambda - a_{22} + a_{12}} \right) \qquad (2.37)$$

which likewise describes *spatial reproductive value*, the contribution of an individual in habitat i to future generations (in both habitats).

Adaptive evolution may occur if genetic variants arise with different values for the a_{ij}. A novel mutant spawns a subpopulation, whose dynamics can also be described by a 2×2 matrix model; this subpopulation settles into its own stable patch distribution and grows at its own asymptotic growth rate. If this growth rate exceeds that of the resident clone, then initially the new clone is favored, and it will (deterministically) increase in frequency. Because favorable mutants can be lost to stochastic birth-death effects when sufficiently rare, the larger the positive effect of the mutational change upon clonal fitness, the more likely it is to increase when rare.

A concept which underlies much of the modern theory of life history evolution (e.g., the evolutionary theory of senescence) is the notion of the *force of selection*. If we have a measure of fitness F which is a function of parameters q_i, the force of selection on parameter i is $\partial F / \partial q_i$. If $\partial F / \partial q_i > 0$, a clonal variant with slightly higher q_i increases when rare; the larger this quantity is, the more rapidly the mutant spreads, and the less likely it is to be lost due to demographic stochasticity.

In the above matrix model, an appropriate fitness measure is the dominant eigenvalue of A. This measure is a function of all the a_{ij}. Caswell (1989) formalized the notion of force of selection for transition matrices using eigenvalue sensitivity analysis. He showed that if one tweaks only element a_{ij} in a transition matrix, the effect on the dominant eigenvalue is

$$\frac{\partial \lambda}{\partial a_{ij}} = \frac{v_i w_j}{<v, w>} \qquad (2.38)$$

where $<v, w>$ is the inner product of v and w. If mutations arise which slightly alter single matrix elements, the above expression can be used to evaluate the relative strength of selection favoring, or disfavoring, them. Selection should be strongest for transitions from classes that numerically dominate the population (large w_j) into classes with a high reproductive value (large v_i). The quantity $v_i w_j$ in essence describes a demographic "weight" accorded by selection to favor (or disfavor) some transitions, over others.

More generally, mutations may affect multiple transition elements. Assume all the matrix elements are functions of a single parameter, q. The strength of selection favoring mutations increasing q is

$$\frac{\partial \lambda}{\partial q} = \sum_{i,j} \frac{v_i w_j}{\langle v, w \rangle} \frac{\partial a_{ij}}{\partial q}. \tag{2.39}$$

We now make the model a bit more concrete. We will first describe an ecological realization of the above matrix model, then return to the evolutionary question. Imagine that in generation t, the growth rate in habitat i before dispersal is $R_i(t)$. A fraction e of individuals in habitat 1 disperse to habitat 2; a fraction e' of individuals disperse from habitat 2 to 1. The following matrix model describes population growth in the two coupled habitats:

$$\begin{pmatrix} N_1(t+1) \\ N_2(t+1) \end{pmatrix} = \begin{pmatrix} (1-e)R_1 & e'R_2 \\ eR_1 & (1-e')R_2 \end{pmatrix} \begin{pmatrix} N_1(t) \\ N_2(t) \end{pmatrix}. \tag{2.40}$$

Now assume that the external environment is constant and that habitat 2 is a sink with R_2 less than 1. In habitat 1, growth rates are locally density dependent, such that R_1 declines monotonically as a function of density. Denote the density at which $R_1(N_1) = 1$ as K_1 (the carrying capacity of habitat 1) and let the maximal growth rate as N_1 approaches 0 be R_1'.

If the population persists at a stable equilibrium, $\lambda = 1$. After substitution and some algebraic manipulation, we find

$$R_1(N_1^*) = \frac{1 - (1-e')R_2}{(1-e) - (1-e-e')R_2} > 1 > R_2 \tag{2.41}$$

where N_1^* is equilibrium density in habitat 1. Thus, $N_1^* < K_1$; coupling to a sink depresses source density. This increases source growth rates, compensates for decline in the sink, and permits landscape-level equilibrium. The ability of a species to compensate is set by its maximal growth rate in the source, R_1'; if this is less than the left quantity in the above inequality, the species is driven to extinction because dispersal drains away source growth. A sufficient condition for persistence is that $R_1'(1-e) > 1$. We assume source growth rates permit persistence.

At demographic equilibrium, the realized source growth rate is independent of source parameters (e.g., R_1') and depends solely on the rate of decline in the sink, and dispersal. At this equilibrium, the matrix elements have fixed values, and we can apply the machinery of the force of selection to study adaptation. The relative reproductive values and patch abundances of source and sink habitats are as follows: (1) $v_{\text{source}} > v_{\text{sink}}$ if $1 > R_2$ (which is always true for a sink); (2) $w_{\text{source}} > w_{\text{sink}}$ if $1 > (1-e')R_2 + eR_1(N_1^*)$ (which often — but not always — holds).

Now, consider mutations which improve fitness slightly in the sink habitat. The cumulative effect of such mutations as they become fixed may be to transform a sink habitat into a potential source habitat. If so, a species' niche will have evolved: the population can persist in the original sink habitat without immigration. We now have the ingredients needed for predicting the likelihood of such evolution. Holt (1996) describes a number of limiting cases (e.g., involving tradeoffs in fitness in the two habitats). Here we consider two simple examples (the student should work out details as an exercise). Consider mutations whose only effect is a slight increase in sink fitness (with no back-effect on source-fitness parameters).

First, assume $e \approx 0$ (little dispersal, source-to-sink). This implies $R_1 \approx 1$, and $a_{21} \approx 0$. Hence, the stable patch distribution is approximately $(w_1, w_2) = (1, 0)$. After substitution, one finds $\partial\lambda/\partial R_2 \approx 0$. The force of selection for increasing fitness in the sink is negligible, basically because no individuals encounter the sink habitat.

Second, assume $e' \approx 0$ (little back-dispersal, sink-to-source). In this case, $R_1 \approx 1/(1-e)$, and $a_{12} \approx 0$, so the vector of spatial reproductive values is $(v_1, v_2) \approx (1, 0)$. Again one finds that $\partial\lambda/\partial R_2 \approx 0$. Because individuals in the sink make no long-term contribution to the overall population, small improvements in their fitness are of negligible evolutionary importance. Moreover, if mutations arise which have deleterious effects in the sink (but not in the source), selection is weak for removing such mutants from the population. With recurrent mutation, the load of deleterious mutations is likely to be heavier in the sink, than in the source (Kawecki 1995; Holt 1996).

This result meshes with the explicit genetic models for a "black-hole" sink in the last section. The two-patch model should converge on a black-hole-sink model when $e' \to 0$. We showed above that mutations of very small effect on fitness are unlikely to be selected in black-hole sink at demographic equilibrium. Because the expression for the force of selection aims at characterizing the fate of mutations of small effect, the two results are equivalent.

Drawing these examples together, they suggest that niche evolution is less likely when dispersal rates are low, or are asymmetrical, with little dispersal from sink to source. Conversely, niche evolution may be more likely if dispersal rates are high and symmetrical (Kawecki 1995; Holt 1996).

2.6   EVOLUTION OF DISPERSAL AND TEMPORAL HETEROGENEITY

2.6.1   Implications for niche conservatism

This model, explored in the last section, highlighted the importance of dispersal in defining how selection averages over different environments in determining the evolutionary trajectory of a species. There is an enormous literature on the ecology and evolution of dispersal. This is an entire topic on its own, beyond the scope of this chapter. However, the above simple matrix model can be used to illustrate a few basic points about the evolution of dispersal relevant to niche evolution in heterogeneous landscapes.

Consider genetic variants which have the same fitness within patches, but differ in their rates of movement. Evolution in dispersal occurs because different dispersal syndromes define how a given variant experiences environmental heterogeneity in determining its overall fitness. Let habitat 1 have higher fitness than habitat 2. In the two-habitat matrix model, for the moment assume that local fitnesses R_i are fixed, and that dispersal is symmetrical among patches (i.e., $e = e'$). Overall fitness λ is a function of dispersal rate e, $\lambda(e)$; fitness decreases monotonically with e, declining from $\lambda(0) = R_1$ to $\lambda(.5) = (R_1 + R_2)/2$ to $\lambda(1) = \sqrt{R_1 R_2}$.

Now assume the population is initially fixed for a particular dispersal rate, e'', that fitness in the source habitat is density dependent, and that the resident population is at demographic equilibrium (viz., $\lambda(e'') = 1$). An invading clone with a different dispersal rate, when rare, experiences density dependence in the source from the resident. When the invader is rare, the resident's abundance can be assumed fixed during the initial stages of invasion; the invader thus experiences constant habitat-specific growth rates and settles into its stable patch distribution and asymptotic growth rate. Because of the monotonic relationship between overall growth rate and dispersal rates, a rare clone with lower e than the resident always increases when rare; a clone with higher e is excluded. A fuller analysis shows that a polymorphic equilibrium is not feasible, provided one habitat has a fixed fitness less than one, and the system moves toward demographic equilibrium.

This suggests that dispersal rates should evolve toward lower values in spatially heterogeneous (but temporally constant) environments. Given unlimited flexibility in dispersal, the evolutionarily stable state of the system described by our two-patch model is zero dispersal, with all individuals occupying habitat 1, and none in habitat 2. (This conclusion depends on the assumption that abundances are sufficiently large to be treated as continuous variables, rather than discrete integers (Holt 1985; Holt and McPeek 1996).) As dispersal becomes lower, so does the exposure of individuals to the sink habitat. We earlier saw that the force of selection favoring improved adaptation to the sink becomes negligible at low dispersal rates. Thus, evolution of dispersal in a spatially heterogeneous landscape indirectly strengthens the tendency toward niche conservatism.

McPeek and Holt (1992) and Holt and McPeek (1996) have explored the evolution of dispersal in two-habitat models, in which local fitness in each habitat is a monotonically declining function of density, $R_i(N_i)$, and there is some $N_i = K_i > 0$, where $R_i(K_i) = 1$, but $K_1 > K_2$. In this case, no habitat is inevitably a sink. However, given dispersal, if more individuals leave a patch than enter it, densities there decline, leading to an increase in local fitness. Conversely, more individuals must enter than leave the low-K patch, pushing numbers up and depressing local fitness. The net effect of dispersal is to create gradients in local fitness, down which individuals on average tend to move. This is clearly disadvantageous.

In one special circumstance, however, dispersal can occur without this deleterious fitness effect. If $eK_1 = e'K_2$, as many individuals leave as enter each habitat. Thus, each habitat equilibrates at its respective carrying capacity, such that fitnesses are equalized across space (the "ideal free distribution" of habitat selection theory (Fretwell 1972)). McPeek and Holt (1992) showed that this fitness equilibration could be generated in two distinct ways: (1) phenotypic plasticity, in which each individual disperses at different rates in different habitats, or (2) a mixture of fixed dispersal types, one low and one high.

An extreme but illuminating case is for all individuals to leave their natal habitat each generation, but then to resettle into the ideal free distribution. The probability that an individual will end up in habitat i is thus $K_i/(K_1 + K_2)$. Individuals carrying novel mutations with habitat-specific effects on fitness in habitat i are likely to experience this change in fitness a fraction $K_i/(K_1 + K_2)$ of generations. This implies that adaptive evolution is "skewed" toward the habitat with the greater K. For a given allelic change in local fitness, positive selection should be greater in the habitat with higher K, and negative selection weeding out deleterious mutants should likewise be stronger. This may imply that low-K habitats, initially within the niche of the species, might be lost over evolutionary time.

Drawing together the various strands of theory presented above, we see that spatial heterogeneity alone tends to foster niche conservatism. Given limited dispersal, selection tends to be weighted against adaptive improvement in sink habitats, outside a species' niche. This tendency is weakened if dispersal forces individuals to experience the sink habitat. Yet selection acts against dispersal if there is spatial heterogeneity in fitness, which is ensured if some habitats are permanent sinks. Considering the coevolution of dispersal and local adaptation suggests an overall tendency toward increased habitat specialization or niche conservatism.

2.6.2   Temporal variation and niche evolution

All the above models assumed environments in which fitness parameters varied in space, but not in time. Introducing temporal variation raises challenging, unsolved research problems in evolutionary ecology. In this section we will touch on several distinct issues, indicating the range of effects of temporal variation to be expected.

Consider again the haploid black-hole-sink model. Let local absolute fitnesses and immigration rates vary with time, as follows

$$N_1(t+1) = N_1(t)W_1(t) + I(t) \tag{2.42}$$
$$N_2(t+1) = N_2(t)W_2(t) \tag{2.43}$$

where allele 1 is the ancestral immigrant type, and allele 2 is a new mutation. Further, imagine that temporal variation is cyclic, with period T. In the absence of density dependence, the fate of allele 2 is clearly independent of the fitness or rate of immigration of allele 1. Allele 2 increases, provided its geometric mean rate of increase over T generations exceeds 1; if the geometric mean fitness is less than 1, allele 2 will decline toward extinction. Because the geometric mean is always less than the arithmetic mean, temporal variation in fitness makes it harder for an allele to increase when rare.

This simple model suggests that temporal environmental variation tends to hamper niche evolution in sink environments. Extending this to the coupled source-sink environment quickly leads to models which are analytically intractable. There are two sources of difficulty.

First, even in the absence of density-dependence, temporal variation in fitness parameters can confound expectations. Tuljapurkar (1991) [p. 82] provides an interesting example for a 2×2 matrix model. There is cyclic variation between matrix A in one generation, with dominant eigenvalue $\lambda_A < 1$, and matrix B in the next, with dominant eigenvalue $\lambda_B < 1$. Yet the overall eigenvalue for the compound matrix AB, which describes growth over successive generations, exceeds 1. Unraveling the effects of different patterns of temporal variation in fitness parameters on the overall course of selection in a spatially heterogeneous environment is an important, challenging problem.

Second, given density dependence, temporal variation in fitness parameters implies variation in local densities. Even in the simplest models (e.g., logistic growth in two habitats in a cyclic environment), this leads to nonlinear expressions for local density that cannot be solved explicitly; thus the temporal pattern of variation in fitness cannot be expressed analytically.

One indirect consequence of temporal variation which may have profound implications for niche evolution is its effect on dispersal. If local fitness varies through time, dispersal can become advantageous. Theoretical studies have highlighted the importance of asynchronous temporal variation in fitness in promoting the evolution of dispersal in heterogeneous landscapes. Such temporal variation can arise from extrinsic environmental factors or endogenously. For instance, Holt and McPeek (1996) examined a two-habitat model in which local fitnesses were defined by $R_i(N_i) = \exp[r_i(1 - N_i/K_i)]$. As is well known, at low r_i, in the absence of dispersal populations settle into stable equilibria, whereas at high r_i, chaotic dynamics emerges. At low r, dispersal is strongly selected against. At high r, the population exhibits chaotic dynamics; a dispersing clone can increase when rare, and persists in a stable equilibrium with a low-dispersal clone. If individuals have habitat-specific dispersal rates, a chaotic population evolves toward a state with persistent dispersal. Similarly, extrinsic temporal variation in fitness parameters that is uncorrelated among habitats favors persistent dispersal (McPeek and Holt 1992). Temporal variation in local fitness parameters thus favors dispersal. High rates of dispersal in turn tend to weaken forces favoring niche conservatism. Thus, temporal environmental variation may indirectly lead to niche evolution via its influence on the evolution of dispersal.

2.7   NICHE EVOLUTION IN METAPOPULATIONS

So far, we have concentrated on the details of adaptive evolution in very simple landscapes: single habitat patches coupled to an external source, and pairs of habitat patches with reciprocal dispersal. Most species exist in much more complex landscapes, with mosaics of discrete habitat types and smooth gradients. A useful way station between simple one- and two-habitat landscapes and realistic landscapes is provided by metapopulation models which ignore the details of localized dispersal but do capture some important aspects of patchiness. In this final section of the chapter, we explore the interplay of niche evolution and metapopulation dynamics.

There are two canonical metapopulation structures: (1) island-mainland, and (2) multiple identical patches (Hanski 1991). We consider each of these in turn:

2.7.1 Island biogeography

The black-hole-sink model considered a species in a single habitat patch, coupled to a source which did not experience reciprocal effects. Now imagine there are many such habitats, which like islands in a sea may vary in size and distance from the source.

Rather than consider the detailed population and evolutionary dynamics of the species in each patch, we can attempt to abstract the essence of the microevolutionary and population dynamic processes as follows (where, for simplicity, we refer to "islands" rather than patches):

1. All immigration is from the mainland, which is assumed to contain a species at its evolutionary equilibrium. Colonization is defined as immigration onto unoccupied islands; the rate parameter c_m defines the rate of colonization per empty island.

2. All colonizing propagules are initially maladapted to the local environment and therefore will inevitably go extinct, in the absence of evolution (see Section 2.2). We assume that such extinctions are described by an exponential distribution with rate parameter e_m. All the rate parameters may vary with island area or distance (see below).

3. Given appropriate genetic variation, a local population may evolve so as to increase its mean fitness or carrying capacity, which enhances population persistence. Catastrophes can still occur, however, leading to local extinctions even for well-adapted populations. We assume that such extinctions occur at rate $e_a < e_m$.

4. The final ingredient we need is the rate of evolution. We assume that local populations exist in just two states: "maladapted" (their original state, just after colonization) and "adapted" (their final state, conditional on local persistence, after selection has pushed the population to its new, local optimum). For simplicity, we ignore intermediate states and assume that an exponential distribution with rate parameter E describes evolutionary transitions from maladapted to adapted states.

Any given island can occur in three states: empty; occupied but maladapted; occupied and adapted. Let P_m be the fraction of islands in which populations are maladapted, and P_a be the fraction in which local adaptation has occurred. The fraction of empty islands available for colonization is $1 - P_m - P_a$. The dynamics of the system are described by a coupled pair of differential equations:

$$\frac{dP_m}{dt} = c_m(1 - P_a - P_m) - EP_m - e_m P_m \qquad (2.44)$$

$$\frac{dP_a}{dt} = EP_m - e_a P_a \qquad (2.45)$$

At equilibrium,

$$P_a^* = \frac{E}{e_a}P_m^*, \qquad \frac{P_a^*}{P_a^* + P_m^*} = \frac{E}{E + e_a} \equiv \epsilon \qquad (2.46)$$

The quantity ϵ describes the fraction of island populations which have become adapted to their environments. Because we assume no interisland migration, each such population is a distinct taxon, so ϵ also measures the fraction of island populations which might be viewed as endemic by a taxonomist. By writing down the above set of equations, we implicitly assume that the number of islands is sufficiently great that ensemble dynamics can be treated deterministically.

Island area, and distance to the source, can influence both the ecological rate parameters of colonization and extinction, and the rate of evolution from a maladapted to adapted state. Somewhat surprisingly, the average equilibrial evolutionary state does not depend upon the rate of colonization from the source, or the rate of extinction of maladapted populations, but only upon the rate of local evolution and extinctions of adapted populations.

Consider first purely ecological effects upon extinction. Owing to demographic stochasticity, adapted populations on large islands are likely to persist longer, per population, than adapted populations on small islands. If dispersal from the source is rare, it is not clear that distance should have any systematic effect upon extinction. However, there may be indirect effects of distance on extinction rates, such as fewer competing or predatory species on more distant islands, leading to reduced extinctions at greater island distance from the source. Considering just these ecological effects, one expects e_a to increase with island area, and possibly to increase as well with island distance from the source (see Holt (1997) for examples and further discussion).

Island area and distance could also indirectly determine the rate of microevolution, E. For instance, if evolution is limited by the pool of variation, then all else being equal, larger islands will harbor larger populations, which can generate more variation via mutation and sustain such variation in the face of genetic drift. The rate of evolution could thus increase with island area, augmenting the ecological effect of island area on endemism.

2.7.2 "Proper" metapopulation

In the long run, even source populations can go extinct, or evolve. Consider a species which occupies an ensemble of habitat patches, where each occupied patch is a potential source of colonists for empty patches. Let P be the fraction of patches occupied. The canonical metapopulation model (Hanski 1991) is

$$\frac{dP}{dt} = cP(1 - P) - eP \qquad \Rightarrow \qquad P^* = 1 - \frac{e}{c} \qquad (2.47)$$

where e is the rate of local extinction, per patch, c describes the rate of colonization of empty patches, per occupied patch, and P^* is patch occupancy at equilibrium.

Now, as before, assume that local populations can be either adapted or maladapted, and that their evolutionary state is made manifest in local extinction rates. One can imagine various scenarios regarding colonization (Holt and Gomulkiewicz, in prep.). The simplest is to assume that only empty patches are colonized, and that only adapted populations are sufficiently vigorous to send out colonists. Assuming again that the rate of local evolution is characterized by a constant transition-rate parameter, E, the metapopulation model is

$$\frac{dP_m}{dt} = -e_m P_m - E P_m \tag{2.48}$$

$$\frac{dP_a}{dt} = c_a(1 - P_a - P_m)P_a + E P_m - e_a P_a. \tag{2.49}$$

There are two possible equilibria: (1) global extinction, if $e_a > c_a$; (2) $P_m^* = 0$, and $P_a^* = 1 - e_a/c_a$ if $e_a < c_a$. The rate of evolution, E, is thus irrelevant to the long-term state of the system (in sharp contrast to the above island model). However, evolution can matter crucially in determining transient dynamics in the metapopulation. Consider a system in which there has been an abrupt change in climate, such that all initial populations are maladapted. Numerical integration of the above model reveals that a species may display an excursion to very low values of occupancy $P_a + P_m$ before the population increases to its eventual equilibrium. As noted above, a key assumption in patch occupancy models is that the number of patches in question is very large, so that deterministic approximations of stochastic processes are sensible. In real metapopulations, when occupancy gets very low, there is likely to be a small number of actual populations in question, and a metapopulation equivalent of demographic stochasticity can arise (Hanski et al. 1996). Analogous to our initial model of adaptation in a changed environment in a single patch, a metapopulation may suffer extinction due to chance events as it passes through a phase of transient maladaptation in a novel environment.

The above metapopulation models, of course, provide a caricature of population and evolutionary dynamics. Future work will address the adequacy of some of their key assumptions (e.g., constant evolutionary rates). Moreover, an obvious next step will be to assume localized dispersal, and to examine evolutionary dynamics in environments with different patterns of heterogeneity. Environments are heterogeneous in many ways, ranging from gentle spatial gradients, fixed in time, to ephemeral habitat patches winking on and off during succession, to landscapes with complex fractal spatial structure. The "texture" of the selective environment is likely to be a key determinant of whether or not a species exhibits niche conservatism, or evolution.

The basic message of this chapter is that population dynamics acts as a kind of constraint within which evolutionary dynamics occur, and that an understanding of niche conservatism and evolution, in particular, mandates analyzing evolutionary processes within an appropriate landscape and population dynamical context.

REFERENCES

Bradshaw, A. D. 1991. The Croonian Lecture 1991: Genostasis and the limits to evolution. *Philosophical Transactions of the Royal Society of London Series B, Biological sciences*, **333**, 289–305. {*25, 48*}

Brown, J. S. and Pavlovic, N. B. 1992. Evolution in heterogeneous environments: Effect of migration on habitat specialization. *Evolutionary Ecology*, **6**, 360–382. {*26, 48*}

Burger, R. and Lynch, M. 1995. Evolution and extinction in a changing environment: a quantitative-genetic analysis. *Evolution*, **49**, 151–163. {*26, 48*}

Burt, A. 1995. Perspective: The evolution of fitness. *Evolution*, **49**, 1–8. {*27, 48*}

Caswell, H. 1989. *Matrix Population Models: Construction, Analysis, and Interpretation*. Sunderland, MA: Sinauer Associates, Inc. ISBN 0-87893-094-9 (hardcover), 0-87893-093-0 (paperback). Pages xiv + 328. {*39, 40, 49*}

Crow, J. F. and Kimura, M. 1970. *Introduction to Population Genetics Theory*. Minneapolis, MN: Burgess Publishing Company. ISBN 0-06-356128-X (Harper and Row paperback, 1972 reprint), 0-8087-2910-2. Pages xiv + 591. {*33, 49*}

Falconer, D. S. 1989. *Introduction to Quantitative Genetics*. Third edn. New York, NY: John Wiley and Sons. ISBN 0-470-21162-8 (Wiley), 0-582-01642-8 (pbk). Pages xii + 438. {*27, 28, 49*}

Fisher, R. A. 1958. *The Genetical Theory of Natural Selection*. New York, NY: Dover Publications, Inc. Page 291. {*27, 49*}

Fretwell, S. D. 1972. *Populations in a Seasonal Environment*. Princeton, NJ: Princeton University Press. ISBN 0-691-08105-0, 0-691-08106-9 (paperback). Pages xxiii + 217. {*43, 49*}

Gomulkiewicz, R. and Holt, R. D. 1995. When does evolution by natural selection prevent extinction? *Evolution*, **49**, 201–207. {*30, 31, 49*}

Hanski, I. 1991. Single-species metapopulation dynamics: Concepts, models, and observations. *Pages 89–103 of:* Gilpin, Michael E. and Hanski, Ilkka (eds), *Metapopulation Dynamics: Empirical and Theoretical Investigations*. Biological journal of the Linnean Society, vol. 42(1–2). New York, NY: Academic Press. {*45, 47, 49*}

Hanski, I., Moilanen, A., and Gyllenberg, M. 1996. Minimum viable population size. *American Naturalist*, **147**, 527–541. {*48, 49*}

Holt, R. D. 1985. Population dynamics in two-patch environments; some anomalous consequences of an optimal habitat distribution. *Theoretical Population Biology*, **28**, 181–208. {*43, 49*}

Holt, R. D. 1996. Demographic constraints in evolution: Towards unifying the evolutionary theories on senescence and niche conservatism. *Evolutionary Ecology*, **10**, 1–11. {*39, 42, 49*}

Holt, R. D. 1997. Rarity and evolution; some theoretical considerations. *Pages 209–234 of:* Kunin, W. and Gaston, K. (eds), *The Biology of Rarity*. London, UK: Chapman and Hall, Ltd. In press. {*47, 49*}

Holt, R. D. and Gaines, M. S. 1992. Analysis of adaptation in heterogeneous landscapes: Implications for the evolution of fundamental niches. *Evolutionary Ecology*, **7**, 433–447. {*25, 49*}

Holt, R. D. and Gomulkiewicz, R. 1996. How does immigration influence local adaptation? A re-examination of a familiar paradigm. *American Naturalist*. In press. {*35, 38, 49*}

Holt, R. D. and McPeek, M. A. 1996. Chaotic population dynamics favors dispersal. *American Naturalist*. In press. {*43, 45, 49*}

Karlin, S. and Taylor, H. M. 1975. *A First Course in Stochastic Processes*. Second edn. New York, NY: Academic Press. ISBN 0-12-398552-8. Pages xvi + 557. {*32, 34, 49*}

Kawecki, T. J. 1995. Demography of source-sink populations and the evolution of ecological niches. *Evolutionary Ecology*, **7**, 155–174. {*26, 42, 49*}

Lande, R. 1976. Natural selection and random genetic drift in phenotypic evolution. *Evolution*, **30**, 314–334. {*28, 49*}

Lenski, R. E. and Bennett, A. F. 1993. Evolutionary response of *Escherichia coli* to thermal stress. *American Naturalist*, **142**, S47–S64. {*32, 49*}

Lin, C. C. and Segel, L. A. 1988. *Mathematics Applied to Deterministic Problems in the Natural Sciences*. Philadelphia, PA: SIAM Press. ISBN 0-89871-229-7. Pages xxi + 609. {*32, 49*}

Lynch, M. and Lande, R. 1993. Evolution and extinction in response to environmental change. *Pages 234–250 of:* Kareiva, P. M., Kingsolver, J. G., and Huey, R. B.

(eds), *Biotic Interactions and Global Climate Change*. Sunderland, MA: Sinauer Associates, Inc. {*26, 49*}

MacArthur, R. H. and Wilson, E. O. 1967. *The Theory of Island Biogeography*. Monographs in Population Biology, vol. 1. Princeton, NJ: Princeton University Press. Pages xi + 203. {*27, 50*}

McPeek, M. A. and Holt, R. D. 1992. The evolution of dispersal in spatially and temporally varying environments. *American Naturalist*, **140**, 1010–1027. {*43, 45, 50*}

Nagylaki, T. 1977. *Selection in One- and Two-locus Systems*. Lecture notes in biomathematics, vol. 15. Berlin: Springer Verlag. ISBN 0-387-08247-6. Pages vii + 208. {*35, 50*}

Nagylaki, T. 1992. *Introduction to Theoretical Population Genetics*. Biomathematics, vol. 21. Berlin: Springer Verlag. ISBN 3-540-53344-3 (Berlin), 0-387-53344-3 (New York). Pages xi + 369. {*36, 50*}

Pease, C. M., Lande, R., and Bull, J. J. 1989. A model of population growth, dispersal, and evolution in a changing environment. *Ecology*, **70**, 1657–1664. {*26, 50*}

Renshaw, E. 1991. *Modelling Biological Populations in Space and Time*. Cambridge, UK: Cambridge University Press. ISBN 0-521-30388-5 (hardback), 0-521-44855-7 (paperback). Pages xvii + 403. {*25, 27, 50*}

Tuljapurkar, S. 1991. *Population Dynamics in Variable Environments*. Lecture Notes in Biomathematics, vol. 85. Berlin: Springer Verlag. ISBN 0-387-52482-7. Page 154. {*44, 50*}

3. REFLECTIONS ON MODELS OF EPIDEMICS TRIGGERED BY THE CASE OF PHOCINE DISTEMPER VIRUS AMONG SEALS

Odo Diekmann

3.1 INTRODUCTION

In the spring and summer of 1988 colonies of harbour (*Phoca vitulina*) and grey (*Halichoerus gryphus*) seals in the coastal waters of Northern Europe were struck by an infectious disease that caused the death of about half of the population. A little later a *morbilli* virus was identified as the cause of the disease (Osterhaus and Vedder 1988). It was baptized Phocine Distemper Virus (PDV).

It is thought that the virus is transmitted via aerosols when seals come out of the water to rest or sunbathe on sandbanks at low tide, roughly twice every 24 hours. The gathering at sandbanks generates a contact process, and transmission of the infective agent is a superimposed event. Consequently, as in many other contexts, the modeling of the spread of the infectious disease requires that we model the underlying contact process.

The seals of Northern Europe form a metapopulation, a collection of many local subpopulations or colonies, loosely coupled by incidental migrations (Gilpin and Hanski 1991; Reijnders et al. 1981; Reijnders et al. 1996). Not much is known for sure about the exchange between colonies, and, as we shall see below, the spread of PDV raises questions about the frequency of migrations and/or visits. The colonies differ widely in size, mainly as a result of differences in local habitat quality. Almost all colonies seem to have been infected with PDV, and the fraction of individuals in a colony that died from the disease is roughly the same for all colonies (Heide-Jorgensen and Harkonen 1992).

In this chapter we want to investigate what mathematical modeling can teach us about a case like this. It will turn out that we have to reflect upon the assumptions underlying traditional equations and develop some new variants. Rather than solving all the problems, these insights will help us to think about new questions.

3.2   THE FINAL-SIZE EQUATION

Seals are born and seals die. The characteristic time scale of this demographic turnover is of the order of years. The time scale at which PDV swept through the seal population is of the order of weeks. When considering the spread of PDV, it makes sense to ignore demographic turnover.

Consider a colony without births and deaths and assume that PDV is introduced in some way. What fraction $s(\infty)$ will escape from ever being infected and what fraction $1 - s(\infty)$ will fall victim? A first dichotomy is whether $s(\infty) < 1$ or $s(\infty) = 1$. Let the symbol R_0 denote the **basic reproduction ratio** associated with the disease, defined as the expected number of secondary cases per primary case in the initial phase of the epidemic, when we do not yet have to take into account that the "resource" of susceptibles is being "consumed" by the virus. Roughly speaking, R_0 is the product of three factors: (i) contact intensity, (ii) probability of transmission given a contact between an infective and a susceptible, (iii) the length of the infectious period. The definition of R_0 implies that it has a threshold value 1 (think of growth on a generation basis). In particular, we will have $s(\infty) = 1$ for $R_0 < 1$ and $s(\infty) < 1$ for $R_0 > 1$. But can we actually determine $s(\infty)$ when $R_0 > 1$?

The traditional answer is that $s(\infty)$ can be found as the unique root in $(0, 1)$ of the equation

$$s(\infty) = e^{R_0(s(\infty)-1)}. \tag{3.1}$$

A derivation is presented in the next section. A stochastic fine-tuning of this answer asserts that, even when $R_0 > 1$, there exists a positive probability that the virus goes extinct after a minor outbreak affecting only a negligible fraction of the population, whereas major outbreaks show some variation around the mean value $s(\infty)$.

We claim that for the present case $s(\infty)$ has to be found as the unique root in $(0, 1)$ of the alternative equation

$$s(\infty) = (s(\infty) + f(1 - s(\infty)))^{R_0/(1-f)} \tag{3.2}$$

where f is the probability that an infected individual survives the disease. Underlying the distinction between (3.1) and (3.2) are assumptions concerning the contact process. In case of (3.1), it is assumed that the per capita number of contacts per unit of time is proportional to colony size, and hence this number diminishes as the colony becomes smaller as a result of disease-induced mortality. In case of (3.2), it is assumed that the per capita number of contacts per unit of time is a constant, independent of colony size. The idea is that upon coming out of the water the seals arrange themselves with typical nearest-neighbor distances and thus occupy an area that is proportional to colony size. Clearly this cannot be true when space is a limiting factor. It is thus a crucial empirical issue to determine whether space is indeed limiting.

Hidden in the notation is another distinction between these two cases. When contact intensity is proportional to colony size, R_0 is also. As a consequence, there exists a critical colony size below which the virus cannot spread. In contrast R_0 is independent of colony size when the effective density, and hence contact intensity, is independent of colony size.

Before going into the derivation of (3.1) and (3.2), we analyze the consequences of the difference. We compute, for a fixed value of R_0 but as a function of f, the fraction $(1 - f)(1 - s(\infty))$ that die from the disease, with $s(\infty)$ either defined by (3.1) or by (3.2). The result is depicted in Figure 3.1. The

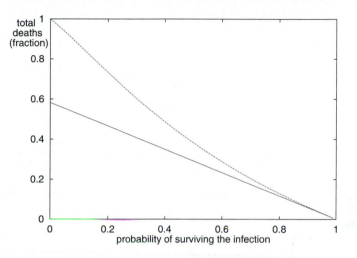

Figure 3.1. The fraction $(1 - f)(1 - s(\infty))$ of the total population that dies as a function of the probability f that an individual survives the infection with $R_0 = 1.5$. The solid straight line corresponds to (3.1) and the dashed curved line to (3.2). With (3.2), the density remains constant as animals die, and all animals are infected and die when $f = 0$. With (3.1), the density decreases as animals die, and some animals always survive the epidemic.

difference results from the fact that immune individuals absorb some of the contacts of infectives, and thus hinder contacts between infectives and susceptibles, unlike dead individuals. When R_0 is large, $s(\infty)$ is very small ($\sim e^{-R_0}$) and this effect is negligible on an absolute scale (but it does matter on a relative scale). Figure 3.1 shows that the effect can be substantial when R_0 exceeds the threshold value only slightly and the survival probability f is small. In the case of seals infected by Phocine Distemper Virus the value of f may very well depend on their general physiological condition, which, in turn, is determined by environmental conditions like pollution. Parameter estimates for various regions (de Koeijer et al. 1996) support the idea that pollution may lead to a lower value of f and, thereby, to an intensification of the epidemic.

3.3 DERIVATION OF THE FINAL SIZE EQUATION

Given a contact between a susceptible individual and an infective individual, let p denote the probability that the virus is successfully transmitted. There are several difficulties with this "definition." First, it is not clear what "contact" really means. In the present setting we shall interpret it as "coming sufficiently close for aerosol-mediated transmission to be possible," realizing that this is still rather vague. Second, p will not be a constant, but will depend on the time τ elapsed since the infective was infected itself. Third, there may be individual variation, but then we simply take the average as the ingredient for our deterministic description. Figure 3.2 presents an example of how p typically depends on disease age τ.

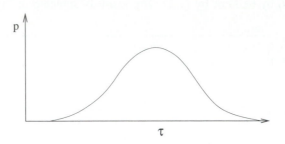

Figure 3.2. A typical dependence of infectivity p on time since infection τ, increasing from zero at the time of infection, peaking at the time of maximum viral load, and decreasing during recovery.

Let $B(\tau)$ denote the probability an individual is alive at time τ after infection. Then $B(0) = 1$ and $B(\infty) = f$. We define $A(\tau) = p(\tau)B(\tau)$. $A(\tau)$ combines the probability of taking part in the contact process with the probability of transmitting the virus, given a contact with a susceptible.

Let $C(t)$ denote the expected number of contacts per unit time of an arbitrary individual at time t. Let $i(t)$ denote the incidence (i.e., the number of new cases per unit of time) at time t. Then our assumptions directly imply that

$$i(t) = C(t)\frac{S(t)}{N(t)} \int_0^\infty A(\tau)i(t-\tau)\,d\tau \tag{3.3}$$

where $S(t)$ denotes the number of susceptibles and $N(t)$ the total number of individuals (out of the $C(t)$ contacts/time of an infective, a fraction $S(t)/N(t)$ will be with susceptibles). The two variants differ in the assumptions concerning $C(t)$. If space is a limiting factor the density will be proportional to $N(t)$ and as $C(t)$ is proportional to density, we end up with the conclusion that $C(t)$ is proportional to $N(t)$. If space is not a limiting factor, $C(t)$ is a constant, independent of t.

To proceed with the derivation we observe that, because there is no demographic turnover, $i(t) = -\dot{S}(t)$. Both give the number of infections per unit time, one as the entry rate into the infected category and the other as the exit rate from the susceptible category. When $C(t) = \alpha N(t)$, (3.3) can be rewritten as

$$\dot{S}(t) = \alpha S(t) \int_0^\infty A(\tau)\dot{S}(t-\tau)\,d\tau. \tag{3.4}$$

Dividing by $S(t)$ and integrating from $-\infty$ to $+\infty$, we find

$$\ln \frac{S(\infty)}{S(-\infty)} = \alpha \int_0^\infty A(\tau)\,d\tau\,(S(\infty) - S(-\infty)), \tag{3.5}$$

which, with the identifications

$$s(\infty) = \frac{S(\infty)}{S(-\infty)} \tag{3.6}$$

and

$$R_0 = \alpha \int_0^\infty A(\tau)\, d\tau\, S(-\infty), \tag{3.7}$$

is just another way of writing (3.1).

As a derivation of this expression for R_0, consider the following argument. Far into the past everybody was susceptible, so $S(-\infty) = N(-\infty)$ and $C(-\infty) = \alpha N(-\infty) = \alpha S(-\infty)$. Consider a newly infected individual during the initial phase of the epidemic, far in the past. After τ units of time, it will be alive with probability $B(\tau)$. If alive, it makes contacts with rate $C(-\infty)$. As these contacts must be with susceptibles, a contact leads to transmission with probability $p(\tau)$. Hence the probability per unit time to generate a secondary case is equal to $\alpha B(\tau) S(-\infty) p(\tau)$. Integration over all τ gives R_0.

In the case just discussed it does not matter (for contacts between susceptibles and infectives) whether individuals become immune at the end of the infectious period and live happily on, or instead die as a result of the disease. But when the contact intensity is a fixed constant, it does matter. To deal with that situation we therefore need B as a separate model ingredient in order to express $N(t)$ in terms of S and $\dot{S}$:

$$N(t) = S(t) - \int_0^\infty B(\tau)\dot{S}(t - \tau)\, d\tau. \tag{3.8}$$

This equation expresses the fact that there are two categories of living individuals: (i) susceptibles and (ii) those infected sometime in the past that survive until the present. (If you are puzzled by the minus sign, recall that $\dot{S}(t) < 0$.) Our task is now to derive the final-size equation (3.2) from (3.8) and

$$\dot{S}(t) = C \frac{S(t)}{N(t)} \int_0^\infty A(\tau)\dot{S}(t - \tau)\, d\tau. \tag{3.9}$$

This will not work unless there is a special relationship between A and B. It is an **open problem** to derive a characterization of the final size from (3.8) and (3.9) for general A and B.

We first explain how one arrives at the appropriate special relationship by formal mathematical manipulation and then explain that it allows for a clear biological interpretation. When we divide (3.9) by $S(t)$ and integrate, we need an integral of

$$\int_0^\infty \frac{A(\tau)\dot{S}(t - \tau)}{N(t)}\, d\tau. \tag{3.10}$$

This is possible when the numerator is a multiple of $\dot{N}(t)$. If we write (3.8) as

$$N(t) = S(t) - \int_{-\infty}^t B(t - \tau)\dot{S}(\tau)\, d\tau \tag{3.11}$$

and differentiate, we find that

$$\dot{N}(t) = - \int_{-\infty}^t \dot{B}(t - \tau)\dot{S}(\tau)\, d\tau \tag{3.12}$$

(note that the definition of B requires that $B(0) = 1$). If $\dot{B}(\tau) = -\theta A(\tau)$ for some θ, we are in business.

Recall that $A(\tau) = p(\tau)B(\tau)$, where p is expected infectivity at disease-age τ. Therefore $\dot{B}(\tau) = -\theta p(\tau)B(\tau)$ means that the hazard rate for death is proportional to infectivity (which we can interpret as the rate at which copies of the infective agent are produced). This assumption seems warranted when replication of the infective agent in organs from which it is excreted to the outside world is responsible for the morbidity. For some agents excretion and disease may relate to replication in different organs and then the assumption is less defensible. In these cases, the open problem stated above is relevant.

Let us assume that

$$\dot{B}(\tau) = -\theta A(\tau). \tag{3.13}$$

With the notation

$$
\begin{aligned}
f &= B(\infty) \\
R_0 &= C \int_0^\infty A(\tau)\,d\tau = -\frac{C}{\theta} \int_0^\infty \dot{B}(\tau)\,d\tau = C\frac{1-f}{\theta} \\
n(\infty) &= N(\infty)/N(-\infty) \\
s(\infty) &= S(\infty)/N(-\infty),
\end{aligned}
$$

we can now write the result of the integration of (3.9) as

$$s(\infty) = n(\infty)^{R_0/1-f}. \tag{3.14}$$

We rewrite (3.8) as

$$N(t) - S(t) = -\int_{-\infty}^t B(t-\sigma)\dot{S}(\sigma)\,d\sigma \tag{3.15}$$

and take the limit $t \to \infty$ to obtain

$$
\begin{aligned}
N(\infty) - S(\infty) &= -f \int_{-\infty}^\infty \dot{S}(\sigma)\,d\sigma \\
&= -f(S(-\infty) - S(\infty)). \tag{3.16}
\end{aligned}
$$

Dividing by $N(-\infty) = S(-\infty)$, we find

$$n(\infty) - s(\infty) = f(1 - s(\infty)). \tag{3.17}$$

This identity repeats the definition of f: out of those who contracted the infection, a fraction f survived. We can use (3.17) to express $n(\infty)$ in terms of $s(\infty)$ and f. Substitution of the result in (3.14) yields (3.2).

When the disease does not cause deaths, the two models are indistinguishable. And indeed, (3.2) converges to (3.1) as f increases to 1. To prove this is a nice calculus exercise involving the characterization of the exponential function as the limit of products

$$e^x = \lim_{\theta \to \infty} \left(1 + \frac{x}{\theta}\right)^\theta. \tag{3.18}$$

3.4 REMARKS AND OBSERVATIONS

We can rewrite (3.9) and (3.13) in the form

$$\begin{cases} f = \dfrac{n(\infty) - s(\infty)}{1 - s(\infty)} \\ R_0 = (1 - f)\dfrac{\ln s(\infty)}{\ln n(\infty)} \end{cases} \qquad (3.19)$$

This is especially useful when we want to estimate the unknown parameters f and R_0 from field estimates of $n(\infty)$ and $s(\infty)$ (de Koeijer et al. 1996).

As we noted before, there is a positive probability that the virus goes extinct before a major epidemic develops. One can use a branching process approximation for the initial phase to calculate this probability. Even though the outcome depends on details of the submodel for infectivity $p(\tau)$, we think that the estimates for R_0 and $s(\infty)$ suggest that this probability is not negligibly small, perhaps even as large as $1/3$. The fact that, except for remote colonies, all colonies were actually suffering from a major outbreak, strongly suggests that the virus was introduced repeatedly in every colony from the outside. In other words, contacts between individuals of neighboring colonies cannot be very rare.

The assumption of constant effective density leads to the conclusion that the final size, expressed as a fraction, is the same for all colonies. This fits in with observations (Heide-Jorgensen and Harkonen 1992; de Koeijer et al. 1996). It implies that the metapopulation structure has had no influence on the final size at the scale of the entire population.

The data do not show a clear traveling wave along the coast. The virus seems to have made several big jumps, presumably by incidental long-distance migration. To model the spread at the metapopulation level, the approach by Ball et al. (1996) seems most appropriate.

3.5 PERSISTENCE

Why did the virus go extinct? In this case, it seems likely that it ran out of susceptible hosts after it went through the major outbreak. In general, we can distinguish several phases in which infective agents can go extinct. The theory of branching processes tells us how to compute the probability that, even though $R_0 > 1$, introduction of the agent leads only to a minor outbreak. At the end of the first outbreak there is a relatively high probability of going extinct, owing to depletion of susceptibles. I know only one reference (van Herwaarden 1996) in which such a probability is computed. When the agent survives, it may cause a less pronounced second outbreak after newborn susceptibles have accumulated, leading once more to an increased risk of extinction. From a stochastic point of view, an endemic state can only be a quasi-stationary state, with an associated probability per unit of time of becoming absorbed in the state of extinction. Hence the probability density function for the extinction time has an exponential tail.

What is the likelihood of extinction after the first outbreak? How does that depend on the time-scale differences between demographic turnover and disease transmission? How is population size involved? Is there a critical community/colony size for persistence? Does a metapopulation structure, allowing for phase differences between outbreaks, promote persistence?

Such questions have been addressed within the context of specific modeling frameworks (Mollison 1995; Grenfell and Dobson 1995; van Herwaarden and Grasman 1995; Diekmann et al. 1995; Diekmann et al. 1996) but, in my opinion, a general understanding has yet to emerge. These questions therefore deserve attention.

An important point that is usually ignored in analytical formulations of models (and sometimes even in simulation models) is that seals (and many other organisms) don't reproduce year long, but instead only during a rather short period. Thus one is led to consider hybrid models, combining continuous time features (transmission) with discrete time features (reproduction). A preliminary study concerning seals and PDV was made by Korthals Altes (1994). The concentrated character of reproduction may make virus persistence even less likely. The class of hybrid population models is certainly of ecological and epidemiological interest and deserves much more attention.

3.6 CONCLUSION

The spread of PDV among seals stimulates us to think about the contact process that serves as the transmission vehicle for the virus. In general, modeling ecological processes in which more than one individual is involved is far more difficult than the modeling of those processes that involve only one individual (often scale differences help to make the former look like the latter, but in the epidemic context that does not work). We argued that when there are no limitations of space, the final-size equation of epidemic theory needs to be modified. The modification makes clear that no critical colony size exists and that the severity of the outbreak, expressed as a fraction, does not depend on colony size either. These findings fit in very well with field observations that may seem puzzling when standard theory, which is concerned with population **densities**, is applied to **numbers** of individuals.

A major epidemic outbreak is conveniently described deterministically. But by its very dynamics it will bring the population to a state that requires a stochastic description again, to deal with the issue of persistence as opposed to extinction. To the best of my knowledge there does not yet exist a systematic method to treat this switch from a deterministic to a stochastic description. Such a method would greatly facilitate development of further understanding of the population dynamics of infective agents.

REFERENCES

Ball, F., Mollison, D., and Scalia-Tomba, G. 1996. *Epidemics with two levels of mixing*. Submitted. {*57, 58*}

de Koeijer, A. A., Diekmann, O., and Reijnders, P. 1996. *Modelling the spread of Phocine Distemper Virus (PDV) among harbor seals*. Submitted. {*53, 57, 58*}

Diekmann, O., de Jong, M. C. M., de Koeijer, A. A., and Reijnders, P. 1995. The force of infection in populations of varying size: A modelling problem. *Journal of Biological Systems*, **3**, 519–529. {*58*}

Diekmann, O., de Koeijer, A. A., and Metz, J. A. J. 1996. On the final size of epidemics within herds. *Canadian Applied Mathematics Quarterly*, **4**(1), 21–30. {*58, 59*}

Gilpin, M. and Hanski, I. (eds). 1991. *Metapopulation Dynamics: Empirical and Theoretical Investigations*. Biological Journal of the Linnean Society, vol. 42. New York, NY: Academic Press. ISBN 0-12-284120-4. Page 336. {*51, 59*}

Grenfell, B. T. and Dobson, A. (eds). 1995. *Ecology of Infectious Diseases in Natural Populations*. Cambridge, UK: Cambridge University Press. ISBN 0-521-46502-8. Pages xii + 521. {*58, 59*}

Heide-Jorgensen, M. P. and Harkonen, T. 1992. Epizootiology of the seal disease in the eastern North Sea. *Journal of Applied Ecology*, **29**, 99–107. {*51, 57, 59*}

Korthals Altes, H. 1994. *Effects of PDV on the population dynamics of seals*. M.Phil. thesis, Leiden University, Leiden, The Netherlands. {*58, 59*}

Mollison, D. (ed). 1995. *Epidemic Models: Their Structure and Relation to Data*. Cambridge, UK: Cambridge University Press. ISBN 0-521-47536-8. Pages xvii + 424. {*58, 59*}

Osterhaus, A. D. M. E. and Vedder, E. J. 1988. Identification of virus causing recent seal deaths. *Nature*, **338**, 20. {*51, 59*}

Reijnders, P. J. H., Drescher, H. E., van Haaften, J. L., Hansen, E. Bogebjerg, and Tougaard, S. 1981. Population dynamics of the harbour seal in the Wadden Sea. *Pages 19–32 of:* Reijnders, P. J. H. and Wolff, W. J. (eds), *Marine Mammals of the Wadden Sea: final report of the section 'marine mammals' of the Wadden Sea Working Group*. Rotterdam: Balkema. {*51, 59*}

Reijnders, P. J. H., Brasseur, S. M. J. M., and Ries, E. H. 1996. The release of seals from captive breeding and rehabilitation programmes: a useful conservation management tool? *Pages 54–65 of:* Aubin, David J. St., Geraci, Joseph R., and Lounsbury, Valerie J. (eds), *Rescue, Rehabilitation and Release of Marine Mammals: An Analysis of Current Views and Practices: Proceedings of a workshop held in Des Plaines, Illinois, 3–5 December 1991*. NOAA Technical Memorandum NMPS-OPR-8, vol. NMFS-OPR 8. Washington, DC: NOAA. {*51, 59*}

van Herwaarden, O. A. 1996. Stochastic epidemics: The probability of extinction of an infectious disease at the end of a major outbreak. *Journal of Mathematical Biology*. In press. {*57, 59*}

van Herwaarden, O. A. and Grasman, J. 1995. Stochastic epidemics: Major outbreaks and the duration of the epidemic. *Journal of Mathematical Biology*, **33**, 581–601. {*58, 59*}

4. SIMPLE REPRESENTATIONS OF BIOMASS DYNAMICS IN STRUCTURED POPULATIONS

R. M. Nisbet, E. McCauley, W. S. C. Gurney, W. W. Murdoch, and A. M. de Roos

4.1 INTRODUCTION

The research described in this chapter represents part of a larger program whose aim is to provide the knowledge and tools to better understand the dynamics of natural and managed ecosystems. More specifically, we aim to produce mathematical models that can translate the effects of environmental stress on individual aquatic organisms to the dynamics of populations. One very practical concern motivates this work: while environmental management demands understanding of long-term effects of stress on populations of plants and animals, much experimental information relates only to short-term effects on individuals. We are addressing this concern by establishing how to develop testable, individual-based models (DeAngelis and Gross 1992) capable of predicting population responses to environmental change. For example, the physiological response of many animals to certain forms of environmental stress (e.g., eutrophication, toxicants, lake acidification) involves changes in the rates of assimilation and utilization of food. Our overall aim is to predict the consequences for the dynamics of natural populations of these changes in individual energy acquisition and use.

The system for which these questions will be explored is the zooplankton *Daphnia* and its algal food supply. This system is appropriate for many reasons. *Daphnia* is an important genus in many natural ponds and lakes, and previous studies (Murdoch and McCauley 1985; McCauley and Murdoch 1987; McCauley et al. 1988) have identified common dynamic patterns across many populations; thus any understanding gained from population models is likely to be ecologically important. The physiology of individuals of the commoner *Daphnia* species is well documented (McCauley et al. 1990; McCauley and Murdoch 1990; Glazier and Calow 1992). Several investigators (Slobodkin 1954; Marshall 1978; Frank 1960; Goulden et al. 1982a) studied food-limited laboratory populations of *Daphnia*. *Daphnia* populations have been studied in stock tanks (McCauley et al. 1988; McCauley and Murdoch 1990) and in enclosures in lakes (Lynch 1979; Leibold 1989). There are many time series obtained by sampling natural *Daphnia* populations (McCauley and Murdoch

(1987) and references therein), and *Daphnia* are an important component of the zooplankton in lakes that have been subjected to experimental manipulation (Carpenter and Kitchell 1993). Thus this system affords the possibility of studying the individual-to-population link in systems of increasing complexity.

This hierarchy of experimental data is of particular value since all individual-based modeling involves judgments on the selection of variables and interactions for inclusion. For this reason, it is of particular interest to model carefully systems where we know we can omit many complicating factors. Thus, for example, although natural *Daphnia* populations in a lake (a) are made up of many clones (Herbert and Crease 1980; Lynch 1979; Carvalho and Crisp 1987), (b) exhibit diurnal vertical migration and thereby experience a temperature environment at least partly under their control (Lampert 1989), (c) take food from a complex phytoplankton assemblage of "edible" and "inedible" species (Kretzschmar et al. (1993) and references therein and Grover (1995)) and (d) have predators and competitors, none of these complications affects the link from individual to laboratory populations of a single clone. Complications (a), (c) and (d) are potentially involved in the next "step up" (to stock tanks: volume around 1000 L), but all are under the control of the experimenter. This means we can approach systematically the eventual task of modeling truly natural populations through a series of steps corresponding to increasing spatial scale and biological or environmental complexity.

The key dilemma in an individual-based approach to modeling natural systems is that the large number of species and time scales involved implies that models with detailed descriptions of them all would be impractical and probably useless. We have argued elsewhere (Murdoch et al. 1992; Murdoch and Nisbet 1996) that the way forward is to analyze the system's dynamics from different focal points, in each case reaching out along the web of interactions only as far as is needed and retaining detail on nonfocal components only when it is important. In previous work on the population dynamics of *Daphnia* (Nisbet et al. 1989; McCauley et al. 1996), we incorporated a large amount of detail on this important zooplankter, while opting for maximal simplicity in modeling other components of the system. However, other workers with different focal points may require a simple description of the phenomena we are modeling; e.g., an ecosystem model might have a single component "zooplankton." This leads to an additional requirement for our modeling work — understanding model simplification.

The theory of structured population dynamics (Metz and Diekmann 1986; Tuljapurkar and Caswell 1996) is the natural mathematical context in which to study the individual-to-population link. One route to simplification of structured population models is to identify situations where the dynamics of lumped variables (e.g., total numbers or biomass) can be fully described in terms of ordinary differential equations (ODEs). Metz and Diekmann (1991) have called this approach "linear chain trickery." A key requirement for successful reduction of structured population models to a set of ODEs appears to be age and size-independent mortality. In this chapter we explore one simple example of such model simplification and make assumptions that lead to ODE models of the *biomass* dynamics of *Daphnia* populations. We challenge these models with experimental data on single-species and two-species laboratory populations and use the results to discuss tests against field data on *Daphnia* in lakes differing in enrichment (Murdoch and Nisbet 1996).

4.2 BIOMASS DYNAMICS

4.2.1 Basic model

The simplest model of a food-limited population has Lotka-Volterra dynamics. We consider two species, "producers" or "food" with biomass density $F(t)$ and "herbivores" or "consumers" (*Daphnia*) with biomass density $C(t)$. We assume:

1. Consumers search randomly for food at a rate proportional to their biomass. This is broadly consistent with data on the allometry of feeding for *Daphnia pulex* (McCauley et al. (1990) and references therein), as both feeding rates and dry weight vary in the same manner with size (approximately as length$^{2.4}$). The assumption does not take account of a more detailed analysis in the same paper where statistically significant differences in the allometry for juveniles and adults are detected. The effect of consumer satiation is described by a "type II functional response" (individual feeding rate is a saturating function of food density).

2. All food encountered by nonsatiated consumers is eaten and converted into consumer biomass with a constant efficiency ε. The evidence for and against this assumption is reviewed in McCauley and Murdoch (1990).

3. Consumers have a constant age- and food-independent per capita death rate m. This is by far the weakest assumption of the model, being inconsistent with all life-table data on *Daphnia* (e.g., Porter et al. (1983) for *D. magna*, McCauley et al. (1990) for *D. pulex*, and many others). However as noted in the introduction, it is the key to the mathematical simplifications that allow representation of biomass dynamics in terms of ODEs.

4. Basal maintenance requirements cause all consumers to lose biomass at a constant rate b per unit biomass. This assumption is consistent with data that relates respiration rate linearly to assimilation rate, and thereby to food density (Bohrer and Lampert 1988), provided we take care in the definition of assimilation efficiency (Nisbet et al. 1991).

The dynamics of producer biomass involves the combined effects of primary production, external food supply rate, and grazing losses. We represent the net rate of food input by Φ, and in this paper consider three forms for Φ. First, we consider lab experiments performed in *transfer culture* where the consumers are transferred to a fresh container with a known food density, F_R, at regular intervals T. If, as we believe to be the case with most experiments on *Daphnia*, the food is depleted effectively to zero in each transfer, the simplest approximation to this situation is to work with the time-averaged food supply, and assume

$$\Phi = \frac{F_R}{T}. \tag{4.1}$$

Second, there is at least one situation (see Section 4.4) where the short-term fluctuations caused by transfer culture conditions seriously affect the population dynamics; in this case we set $\Phi = 0$ but reset the food density to F_R at times $t = nT$, where n is an integer. Third, when modeling field populations, we recognize that the food supply comes from primary production, not external inputs, and assume logistic growth, so that

$$\Phi = rF\left(1 - \frac{F}{K}\right), \tag{4.2}$$

where r and K represent intrinsic growth rate and carrying capacity respectively.

Our basic model involves two ODEs. The food dynamic equation, obtained by balancing net input with consumption, is

$$\frac{dF}{dt} = \Phi - \frac{I_{max}FC}{F + F_h} \tag{4.3}$$

where I_{max} is the maximum feeding rate per unit of consumer biomass, F_h is the half saturation constant in the type II functional response, and we make an appropriate choice of Φ as discussed above. The consumer equation may be derived from the standard partial differential equations for structured populations (see the Appendix), or heuristically by considering the biomass balance: new biomass is produced by conversion of food, and biomass is lost through death m and respiration b. Thus

$$\frac{dC}{dt} = \frac{\varepsilon I_{max}FC}{F + F_h} - (m + b)C. \tag{4.4}$$

4.2.2 Equilibrium demography

The equation for consumer biomass dynamics was derived without specifying the precise rules for partitioning energy from assimilated food between growth and reproduction. This is possible because of the assumptions of a constant feeding rate per unit biomass of consumer, and of a constant assimilation efficiency, irrespective of the ultimate physiological use of the assimilate. Thus both the immediate contribution to biomass, and the subsequent effects on grazing rates, are identical, whether the assimilate is allocated to somatic growth or to eggs. In order to model *demographic* properties of the population such as fecundity (i.e., per capita egg-production rate) or age structure, it is necessary to make further assumptions.

One particularly simple group of models is obtained if we assume that all individuals have a biomass w_b at birth, and mature (i.e., start reproduction) when they attain a critical biomass w_m, and that juveniles assign 100% of net production (assimilation minus maintenance) to biomass. Then at a constant food density $\overline{F}$, the time to grow from birth to maturity is

$$T_J = \left(\frac{e I_{max}\overline{F}}{\overline{F} + F_h} - b\right)^{-1} \ln\left[\frac{w_m}{w_b}\right]. \tag{4.5}$$

However, if the consumer population is food limited and at equilibrium, the food density cannot have an arbitrary value but must take the value that makes consumer birth and death rates equal. General theory for equilibrium demography of food-limited populations was developed by Gurney et al. (1996), but for the current simple model the results are intuitive. The time T_J taken to grow through the juvenile stage, and the through-stage survival, S_J for juveniles, can be derived by replacing $\overline{F}$ in (4.5) with the equilibrium food density F^* from (4.4) to obtain

$$T_J = \frac{1}{m} \ln \left[\frac{w_m}{w_b} \right], \qquad S_J = \exp(-mT_J). \qquad (4.6)$$

The average adult fecundity, $\overline{\beta}$, and the proportion of the equilibrium population that is juvenile, P_J, are related to the through-stage-survival for juveniles, S_J, and thereby to T_J and to the attributes of individual consumers by

$$\overline{\beta} = m/S_J; \qquad P_J = 1 - S_J. \qquad (4.7)$$

4.3 SINGLE-SPECIES LABORATORY POPULATIONS

The simple model of biomass dynamics developed in the previous section has been tested against two sets of data from small laboratory populations in *transfer culture*, i.e., the populations are kept in fixed volumes of water and transferred at regular intervals to fresh containers with a known amount of food:

- A set of replicated experiments by McCauley (unpublished) with small laboratory populations (275 mL volume) of *D. pulex*. The populations were limited by a single food type (*Chlamydomonas reinhardii*, a small alga edible by all sizes of *D. pulex*). Some results are shown in Figure 4.1.

- Similar experiments by Goulden et al. (1982b) with *D. galeata mendotae*, a smaller species of *Daphnia*. Two experimental regimes were used: a "medium food" regime in which individuals were transferred to containers with a food density of 0.25 mgC L^{-1} every two days, and a second set of studies with transfers every four days to "high food" with a concentration of 2.5 mgC L^{-1}.

Table 4.1 contains our estimates of parameter values for the two *Daphnia* species under the appropriate experimental conditions. All parameters are estimated from experiments on individuals, so there are *no* adjustable parameters when we come to model tests against populations. Two problems with the parameter estimates merit comment. First, in view of our (necessary) assumption of age-independent mortality, our estimates of m are only "guesstimates". Second, there are problems with estimation of the parameter I_{max}, discussed in the Appendix to Nisbet et al. (1991). This parameter can be estimated in two ways — from direct measurements of individual feeding rates, and from the exponential growth rate of small populations with ample food. Typically these approaches give very different values, the discrepancy being at least a factor

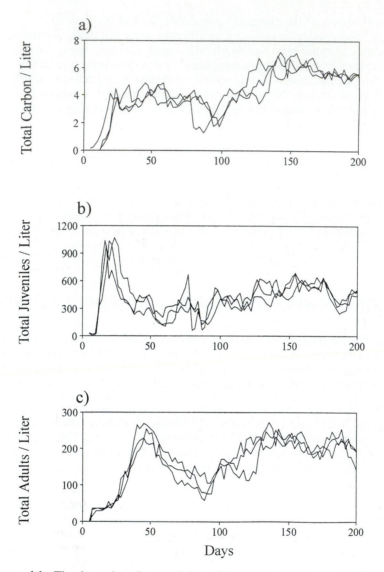

Figure 4.1. The dynamics of a population of *D. pulex* in laboratory transfer culture. The population was counted at transfer, so there is no sampling error. Length was estimated at transfer, with individuals being assigned to one of the following size classes: neonates (< 0.7 mm); small juveniles (< 1.0 mm); large juveniles (1.0–1.6mm); small adults (1.6–2.0 mm); large adults (over 2.0 mm). Carbon was calculated using a formula of Paloheimo et al. (1982) relating dry weight to length and assuming carbon content equal to 42% of dry weight. (a) Total carbon density; (b) Juvenile density; (c) Adult density.

of 2 for *D. pulex*. Thus we regard the values of this parameter in Table 4.1 as particularly unreliable.

Table 4.2 summarizes the equilibrium dynamics. There are at least two sources of uncertainty in the quoted biomasses. First, it is not certain

Table 4.1. Parameters of the basic model for two species of *Daphnia* and one species of *Bosmina*. Details of calculations: (a) neonate length = 0.7 mm, length at maturation = 1.5 mm; biomasses calculated using dry-weight/length relationship of Paloheimo et al. (1982) and assuming carbon/dry-weight ratio of 0.42; (b) from Nisbet et al. (1991); (c) length at birth = 0.38 mm, and length at maturation = 1.08 mm (Urabe 1988)(Lynch, 1980); biomasses calculated using length-carbon relationship of Urabe and Watanabe (1991); (d) calculated from information in McCauley et al. (1996); (e) length at birth = 0.18 mm and length at maturation = 0.27 mm; biomasses from length/carbon relationship in McCauley et al. (1996).

Quantity	Variable	Units	*D. pulex*	*D. galeata*	*Bosmina*
Neonate carbon mass	$0.42w_b$	mgC	0.0011^a	0.00023^c	0.00012^e
Maturation carbon mass	$0.42w_m$	mgC	0.0069^a	0.0033^c	0.00033^e
Respiration rate	b	day^{-1}	0.12^b	0.23^d	0.23^d
Death rate	m	day^{-1}	0.03	0.04	0.04
Assimilation efficiency	ϵ	-	0.5^b	0.75^d	0.92^d
Maximum specific ingestion rate	I_{max}	day^{-1}	1.0^b	6.5^d	2.0^d
Half-saturation constant	F_h	mgC L^{-1}	0.16^b	0.98^d	0.18^d

that the populations have in fact achieved equilibrium. This problem is apparent from Figure 4.1 for our *D. pulex* experiments. A similar doubt can be raised with the *D. galeata mendotae* experiments, owing to their short duration (around 70 days). Second, there is some variability among replicates. This variability is small in our experiments (Figure 4.1) but is much larger in those of Goulden et al. (1982a) — see Figures 1 and 2 of McCauley et al. (1996). Given the nature of the sources of uncertainty, any estimate of likely error in equilibrium biomasses must be subjective; ±25% is probably reasonable for our experiments, with a somewhat higher uncertainty in those of Goulden et al. (1982a). The uncertainty in the proportion juvenile is lower — perhaps ±15%.

The simplest model representation of this type of population is obtained by neglecting any fluctuations in food density and assuming a constant food supply rate (4.1). The solution to the differential equation for *D. galeata mendotae* is plotted in Figure 4.2b, the main feature being that the *Daphnia* biomass density is predicted to approach monotonically an equilibrium level given by

$$C^* = \frac{\varepsilon F_R}{(m+b)T}. \tag{4.8}$$

Table 4.2. Tests of basic model for small laboratory populations of *Daphnia*. The biomass density for *D. pulex* was obtained by using the observed length distribution, together with the weight/length relationship of Paloheimo et al. (1982) and an assumed carbon/dry-weight ratio of 0.42. The biomass density for *D. galeata* was estimated from numerical densities reported by Goulden et al. (1982a) by choosing nominal sizes of 0.7mm and 1.1mm for "typical" juveniles and adults, respectively, and then using the length-carbon relationship of Urabe and Watanabe (1991).

	D. pulex	*D. galeta*	
		Low food	**High food**
Experimental conditions			
Food density F_R (mgC/L)	3.5	0.25	2.5
Mean transfer interval T (day)	2.33	2	4
Equilibrium observations			
Biomass density (mgC/L)	5.5	0.27	1.45
Proportion juvenile	0.71	0.9	0.7
Model predictions			
Biomass density (mgC/L) (Equation (4.8))	5.10	0.35	1.73
Biomass density (mgC/L) (Equation (4.9))	4.26	0.22	1.30
Proportion juvenile	0.84	0.93	0.93
Juvenile stage duration (day)	61	66	66

This formula has two encouraging features: the parameter I_{max} in which we have least confidence does not appear, and the other uncertain parameter m appears only in combination with b (whose magnitude is much greater), and hence its impact on the predicted biomass is lessened.

If we take account of the fluctuations in food supply in the manner described in the text following equation (4.1), the food and *Daphnia* trajectories vary in a manner also shown in Figure 4.2. The food density drops rapidly toward zero after each transfer (Figure 4.2a), and the *Daphnia* biomass density eventually fluctuates around a mean level given by (4.8). An approximation to the *Daphnia* biomass density immediately before the transfers is obtained by assuming all food is consumed immediately on transfer to the new container, and followed by exponential decline in biomass due to death and respiration. In this approximate solution, the equilibrium biomass just before transfer is

$$C^* = \frac{\varepsilon F_R}{e^{(m+b)T} - 1}. \tag{4.9}$$

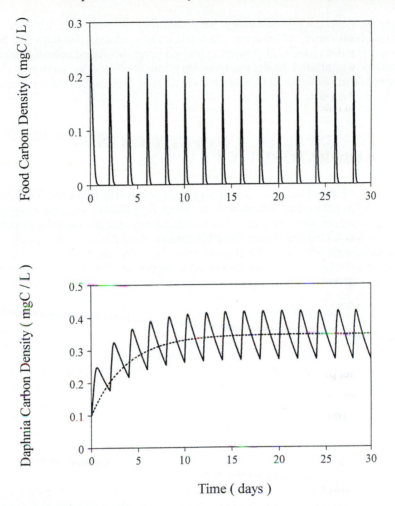

Figure 4.2. Model predictions for the food and biomass dynamics of laboratory populations of *Daphnia galeata mendota*: the experiments of Goulden et al. (1982b). The top panel shows the fluctuations in food density caused by the transfer culture protocol. The bottom panel shows the *Daphnia* biomass. The broken curve assumes a constant food supply rate; the continuous curve incorporates the two-day transfers. Parameter values are from Tables 4.1 and 4.2.

Table 4.2 confronts both predictions of equilibrium biomass with results from the three experiments. The agreement is very good, given that all parameters characterizing the *Daphnia* were estimated independently from the population experiments. This implies that, at least in terms of biomass balance at equilibrium, individuals in populations are exhibiting similar physiology to individuals in isolation.

The model does not do well in predicting equilibrium demography. Data suggest that at higher food densities, the proportion of juvenile *D. galeata* is significantly reduced; the model predicts no effect. However, the case against the model becomes overwhelming when we note that en route to computing the proportion of juveniles, we obtain the juvenile stage duration (see (4.6)), which turns out to have predicted values of over 60 days for both species. For our clone of *D. pulex*, the longest juvenile stage duration observed consistently is around 20 days even at very low food, and we have estimated a similar upper limit for *D. galeata mendotae* (see McCauley et al. (1996) and references therein).

4.4   INTERSPECIFIC COMPETITION IN LABORATORY POPULATIONS

For each of the two transfer regimes discussed in Section 4.3, Goulden et al. (1982b) performed experiments in which *D. galeata mendotae* competed with a smaller, freshwater herbivore of a different genus, *Bosmina longirostris*. In the two-day transfer, lower-food treatments, *Bosmina* and *Daphnia* coexisted for around 70 days, with *Bosmina* the more abundant species at the end of the experiment. In the four-day-transfer, higher-food treatments, there was coexistence again throughout the experiment, but here *Daphnia* dominated the community by the end of the experiment.

The theory of competition between two species living on a single resource is well studied. The analysis is straightforward for situations where each species would achieve a stable equilibrium in the absence of the other (e.g., Smith and Waltman (1995) and references therein). There, coexistence is impossible, the "winner" being the species capable of sustaining its biomass at the lower food density. In our models, this single-species, equilibrium food density is calculated from equation (4.4) and thus has a value independent of Φ, the rate of external food-supply or of primary production. This would imply that experimental manipulations such as those performed by Goulden et al. (1982a) should not affect the outcome of competition.

Goulden's experiments did not run for long enough to falsify unambiguously the equilibrium theory, but the change in numerically dominant species strongly points to the possibility that the outcome of competition is influenced by the food-supply regime, and it is certainly plausible that coexistence, or at least very slow competitive exclusion, would have been observed if the experiments had run for much longer periods. We now show that both the change in dominant species and the possibility of coexistence are predicted by our basic model, provided we include the rapid fluctuations in food density induced by the transfer regime. It is well known that internally generated fluctuations in resource density may increase the likelihood of coexistence of consumers subsisting on a single resource (Levins 1979; Armstrong and McGehee 1980); the analysis here sets out in more detail how this mechanism might operate in transfer culture and makes predictions concerning competition between *Daphnia* and *Bosmina*.

Our model of competing herbivores is a natural extension of the basic model in Section 4.2.1:

$$\frac{dF}{dt} = \Phi - I_1(F)C_1 - I_2(F)C_2 \tag{4.10}$$

$$\frac{dC_1}{dt} = C_1\left[\varepsilon_1 I_1(F) - m_1 - b_1\right] \tag{4.11}$$

$$\frac{dC_2}{dt} = C_2 \left[\varepsilon_2 I_2(F) - m_2 - b_2 \right] \qquad (4.12)$$

with

$$I_1(F) = \frac{I_{m1}F}{F + F_{h1}} \quad \text{and} \quad I_2(F) = \frac{I_{m2}F}{F + F_{h2}}. \qquad (4.13)$$

We investigate coexistence by determining conditions under which a small population of species 2 can successfully "invade" when species 1 is "resident" and vice versa. Coexistence is deemed possible if each species can invade when its competitor is resident.

First suppose species 1 to be the resident. Then as $t \to \infty$, food density approaches some asymptotic trajectory $F_1^*(t)$ with period T. The time-average specific growth rate of the resident consumer on this asymptotic trajectory must be zero; i.e., for any t,

$$\varepsilon_1 \int_{t-T}^{t} I_1(F_1^*(s)) \, ; ds = m_1 + b_1. \qquad (4.14)$$

Species 2 can invade if, over a time interval of duration T, its biomass grows in the environment $F_1^*(t)$, i.e., if

$$\varepsilon_2 \int_{t-T}^{t} I_2(F_1^*(s)) \, ds > m_2 + b_2. \qquad (4.15)$$

Now let $F_2^*(t)$ be the asymptotic trajectory when species 2 is resident. Then, by analogy with (4.14),

$$\varepsilon_2 \int_{t-T}^{t} I_2(F_2^*(s)) \, ds = m_2 + b_2, \qquad (4.16)$$

so species 2 can invade species 1 if

$$\int_{t-T}^{t} I_2(F_1^*(s)) \, ds > \int_{t-T}^{t} I_2(F_2^*(s)) \, ds. \qquad (4.17)$$

Similarly, species 1 can invade species 2 if

$$\int_{t-T}^{t} I_1(F_2^*(s)) \, ds > \int_{t-T}^{t} I_1(F_1^*(s)) \, ds. \qquad (4.18)$$

For coexistence, we require (4.17) and (4.18) to be valid simultaneously.

The coexistence conditions become tractable with two further assumptions. First, we assume that all food supplied at one transfer is consumed before the next transfer. Then the average biomass densities of each species when "resident" are obtained by equating total biomass gains and losses over the transfer interval to obtain

$$C_1^* = \frac{\varepsilon_1 F_R}{(m_1 + b_1)T} \quad \text{and} \quad C_2^* = \frac{\varepsilon_2 F_R}{(m_2 + b_2)T}. \qquad (4.19)$$

Second, we assume that the fluctuations in consumer populations about this average value are small, implying that (for example) when species 1 is resident, the food dynamics (that define the environment for the invader) are given by

$$\frac{dF_1^*}{dt} = -C_1^* I_1(F_1^*).$$ (4.20)

This result, and its analog for F_2^*, can be used to change variables in the integrals in inequalities (4.17) and (4.18). Some tedious algebra then leads to the result that coexistence requires

$$\int_0^{F_R} \frac{I_2(s)\,ds}{I_1(s)} \geq \frac{(m_2 + b_2)\varepsilon_1 F_R}{(m_1 + b_1)\varepsilon_2}$$ (4.21)

and

$$\int_0^{F_R} \frac{I_1(s)\,ds}{I_2(s)} \geq \frac{(m_1 + b_1)\varepsilon_2 F_R}{(m_2 + b_2)\varepsilon_1}.$$

The conditions can be expressed analytically and graphically for the particular case where the consumers have a type 2 functional response. We define three dimensionless parameter groups:

$$\phi_1 = \frac{F_{h1}}{F_R} \quad \phi_2 = \frac{F_{h2}}{F_R} \quad \theta = \frac{(m_1 + b_1)\varepsilon_2 I_{m2}}{(m_2 + b_2)\varepsilon_1 I_{m1}}.$$ (4.22)

With some algebra, it can then be shown that the coexistence conditions take the form

$$1 + (\phi_2 - \phi_1)\ln(1 + 1/\phi_1) \geq \theta$$ (4.23)

$$1 + (\phi_1 - \phi_2)\ln(1 + 1/\phi_2) \geq 1/\theta.$$ (4.24)

Exact numerical solutions of (4.10)-(4.13) confirm the possibility of coexistence; indeed the coexistence conditions (4.21) are remarkably accurate provided the food density drops very close to zero between transfers. Figure 4.3 shows graphically the conditions for coexistence with a value of θ estimated to be appropriate to the *Daphnia-Bosmina* system. Also shown are exact numerical solutions of the differential equations for three values of F_R. These confirm that *Bosmina* "wins" at low food, there is coexistence at intermediate food densities, and *Daphnia* wins at the highest food levels.

4.5   DISCUSSION

Our primary conclusion is a positive one: simple biomass models, involving parameters that can be estimated from measurements on individuals, are likely to produce reliable predictions of biomass density. Not only is numerical agreement with observed equilibria reasonable, but also a simple ODE-based model is capable of elucidating the subtle dynamics of competing

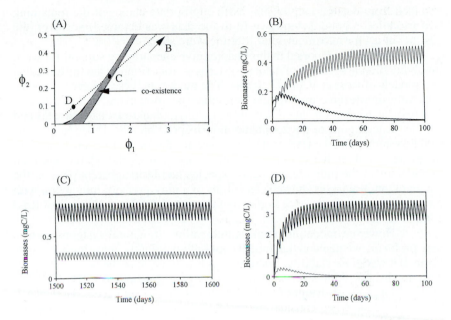

Figure 4.3. Competitive exclusion and coexistence as predicted by the model in Section 4.4. Parameter values are from Table 4.1; species 1 is *Daphnia galeata mendota*, species 2 is *Bosmina longirostris*. (A): Graphical representation in the ϕ_1–ϕ_2 plane of the coexistence conditions in (4.17) and (4.18). The points B–D on the line correspond to F_R of 0.25, 0.65 and 2.00 mgC/L respectively. (B): Exact biomasses against time (obtained by numerical solution of (4.10)–(4.13)) for *Daphnia* (fine line) and *Bosmina* (bold line) for simulated two-day experiments and $F_R = 0.25$ mgC/L. (C) and (D): As (B) with $F_R = 0.65$ and 2.00 mgC/L. The exact solutions shown in graphs (B)–(D) have dynamics consistent with predictions from the approximate analysis shown in (A).

consumer populations in transfer culture and of making predictions consistent with observations.

The main qualification is that if we require predictions of size structure or of the demography of populations at equilibrium, a more detailed model is required. This conclusion is consistent with the more general work of Gurney et al. (1996), who showed that the equilibrium demography of food-limited populations is highly sensitive to details of the functions describing size dependence of assimilation and feeding rates, the rules for partitioning assimilate among growth, reproduction, and maintenance, and the assumptions concerning mortality. Any one of these, or a combination, may be responsible for the problems in understanding the demography of laboratory *Daphnia* populations at equilibrium. We have work in progress (de Roos et al. 1996) which aims to determine the role played by each of these by using a detailed model of growth and reproduction in individual *Daphnia* (Gurney et al. 1990), and a new model of mortality rates. However, we already have strong indications that one key component of any resolution of the problem will be the maximum lifetime of the adult *Daphnia*. This comes from a study of the *D. galeata mendotae* populations using a stage-structured model (McCauley et al. 1996), where all parameters but two (assimilation efficiency and maximum adult life-time) were obtained from data on individuals. With appropriate choice of the remaining two variables, a good quantitative fit to the data was obtained; however, the best fit values for maximum adult lifetime were short.

The results reported in this chapter give encouraging support for work involving biomass models of natural systems — for example, the fjord ecosystem models of Ross et al. (1993; 1994). The models in these papers assumed a relatively simple three-trophic-level structure (phytoplankton, herbivores, carnivores), an assumption that is supported by phytoplankton measurements and detailed calculations of the zooplankton (copepod) biomass dynamics in one of the systems (Killary Harbor, Ireland); see Rodhouse and Rosen (1987). In the present context, the key feature of this work is that although the observed fluctuations in the individual copepod species had little apparent pattern, the zooplankton biomasses showed clear signs of a prey-predator interaction with the gelatinous carnivores (see Figure 3 of Ross et al. (1994) or Figure 5 of Ross et al. (1993).)

However, modeling biomass fluctuations in natural systems opens a new problem: we demand that models not only predict the magnitude of equilibria, but also their stability. Stability was not an issue in the laboratory populations discussed here, nor is it important in the fjord models, where stability is ensured by exchanges of population with the open ocean. However it is a major issue when considering the dynamics of natural *Daphnia* populations. We have analyzed over 30 studies of the *Daphnia*/algal interaction in lakes and ponds spread over much of the northern hemisphere, concentrating on the period between the spring peak *Daphnia* density and the winter decline, and on environments where vertebrate predation was unimportant (Murdoch and McCauley 1985; McCauley and Murdoch 1987; McCauley et al. 1988; Murdoch et al. 1996). These studies identified two general phenomena. (a) Steady populations or small amplitude quasi-cycles with a period of about a *Daphnia* generation occur in a wide range of environments. (b) The seasonal average density of edible algae increases, but very slowly, from oligotrophic to eutrophic lakes and ponds. Thus increasing nutrient levels do not lead to a decrease in stability, nor to the large-amplitude limit cycles associated with the "paradox of enrichment" (Rosenzweig 1969). These cycles are predicted at all but the smallest values of algal carrying capacity K in a model described by our equations (4.2), (4.3), and (4.4).

We recently (Murdoch et al. 1996) completed a study to identify mechanisms capable of reconciling the absence of large-amplitude limit cycles with the very slow rise in equilibrium levels of edible algae with enrichment. By adding new mechanisms to the basic model one at a time, we showed that stability cannot be explained by the following mechanisms:

- A reduction in available inorganic phosphorus through temporary incorporation in *Daphnia* biomass.

- A reduction in *Daphnia*'s feeding on edible algae due to the effects of inedible algae which (a) cause the *Daphnia* to narrow their gape and hence reduce filtering rate, and (b) cause rejection of (edible and inedible) food from the food groove.

- An increase in *Daphnia*'s death rate on moving from oligotrophic to eutrophic lakes.

- Density-dependent mortality (proposed by Gatto (1991)as a stabilizing factor for *Daphnia* populations).

- Ratio-dependent (Arditi and Ginzburg 1989) or density-dependent functional response (Beddington 1975).

We are left with three broad categories of mechanism potentially capable of explaining the observed phenomena. The first invokes details of the individual physiological response of individuals to variations in the food supply, and/or differences among individuals in this response. The second relies on spatial heterogeneity. The third assumes some plasticity in an animal's physiology in different environments. This failure to understand the stability of *Daphnia* populations in the field points to the likelihood that some modification of the formalism in this chapter may be required if we are to use biomass models in situations where dynamic instabilities are possible.

There is, of course, a world beyond *Daphnia* and algae, and developing truly *general* theory is our ultimate goal. The central problem is achieving generality without sacrificing the security that comes from working with testable (and tested) models. We hope the research described here will contribute in two ways. First, and most obviously, we hope the work on *Daphnia* will give insight that will be important in the formulation of future individual-based models for other systems. But we hope the general significance will go further. For example, as noted in the Introduction, the response of many organisms to low levels of toxicants involves changes in the rates of assimilation of food and of respiration by individuals (Donkin et al. 1989; Widdows et al. 1995). Simple biomass models can be used to describe the population level consequences of these changes (Nisbet et al. 1996). As another example, the population-level phenomenon described in the previous paragraph (suppression of a prey or host species to a level far below that set by its resources without loss of stability in spite of high potential growth rates of prey) occurs widely. For example, many insect species under biological control exhibit this phenomenon. We would not necessarily expect a universal mechanism to be operating, but we certainly hope for a finite number of explanations, and that the insight from Murdoch et al. (1996) will be applicable to a number of these systems.

4.6 ACKNOWLEDGMENTS

We thank the participants in the Special Year for Mathematical Biology for discussion and advice. We acknowledge valuable discussions with Andy Brooks, Hans Metz, Erik Muller, Erik Noonburg, Alex Ross, and Will Wilson. This research was supported by the National Science Foundation (Grant DEB-9319301) and the US Environmental Protection Agency (Grant R819433-01-0).

4.7   APPENDIX: DERIVATION OF ODE FOR CONSUMER DYNAMICS

Suppose all individuals have a biomass w_b at birth, and they mature (i.e., start reproduction) when they attain a critical biomass w_m. Juveniles assign 100% of net production (assimilation minus maintenance) to biomass. If net production is positive, adults assign some fraction $\mu(w, F)$ of net production to biomass and the remainder to reproduction. For the purposes of the present discussion it is not necessary to make any further assumptions about the allocation function μ. If net production is negative, adults cease reproduction. With these assumptions, the biomass, w, of an *individual* changes according to the differential equation

$$\frac{dw}{dt} = \xi pw, \qquad \text{with } p = \frac{\varepsilon I_{\max} F}{F + F_h} - b \qquad (4.25)$$

where

$$\begin{cases} \xi = \mu(w, F) & \text{if } p > 0 \text{ and } w > w_m \\ \xi = 1 & \text{otherwise.} \end{cases} \qquad (4.26)$$

The dynamics of a population are described by two partial differential equations (PDEs). An age distribution, $f(\tau, t)$, is defined by specifying that $f(\tau, t)d\tau$ is the number of consumers with ages in an infinitesimal age interval $\tau \to \tau + d\tau$ at time t. As all consumers are assumed to experience the same common food environment $F(t)$, individuals born at the same time grow at the same rate. Then $w(\tau, t)$ can be defined as the weight of an individual aged τ at time t. The population dynamics are then obtained from the solutions of two simultaneous PDEs

$$\frac{\partial f}{\partial t} = -\frac{\partial f}{\partial \tau} - mf \qquad (4.27)$$

$$\frac{\partial w}{\partial t} = -\frac{\partial w}{\partial \tau} + \xi pw \qquad (4.28)$$

which have to be solved with the boundary conditions

$$f(0, t) = w_b^{-1} \int_0^\infty (1 - \xi) pw(\tau, t) f(\tau, t) \, d\tau \qquad (4.29)$$

$$w(0, t) = w_b. \qquad (4.30)$$

To calculate the total consumer biomass density, note that $f(\tau, t)w(\tau, t)d\tau$ represents the biomass density for individuals aged τ to $\tau + d\tau$ at time t. Thus

$$C(t) = \int_0^\infty f(\tau, t)w(\tau, t) \, d\tau. \qquad (4.31)$$

The total rate of change of biomass density of individuals ((4.4)) is obtained by differentiating (4.31), and substituting from (4.27) - (4.30):

$$
\begin{aligned}
\frac{dC}{dt} &= \int_0^\infty \left(f \frac{\partial w}{\partial t} + w \frac{\partial f}{\partial t} \right) d\tau \\
&= -\int_0^\infty \left(\frac{\partial}{\partial \tau} (fw) + mfw - \xi pfw \right) d\tau \\
&= f(0,t)w(0,t) - mC + \int_0^\infty \xi pfw \, d\tau \\
&= (p-m)C = \left(\frac{\varepsilon I_{max} F}{F + F_h} - b - m \right) C.
\end{aligned}
\tag{4.32}
$$

REFERENCES

Arditi, R. and Ginzburg, L. R. 1989. Coupling in predator-prey dynamics: Ratio-dependence. *Journal of Theoretical Biology*, **139**, 311–326. {*75, 77*}

Armstrong, R. A. and McGehee, R. 1980. Competitive exclusion. *American Naturalist*, **115**, 151–170. {*70, 77*}

Beddington, J. R. 1975. Mutual interference between parasites or predators and its effect on searching efficiency. *Journal of Animal Ecology*, **44**, 331–340. {*75, 77*}

Bohrer, R. N. and Lampert, W. 1988. Simultaneous measurement of the effect of food concentration on assimilation and respiration in *Daphnia magna* straus. *Functional Ecology*, **2**, 463–471. {*63, 77*}

Carpenter, S. R. and Kitchell, J. F. (eds). 1993. *The Trophic Cascade in Lakes*. Cambridge, UK: Cambridge University Press. ISBN 0-521-43145-X. Pages xiv + 385. {*62, 77*}

Carvalho, G. R. and Crisp, D. J. 1987. The clonal ecology of *Daphnia magna* (crustacea: cladocera) I. Temporal changes in the clonal structure of a natural population. *Journal of Animal Ecology*, **56**, 453–468. {*62, 77*}

de Roos, A. M., McCauley, E., Nisbet, R. M., Gurney, W. S. C., and Murdoch, W. W. 1996. *Relating individual life-history and population dynamics in* Daphnia pulex: *Is the population more than a collection of individuals?* In preparation. {*73, 77*}

DeAngelis, Donald L. and Gross, Louis J. (eds). 1992. *Individual-based models and approaches in ecology: populations, communities, and ecosystems*. New York, NY: Routledge, Chapman and Hall. ISBN 0-412-03161-2 (hardcover), 0-412-03171-X (paperback). Pages xix + 525. {*61, 77, 79*}

Donkin, P., Widdows, J., Evans, S. V., Worral, C. M., and Carr, M. 1989. Quantitative structure-activity relationships for the effect of hydrophobic organic chemicals on rate of feeding by mussels (*Mytilus edulis*). *Q. Aquatic Toxicology*, **15**, 277–294. {*75, 77*}

Frank, P. W. 1960. Prediction of population growth form in *Daphnia pulex* cultures. *American Naturalist*, **94**, 357–372. {*61, 77*}

Gatto, M. 1991. Some remarks on models of plankton densities in lakes. *American Naturalist*, **137**, 264–267. {*75, 77*}

Glazier, D. S. and Calow, P. 1992. Energy allocation rules in *Daphnia magna* — clonal and age differences in the effects of food limitation. *Oecologia*, **90**(4), 540–549. {*61, 77*}

Goulden, C. E., Henry, L. L., and Tessier, A. J. 1982a. Body size, energy reserves, and competitive ability in three species of cladocera. *Ecology*, **63**, 1780–1789. {*61, 67, 68, 70, 77*}

Goulden, C. E., Henry, L. L., and Tessier, A. J. 1982b. Body size, energy reserves, and competitive ability in three species of cladocera. *Ecology*, **63**, 1780–1789. {*65, 69, 70, 77*}

Grover, J. P. 1995. Competition, herbivory, and enrichment-nutrient-based models for edible and inedible plants. *American Naturalist*, **145**(5), 746–774. {*62, 78*}

Gurney, W. S. C., McCauley, E., Nisbet, R. M., and Murdoch, W. W. 1990. The physiological ecology of *Daphnia*: A dynamic model of growth and reproduction. *Ecology*, **71**(2), 716–732. {*73, 78*}

Gurney, W. S. C., Middleton, D. A. J., Nisbet, R. M., McCauley, E., Murdoch, W. W., and de Roos, A. M. 1996. The equilibrium demography of structured populations. *Theoretical Population Biology*. In press. {*65, 73, 78*}

Herbert, P. D. N. and Crease, T. J. 1980. Clonal coexistence in *Daphnia pulex*: Another planktonic paradox. *Science*, **207**, 1363–1365. {*62, 78*}

Kretzschmar, M., Nisbet, R. M., and McCauley, E. 1993. A predator-prey model for zooplankton grazing on competing algal populations. *Theoretical Population Biology*, **44**(1), 32–66. {*62, 78*}

Lampert, W. 1989. The adaptive significance of vertical migration of zooplankton. *Functional Ecology*, **3**, 21–27. {*62, 78*}

Leibold, M. A. 1989. Resource edibility and the effects of predators and productivity on the outcome of trophic interactions. *American Naturalist*, **134**, 922–949. {*61, 78*}

Levins, R. 1979. Coexistence in a variable environment. *American Naturalist*, **114**, 765–783. {*70, 78*}

Lynch, M. 1979. Predation, competition, and zooplankton community structure: An experimental study. *Limnology and Oceanography*, **24**, 253–272. {*61, 62, 78*}

Marshall, J. S. 1978. Population dynamics of *Daphnia galeata mendotae* as modified by chronic cadmium stress. *Journal of the Fisheries Research Board of Canada*, **35**, 461–469. {*61, 78*}

McCauley, E. and Murdoch, W. W. 1987. Cyclic and stable populations: Plankton as paradigm. *American Naturalist*, **129**, 97–121. {*61, 62, 74, 78*}

McCauley, E. and Murdoch, W. W. 1990. Predator-prey dynamics in environments rich and poor in nutrients. *Nature*, **343**(6257), 455–457. {*61, 63, 78*}

McCauley, E., Murdoch, W. W., and Watson, S. 1988. Simple models and variation in plankton densities among lakes. *American Naturalist*, **132**, 383–403. {*61, 74, 78*}

McCauley, E., Murdoch, W. W., and Nisbet, R. M. 1990. Growth, reproduction, and mortality of *Daphnia pulex*: life at low food. *Functional Ecology*, **4**, 505–514. {*61, 63, 78*}

McCauley, E., Nisbet, R. M., de Roos, A. M., Murdoch, W. W., and Gurney, W. S. C. 1996. Structured population models of herbivorous zooplankton. *Ecology*. In press. {*62, 67, 70, 73, 78*}

Metz, J. A. J. and Diekmann, O. 1986. *The Dynamics of Physiologically Structured Populations*. Lecture Notes in Biomathematics, vol. 68. Berlin: Springer-Verlag. ISBN 0-387-16786-2 (New York), 3-540-16786-2 (Berlin). Pages xii + 511. {*62, 78*}

Metz, J. A. J. and Diekmann, O. 1991. Exact finite dimensional representations of models for physiologically structured populations. I: The abstract foundations of linear chain trickery. *Pages 269–289 of:* Goldstein, Jerome A., Kappel, Franz, and Schappacher, Wwilhelm (eds), *Differential Equations with Applications in Biology, Physics, and Engineering*. New York, NY: Marcel Dekker, Inc. {*62, 78*}

Murdoch, W. W. and McCauley, E. 1985. Three distinct types of dynamic behavior shown by a single planktonic system. *Nature*, **316**, 628–630. {*61, 74, 78*}

Murdoch, W. W. and Nisbet, R. M. 1996. Frontiers of population ecology. *In:* Floyd, R. B. and Sheppard, A. W. (eds), *Frontiers of Population Ecology*. Melbourne, Australia: CSIRO Press. {*62, 75, 78*}

Murdoch, W. W., McCauley, E., Nisbet, R. M., Gurney, W. S. C., and de Roos, A. M. 1992. Individual-based models: Combining testability and generality. *In:* (DeAngelis and Gross 1992). {*62, 79*}

Murdoch, W. W., Nisbet, R. M., McCauley, E., de Roos, A. M., and Gurney, W. S. C. 1996. *Plankton abundance and dynamics across nutrient levels: Tests of hypotheses.* Submitted to Ecology. {*74, 79*}

Nisbet, R. M., Gurney, W. S. C., Murdoch, W. W., and McCauley, E. 1989. Structured population models: A tool for linking effects at individual and population level. *Biological Journal of the Linnean Society*, **37**, 79–99. {*62, 79*}

Nisbet, R. M., McCauley, E., de Roos, A. M., Murdoch, W. W., and Gurney, W. S. C. 1991. Population dynamics and element recycling in an aquatic plant-herbivore system. *Theoretical Population Biology*, **40**(2), 125–147. {*63, 65, 67, 79*}

Nisbet, R. M., Muller, E. B., Brooks, A. J., and Hosseini, P. 1996. *Models relating individual and population response to contaminants.* Submitted to Environmental Modeling and Assessment. {*75, 79*}

Paloheimo, J. E., Crabtree, S. J., and Taylor, W. D. 1982. Growth model of *Daphnia*. *Canadian Journal of Fisheries and Aquatic Sciences*, **39**, 598–606. {*66–68, 79*}

Porter, K. G., Orcutt, J. J. D., and Gerritsen, J. 1983. Functional response and fitness in a generalist filter feeder, *Daphnia magna* (Cladocera: Crustacea). *Ecology*, **64**, 735–742. {*63, 79*}

Rodhouse, P. G. and Roden, C. M. 1987. Carbon budget for a coastal inlet in relation to an intensive cultivation of suspension feeding bivalve molluscs. *Marine Ecology Progress Series*, **36**, 225–236. {*74, 79*}

Rosenzweig, M. 1969. Why the prey curve has a hump. *American Naturalist*, **103**, 81–87. {*74, 79*}

Ross, A. H., Gurney, W. S. C., Heath, M. R., Hay, S. J., and Henderson, E. W. 1993. A strategic simulation model of a fjord ecosystem. *Limnology and Oceanography*, **38**(1), 128–153. {*74, 79*}

Ross, A. H., Gurney, W. S. C., and Heath, M. R. 1994. A comparative study of the ecosystem dynamics of four fjords. *Limnology and Oceanography*, **39**(2), 318–343. {*74, 79*}

Slobodkin, L. B. 1954. Population dynamics in *Daphnia obtusa* kurz. *Ecological Monographs*, **24**, 69–88. {*61, 79*}

Smith, H. L. and Waltman, P. 1995. *The Theory of the Chemostat: Dynamics of Microbial Competition*. Cambridge Studies in Mathematical Biology, vol. 13. Cambridge, UK: Cambridge University Press. ISBN 0-521-47027-7. Pages xvi + 313. {*70, 79*}

Tuljapurkar, Shripad and Caswell, Hal. 1996. *Structured-Population Models in Marine, Terrestrial, and Freshwater Systems*. London, UK: Chapman and Hall, Ltd. ISBN 0-412-07261-0 (hardcover), 0-412-07271-8 (paperback). {*62, 79*}

Urabe, J. 1988. Effect of food conditions on the net production of *Daphnia galeata*: separate effects on growth and reproduction. *Bulletin of Plankton Society of Japan*, **35**, 159–174. {*67, 79*}

Urabe, J. and Watanabe, Y. 1991. Effect of food concentration on the assimilation and production efficiencies of *Daphnia galeata*. *Functional Ecology*, **5**, 635–641. {*67, 68, 79*}

Widdows, J., Donkin, P., Brinsley, M. D., Evans, S. V., Salkeld, P. N., Franklin, A., Law, R. J., and Waldock, M. J. 1995. Scope for growth and contaminant levels in North Sea mussels *Mytilus edulis*. *Marine Ecology Progress Series*, **127**(1–3), 131–148. {*75, 79*}

5. ANCESTRAL INFERENCE FROM DNA SEQUENCE DATA

Simon Tavaré

5.1   INTRODUCTION

After the pioneering paper of Cann et al. (1987), many authors have discussed methods for inferring ancestral history from samples of DNA sequences taken from human populations. Much of this research has focused on the evolution of mitochondrial DNA. These molecules have been exploited in evolutionary studies because of their high mutation rate; this means that DNA sequence differences can be detected between individuals who are quite closely related. In addition, mitochondria are maternally inherited, so these molecules are particularly suited to studying the female lineages in which they arise. One tantalizing problem, usually referred to as "the time to Mitochondrial Eve," is to estimate the time to the most recent common mitochondrial ancestor of the population from which the sample sequences were drawn. The papers of Templeton (1993), Ayala (1995), Wallace (1995) and Wills (1995) provide further background and discussion. More recently, DNA sequence data from the male-specific part of the Y chromosome have begun to appear, along with analyses of the time to "Y Adam." See Dorit et al. (1995), Hammer (1995), Whitfield et al. (1995), and the review of Jobling and Tyler-Smith (1995).

In this paper we describe one approach to drawing inferences about the distribution of the time to the most recent common ancestor (TMRCA) of a population, given data from a sample of DNA sequences taken from that population. In practice we do not know the ancestral history of the DNA sequences in the sample in any detail. Therefore statistical statements about TMRCA have to be based on a stochastic model for this ancestry. We use a model called the *coalescent* (Kingman 1982a; Griffiths 1980; Hudson 1983; Tajima 1983), reviewed briefly in Sections 5.2 and 5.3. The effects of deterministic fluctuations in population size are discussed in Section 5.4.

The sample consists of n individuals from the population of interest. The data $\mathcal{D}$ in the sample are n DNA sequences from a given molecular region. For example, many mitochondrial data sets contain sequences from the control region of the molecule, while Dorit et al. sequenced an intron of the ZFY locus on the Y chromosome. We assume the sequences in the sample are the same length. We can then think of the data as a matrix $X = (x_{ij})$ with n rows, where the entry x_{ij} records the DNA base (either A, C, G, or T) in individual i at site j. A *site*, then, refers to a location in the DNA. It is sometimes convenient to ignore any columns of X that have identical bases in every sequence. The remaining columns are referred to as *segregating sites*; they comprise locations in the DNA sequences where not every individual is identical. The differences observed in the sample sequences arise from the effects of mutation in their ancestry. We suppose that these differences are due to the effects of *substitutions*, the replacement of one base by another when a mutation occurs. We model the locations of the mutations in the ancestral tree of the sample in Section 5.5.

In practice it is often either difficult or uninformative to get explicit mathematical expressions for quantities of interest such as the conditional distribution of TMRCA given the data $\mathcal{D}$. Instead we use a computational approach that simulates observations from the required conditional distribution. Summary statistics such as histograms and moments can then be found from these simulated values in the usual way. In this chapter we summarize the data matrix X in terms of the random quantity S_n, the number of segregating sites in the sample. Conditional distributions of TMRCA given $S_n = k$ can be found by a rejection method, discussed briefly in Section 5.6. Applications of these simulation methods to Y-chromosome data are given in Section 5.7.

5.2   THE COALESCENT

Inferences about the TMRCA of a population are to be made on the basis of a comparison of the DNA sequences from the molecular region of interest from a sample of people in the population. Differences in these sequences come from the effects of mutation in the unknown ancestry of the sample. It follows that to study TMRCA we need a stochastic model for this ancestry. In the molecular regions of interest here (the intron in the ZFY locus on the Y chromosome or the D loop of the mitochondrion for example) there appears to be no recombination. The molecular region is passed on intact, modulo the effects of substitutions, from parent to offspring. As a result, each molecule (or "individual") has a single haploid "parent" in the previous generation (the molecule from which it was copied), that "parent" itself has a single parent in the previous generation, and so on back into time. It is this genealogical process that we have to model.

Population geneticists have modeled such genealogies in a variety of circumstances, in particular when the population size is large. Consider then a particular generation in a large random mating population of constant-size N haploid individuals, and label them $1, 2, \ldots, N$. Population genetics models are often defined by specifying the joint distribution of the numbers $\nu_1, \nu_2, \ldots, \nu_N$ of offspring born to individuals $1, 2, \ldots, N$. For example, the classical Wright-Fisher model specifies that the offspring numbers have a symmetric multinomial distribution:

$$\mathbb{P}(\nu_1 = m_1, \ldots, \nu_N = m_N) = \frac{N! N^{-N}}{m_1! \cdots m_N!} \tag{5.1}$$

where $m_1, \ldots, m_N \in \{0, 1, \ldots, N\}$ satisfy $m_1 + \cdots + m_N = N$ and the offspring numbers in different generations are independent and identically distributed. This prescription shows how to construct the model forwards in time. However, for our inference problem it is much more convenient to study not how parents have offspring, but rather how children "choose" their parents.

The Wright-Fisher model can be described by saying that each individual in a given generation chooses its parent independently of others in its generation, uniformly and at random from the N potential parents in the previous generation. Continuing this process back into the past yields a genealogical tree that relates the individuals in a given generation to their parents, grandparents, and so on.

This genealogy is hard to analyze for a given fixed value of N, but it may be approximated in a simple way when N is large. Notice that the chance that two randomly chosen individuals have distinct parents in the previous generation is $1 - N^{-1}$. It follows that the chance that these two have distinct ancestors in generations $1, 2, \ldots, r$ is $(1 - N^{-1})^r$. If we measure time in units of N generations, so that $r \approx Nt$ for some $t > 0$, we see that the time W_2 during which the sample of two individuals has no common ancestor satisfies

$$\mathbb{P}(W_2 > t) = \left(1 - \frac{1}{N}\right)^{Nt} \approx e^{-t}. \tag{5.2}$$

This shows that in a large population, the time W_2 until two individuals have a common ancestor has (approximately) an exponential distribution with mean 1. What of the genealogy of a sample of size three? Looking back into the past, there will be a first time at which some members of the sample share a common ancestor. At this time, either all three will have a common ancestor, or a particular pair will. In a large population, this last possibility is overwhelmingly the most likely. Furthermore, the time W_3 (measured once more in units of N generations) has approximately an exponential distribution with mean $\binom{3}{2} = 3$, so

$$\mathbb{P}(W_3 > t) \approx e^{-3t}. \tag{5.3}$$

At the time the first pair of individuals has found a common ancestor, the sample of three individuals has two distinct ancestors. The additional time taken for these two to find their common ancestor has the distribution of W_2, independent of W_3.

Thus in a large population we can give a simple description of the genealogy of a sample of n individuals. This stochastic process, known as the *coalescent*, describes the genealogical tree of the sample as time goes back into the past. With time measured in units of N generations, the time W_j during which the sample has j distinct ancestors has an exponential distribution with parameter $\binom{j}{2} = j(j-1)/2$, the times $W_n, W_{n-1}, \ldots, W_2$ being independent for different j. W_j should be thought of as the length of each of the j branches of the genealogical tree when the sample has j distinct ancestors. This tree is bifurcating; at the time W_n, two of the n ancestors are chosen at random and their branches are joined, giving $n - 1$ ancestors for the sample. At the time $W_n + W_{n-1}$, two of these $n-1$ ancestors are chosen at random and their branches are joined, resulting in $n - 2$ distinct ancestors in the sample. This process continues until the time

$$T_n = W_n + \cdots + W_2, \tag{5.4}$$

when all the individuals in the sample have been traced back to their most recent common ancestor (MRCA). A sample path of this process appears in Figure 5.1.

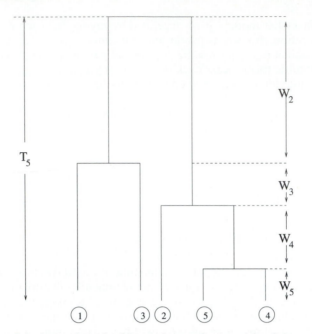

Figure 5.1. A sample path of the coalescent for a sample of size $n = 5$.

The previous discussion was based on the Wright-Fisher model. Remarkably, the same approximation applies to a very wide class of discrete exchangeable reproduction models. Kingman (1982a; 1982b) showed how the coalescent arises as the limiting approximation (as the population size $N \to \infty$) to these underlying discrete genealogies. In this approximation, time is measured in units of $\sigma^{-2} N$ generations, where $\sigma^2 \in (0, \infty)$ is the limiting variance of the number v_1 of offspring born to a typical individual. For ease of exposition, we assume $\sigma^2 = 1$ (as it is for the Wright-Fisher model) in what follows.

The mean time to the MRCA, and so the mean height of the ancestral tree, can be found from (5.4) as

$$
\begin{aligned}
\mathbb{E}T_n &= \mathbb{E}(W_n + \cdots + W_2) & (5.5)\\
&= \mathbb{E}W_n + \cdots + \mathbb{E}W_2 & (5.6)\\
&= \frac{2}{n(n-1)} + \cdots + \frac{2}{2(2-1)} & (5.7)\\
&= 2\left(1 - \frac{1}{n}\right) & (5.8)
\end{aligned}
$$

in coalescent units. The variance of T_n can be computed easily because the W_j are independent and exponentially distributed. We obtain

$$
\mathrm{Var}(T_n) = \sum_{j=2}^{n} \frac{4}{j^2(j-1)^2}. \qquad (5.9)
$$

In large samples this variance is about 1.16, most of which comes from the time W_2 when the sample has just two ancestors. Times are often converted from the coalescent time T_n to years T_n^y via

$$T_n^y = T_n \times N \times G, \tag{5.10}$$

where G is the number of years in a generation.

5.2.1 The ancestral process

In the sequel we make use of the Markov chain $\{A_n(t), t \geq 0\}$ that counts the number of distinct ancestors of the sample of size n at times $t \geq 0$. In Markov chain parlance, this is a *death process*: it starts from $A_n(0) = n$, waits an exponential amount of time W_j in state j, and then moves to state $j - 1$ and so forth. Eventually the process is absorbed in the state 1, at the time T_n. The probability distribution $g_{nj}(t)$ of $A_n(t)$ was found by Griffiths (1979).

$$
\begin{aligned}
g_{nj}(t) &= \mathbb{P}(A_n(t) = j) \\
&= \sum_{k=j}^{n} (-1)^{k-j} e^{-k(k-1)t/2} \frac{(2k-1) j_{(k-1)} n_{[k]}}{j!(k-j)! n_{(k)}},
\end{aligned}
\tag{5.11}
$$

where we have used the notation

$$
\begin{aligned}
a_{(n)} &= a(a+1)\cdots(a+n-1); \quad a_{(0)} = 1; \tag{5.12} \\
a_{[n]} &= a(a-1)\cdots(a-n+1); \quad a_{[0]} = 1. \tag{5.13}
\end{aligned}
$$

Because $\{T_n \leq t\} = \{A_n(t) = 1\}$, the distribution function of T_n follows immediately from (5.11):

$$\mathbb{P}(T_n \leq t) = g_{n1}(t), \quad t \geq 0. \tag{5.14}$$

While this provides an explicit formula for the distribution of T_n, it is harder to find explicit results for other quantities of interest such as the distribution of the total length L_n of the tree, defined by

$$L_n = nW_n + (n-1)W_{n-1} + \cdots + 2W_2. \tag{5.15}$$

Instead we can resort to a Monte Carlo approach, in which observations having the required distribution are simulated. These simulated values can then be used to estimate the probability density of the underlying random variable, together with any required statistics such as percentiles, mean and variance. A convenient introduction to stochastic simulation can be found in Ripley (1987). To illustrate the ideas, we give an algorithm for simulating the times $W_n, W_{n-1}, \ldots, W_2$.

Algorithm 1 *Algorithm to generate $W_n, \ldots, W_2$ for constant population size.* U denotes a random variable with the uniform distribution on (0,1), generated independently at each use.

1. Set $t = 0$, $j = n$.

2. Generate $s = -2 \log(U)/j(j-1)$.

3. Set $w_j = s, t = t + s$.

4. Set $j = j - 1$. If $j \geq 2$, go to 2. Else return $T_n = t$, $W_n = w_n, \ldots, W_2 = w_2$.

Step 2 generates an observation having the exponential distribution with mean $2/j(j-1)$, just as needed for W_j. The value t returned in Step 4 has the distribution of $T_n = W_n + \cdots + W_2$. The algorithm can be modified to generate observations having the distribution of L_n; simply set $l = 0$ at Step 1, $l = l + js$ at Step 3, and return $L_n = l$ at the end of Step 4. Later in this chapter, we exploit this simulation approach in cases where exact results are unobtainable.

Genealogical methods based on variations of the coalescent, using both theoretical and simulation approaches, have proved very powerful for understanding the structure of complex stochastic models in population genetics and as a useful guide to intuition in understanding the evolution of many population genetic phenomena. The recent reviews of Hudson (1991); Hudson (1992) and Donnelly and Tavaré (1995) describe some of these developments.

5.3   THE BIVARIATE ANCESTRAL PROCESS

In order to study TMRCA for a population given sequence data from a sample, we need to understand the *joint* behavior of the genealogy of both population and sample. We make use of the process $\{(A_m(t), A_n(t)), t \geq 0\}$ that counts the number of ancestors in a population of size m and a random sample of size n taken from it. (It is convenient to refer to the set of m individuals as the population, and the subset of size n as the sample. This avoids ambiguity and terms like "sample" and "subsample" or "supersample" and "sample".) This bivariate process is Markovian and it makes transitions from a state of the form (i, j) whenever two individuals in the current population of size i share a common ancestor. If this coalescence event involves the ancestors of two individuals in the sample, then the new state becomes $(i - 1, j - 1)$. In any other case, it is $(i - 1, j)$. From (i, j) we move to

$$
\begin{array}{lll}
(i - 1, j) & \text{at rate} & (i(i - 1) - j(j - 1))/2 & (5.16) \\
(i - 1, j - 1) & \text{at rate} & j(j - 1)/2. & (5.17)
\end{array}
$$

A sample path of the bivariate process for $m = 9, n = 5$ is given in Figure 5.2.

The distribution of $(A_m(t), A_n(t))$ was found by Saunders et al. (1984) as

$$
\mathbb{P}(A_m(t) = l, \quad A_n(t) = k) = g_{ml}(t)q(n, k \,|\, m, l), \qquad (5.18)
$$

where $g_{ml}(t)$ is given in (5.11), and $q(n, k \,|\, m, l)$ is given by

$$
q(n, k \,|\, m, l) = \qquad\qquad\qquad\qquad\qquad\qquad\qquad\qquad (5.19)
$$
$$
\frac{(m - n)!(m - l)!n!(n - 1)!l!(l - 1)!(m + k - 1)!}{(n - k)!(l - k)!m!(m - 1)!k!(k - 1)!(l + n - 1)!(m + k - l - n)!}.
$$

The quantity $q(n, k \,|\, m, l)$ is the probability that the sample of size n taken at any time $t > 0$ has k distinct ancestors given that the population of size m has l ancestors at that time.

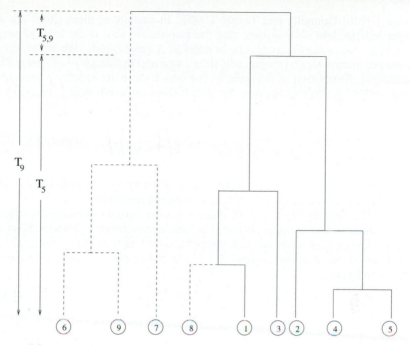

Figure 5.2. A sample path of the bivariate coalescent for a population of size $m = 9$ and a sample of size $n = 5$. The sample individuals are labeled $1, 2, \ldots, 5$.

The sample path in Figure 5.2 shows that at the time T_n when the sample reaches its MRCA, the number $A_m(T_n)$ of distinct ancestors of the population is random. The distribution of the number $A_m(T_n)$ is known. In particular, Watterson (1982) showed that the probability that the population of size m and a sample of size n share a common ancestor is

$$\mathbb{P}(A_m(T_n) = 1) = \frac{(n-1)(m+1)}{(n+1)(m-1)}. \tag{5.20}$$

If the sample is at all large, there is an appreciable chance that the sample and the subsample will share their MRCA, and that the time to the MRCA is thus the same for both sample and subsample. On the other hand, if they do not share a common ancestor then the *extra* time T_{nm} required to reach the MRCA of the sample is stochastically larger than W_2, the time taken for two individuals to be traced back to their common ancestor.

5.4 VARIABLE POPULATION SIZE

In order to apply coalescent methods to human population data, we need to account for the effects of variations in population size through time. Fortunately, this is straightforward in the case of deterministic fluctuations. To keep the presentation simple, we concentrate on the approximation to the Wright-Fisher model once more. The effect of variable population size is to change the joint distribution of the times W_j (Kingman 1982b; Griffiths and

Tavaré 1994b; Donnelly and Tavaré 1995). In particular, these times are no longer independent. We assume that the population size at the time of sampling is N, and again measure time in units of N generations. We write $Np(t)$ for the population size a (coalescent) time t ago and define $\lambda(t) = 1/p(t)$. The conditional distribution of the time W_j for which there are exactly j ancestors of the sample, given that the time for which there are more than j ancestors is s, is

$$\mathbb{P}(W_j > t \mid W_n + \cdots + W_{j+1} = s) = \exp\left(-\binom{j}{2} \int_s^{s+t} \lambda(v)\,dv\right). \tag{5.21}$$

We assume that $\int_0^\infty \lambda(v)\,dv = \infty$ to ensure that any pair of individuals (and thus the sample) can be traced back to a common ancestor.

The process $\{A_n^v(t), t \geq 0\}$ that counts the number of ancestors at time t of a sample of size n taken at time 0 is now a time-inhomogeneous Markov process. Given that $A_n^v(t) = j$, it jumps to $j - 1$ at rate $j(j-1)\lambda(t)/2$. A useful way to think of the process $A_n^v(\cdot)$ is to notice that a realization may be constructed via

$$A_n^v(t) = A_n(\Lambda(t)), \quad t \geq 0, \tag{5.22}$$

where $A_n(\cdot)$ is the corresponding ancestral process for the constant-population-size case, and

$$\Lambda(t) = \int_0^t \lambda(s)\,ds. \tag{5.23}$$

Thus the variable-population-size model is just a deterministic time change of the constant-population-size model. Some of the properties of $A_n^v(\cdot)$ follow immediately from this representation. For example,

$$\mathbb{P}(A_n^v(t) = j) = g_{nj}(\Lambda(t)), \quad j = 1, \dots, n \tag{5.24}$$

where $g_{nj}(t)$ is given in (5.11), and so

$$\mathbb{P}(T_n \leq t) = \mathbb{P}(A_n^v(t) = 1) = g_{n1}(\Lambda(t)), \quad t \geq 0. \tag{5.25}$$

Once more, simulation provides a valuable way to study properties of genealogy when the population size varies. The representation (5.21) gives a direct way to simulate the times $W_n, W_{n-1}, \dots, W_2$.

Algorithm 2 *Algorithm to generate $W_n, \dots, W_2$ with variable population size.* U denotes a random variable with a uniform distribution on $(0,1)$, generated independently at each use.

1. Set $t = 0$, $j = n$.

2. Generate

$$w_j^* = \frac{-2\log(U)}{j(j-1)}.$$

3. Solve for s the equation

$$\Lambda(t+s) - \Lambda(t) = w_j^*. \tag{5.26}$$

4. Set $w_j = s, t = t + s$.

5. Set $j = j - 1$. If $j \geq 2$, go to 2. Else return $T_n = t$, $W_n = w_n, \ldots, W_2 = w_2$.

As noted after Algorithm 1, w_j^* generated in step 2 has an exponential distribution with mean $2/j(j - 1)$. If the population size is constant, then $\Lambda(t) = t$, and Algorithm 2 reduces to Algorithm 1. Observations having the distribution of the tree length L_n can be generated as described after Algorithm 1.

5.4.1 The bivariate process revisited

The analysis of the bivariate ancestral process with variable population size follows immediately from the representation

$$(A_m^v(t), A_n^v(t)) = (A_m(\Lambda(t)), A_n(\Lambda(t))), \quad t \geq 0. \tag{5.27}$$

From this follows the fact that

$$\mathbb{P}(A_m^v(t) = l, A_n^v(t) = k) = g_{ml}(\Lambda(t))q(n, k \mid m, l), \tag{5.28}$$

where $q(n, k \mid m, l)$ is given in (5.19). Note that the combinatorics of the bivariate process remain as they were in the constant-population-size case; only the waiting times between the jumps of the process change. In particular, the probability that the population and the sample share their MRCA is still given by (5.20).

Distributions in the bivariate process can also be simulated easily. Algorithm 3 gives a method for simulating values of the height $T_n = W_n + \cdots + W_2$ of the coalescent tree of the sample of size n, the number $A_m(T_n)$ of ancestors of the sample at the time the subsample reaches its MRCA, and the time T_{nm} from then until the population reaches its MRCA. This extra time may, of course, be 0.

Algorithm 3 *Algorithm for bivariate ancestral process.* U denotes uniform $(0,1)$ random variable, independently generated at each use.

1. Set $a_m = m, a_n = n, t = 0$.

2. Set $w = -2\log(U)/(a_m(a_m - 1))$.

3. Solve for s the equation $\Lambda(t + s) - \Lambda(t) = w$.

4. Set $t = t + s$.

5. Set

$$p = \frac{a_n(a_n - 1)}{a_m(a_m - 1)} \tag{5.29}$$

and $a_m = a_m - 1$.

6. With probability p, set $a_n = a_n - 1$. If $a_n > 1$, go to 2.

7. Set $t_n = t$, $a^* = a_m$. If $a^* = 1$, set $t_{mn} = 0$, and stop. Otherwise, use Algorithm 2 starting from $t = t_n$, $n = a^*$ to generate an observation t_{nm} on the total height of a coalescent tree of a^* individuals, then stop.

The values of t_n, a^*, and t_{nm} returned by a single pass through Algorithm 3 have the joint distribution of the height T_n, the number of ancestors $A_m(T_n)$ of the population at the time the subsample finds its MRCA, and the additional time required to get to the MRCA of the population. Note that $t_m = t_n + t_{nm}$ has the distribution of T_m. More detailed information about the genealogical trees could also be recorded, but this is all we need later on.

5.5   MUTATIONS IN THE GENEALOGICAL TREE

To model the effect of mutations in the genealogy of the sample, we assume that the times at which mutations occur form a Poisson process of constant rate $\theta/2$, independently in each branch of the tree. A branch of length w therefore has a Poisson number of mutations with mean $w\theta/2$. The parameter θ is defined by

$$\theta = 2N\mu, \tag{5.30}$$

where μ is the mutation rate per gene per generation in the underlying discrete model. When the mutation rate μ is of the order of the reciprocal of the population size N, the genealogy and the genetics compete on equal terms; both features are included in the coalescent approximation.

To model the evolution of DNA-sequence data we have to describe how a sequence is changed when a mutation occurs in it. We here use the infinitely-many-sites model of Watterson (1975). Because we are ignoring the effects of recombination (it is not thought to occur in the data at hand), each sequence may be thought of as a completely linked sequence of DNA sites. Whenever a mutation occurs, it occurs at a site that has not had a mutation before.

Observing that each mutation in the coalescent tree introduces a new segregating site into the sample, the number of segregating sites S_n in the sample of n chromosomes is precisely the number of mutations that arise in its genealogical tree. This in turn has a Poisson distribution with mean $\theta L_n/2$, where L_n is the total length of the tree, defined in (5.15) by $L_n = nW_n + (n-1)W_{n-1} + \cdots + 2W_2$. That is,

$$\mathbb{P}(S_n = k \mid L_n = l) = \mathrm{Po}\,(k, \theta l/2), \tag{5.31}$$

where $\mathrm{Po}\,(k, \mu)$ is the Poisson probability

$$\mathrm{Po}\,(k, \mu) = e^{-\mu}\frac{\mu^k}{k!}, \quad k = 0, 1, \dots. \tag{5.32}$$

5.6   CONDITIONING ON THE DATA

Our aim is to find the distribution of the time to the MRCA of a population of size m given data from a sample of n individuals. In this chapter, we summarize the data by taking $\mathcal{D}$ to be the number of segregating sites in the DNA sequences. Prior to sampling, the required probability density is that of T_m, defined as

$$T_m = W_m + \cdots + W_2. \tag{5.33}$$

We call this the *predata* density of TMRCA. The *postdata* density function is that of T_m given $\mathcal{D} = \{S_n = k\}$, which, using Bayes' formula, satisfies

$$f_{T_m}(t \mid \mathcal{D}) \propto f_{T_m}(t) \mathbb{P}(S_n = k \mid T_m = t). \tag{5.34}$$

As in (13) of Tavaré et al. (1996), this can be expressed as follows:

$$
\begin{aligned}
f_{T_m}(t) \mathbb{P}(S_n = k \mid T_m = t) &= \int_0^\infty f_{T_m, L_n}(t, l) \mathbb{P}(S_n = k \mid T_m = t, L_n = l)\, dl \\
&= \int_0^\infty f_{T_m, L_n}(t, l) \mathbb{P}(S_n = k \mid L_n = l)\, dl \\
&= \int_0^\infty f_{T_m, L_n}(t, l) \mathrm{Po}(k, l\theta/2)\, dl. \tag{5.35}
\end{aligned}
$$

In (5.35), $f_{T_m, L_n}(t, l)$ is the joint probability density of T_m and L_n in the bivariate coalescent process. Supposing for the moment that an observation (t, l) could be generated from the joint density of T_m and L_n, we see from (5.35) that the *rejection method* can be used to generate from the conditional distribution in (5.35). The idea is to keep the observation t with probability $u = \mathrm{Po}(k, l\theta/2)$, and reject it otherwise. The rejection step can be improved by noting that t can be accepted with probability u/c for any constant $c > u$. Because $\mathrm{Po}(k, l\theta/2) \leq \mathrm{Po}(k, k)$, we may take $c = \mathrm{Po}(k, k)$, and hence we accept t with probability u given by

$$u = \frac{\mathrm{Po}(k, l\theta/2)}{\mathrm{Po}(k, k)}. \tag{5.36}$$

Ripley (1987)(pg. 60) gives a description of the rejection method.

To generate an observation from the joint distribution of T_m and L_n, we can use Algorithm 3 directly. In addition, it generates observations from the conditional distribution of the number $A_m(T_n)$ of ancestors of the population of size m given $\mathcal{D}$. In summary, we have the following algorithm.

Algorithm 4 *Rejection algorithm for* $f_{T_m}(t \mid S_n = k)$. *U denotes a uniform* (0,1) *random variable, independently generated at each use.*

1. Set $a_m = m$, $a_n = n$, $t = 0$, $l = 0$.

2. Set $w = -2\log(U)/a_m(a_m - 1)$.

3. Solve for s the equation $\Lambda(t + s) - \Lambda(t) = w$.

4. Set $t = t + s$, $l = l + a_n s$.

5. Set

$$p = \frac{a_n(a_n - 1)}{a_m(a_m - 1)}, \tag{5.37}$$

and $a_m = a_m - 1$.

6. With probability p, set $a_n = a_n - 1$. If $a_n > 1$, go to 2.

7. Set $u = \text{Po}(k, l\theta/2)/\text{Po}(k, k)$. Accept (t, a_m) with probability u, else go to 1.

8. Set $t_n = t$, $a^* = a_m$. If $a^* = 1$, set $t_{mn} = 0$, and stop. Otherwise, use Algorithm 2 starting from $t = t_n$, $n = a^*$ to generate an observation t_{nm} on the total height of a coalescent tree of a^* individuals, then stop.

The values of t_n, a^*, t_{nm} generated by a single run through Algorithm 4 have the joint distribution of the sample tree height T_n, the number of ancestors $A_m(T_n)$ of the sample at the time the sample finds its MRCA, and the additional time T_{nm} required to get to the MRCA of the population conditional on having observed k segregating sites in the sample. The value of $t_m = t_n + t_{nm}$ has the distribution of T_m given $S_n = k$.

5.7   APPLICATIONS

Whitfield et al. (1995) sequenced a region of 15,680 base pairs from the Y chromosome of $n = 5$ individuals. They observed just three segregating sites and estimated the coalescence time of the sample to be between 37,000 and 49,000 years. Their analysis was not based on a population genetics model. Tavaré et al. (1996) use coalescent methods to reanalyze these data, using a number of plausible scenarios about variability in the effective population size N and the underlying mutation rate μ. Whitfield et al. (1995) estimated the mutation rate in the region to be $\mu = 3.52 \times 10^{-4}$ substitutions per generation, based on a generation time of $G = 20$ years. Using an estimate of $N = 4,900$, the value used by Hammer (1995), Tavaré et al. (1996) found a 95% credible region for TMRCA of the sample of 30,000–183,000 years.

Here we examine two aspects in more detail: we estimate TMRCA for the population, and we estimate the chance that the sample and the population share their most recent common ancestor, given the data of 3 segregating sites from the sample of 5 individuals. For illustration, we use a model of deterministic fluctuation in population size of the form

$$\lambda(t) = \begin{cases} \alpha^{-1}, & t > V, \\ \alpha^{-t/V}, & 0 \leq t \leq V. \end{cases} \tag{5.38}$$

This corresponds to a model in which the population has constant relative size $\alpha \in (0, 1)$ prior to time V ago, and exponential growth to relative size 1 at the time of sampling. We take a value of 50,000 years for V, and $\alpha = 10^{-4}$. For comparison with our earlier work, we suppose that the effective size in the constant phase is 4900, so that $N = 4.9 \times 10^7$. To apply Algorithm 4 we need a value to use for m. In practice, it is difficult to detect a difference between the distributions of, say, T_{200} and T_{500}, and we choose for this illustration a value of $m = 200$. We use Algorithms 3 and 4 to simulate 10,000 observations from the predata distribution of (T_5, T_{200}) and the corresponding postdata distribution, given that in the sample of size 5 there are 3 segregating sites.

For this demographic model, the predata distribution of T_5 has a mean of 199,000 years and 95% of the distribution of lies in the interval (76,000–464,000) years. (Here and in what follows, all ages are rounded to the nearest 1000 years. We use the shorthand I95 to denote the interval such that 2.5% of the mass of a distribution is to the left of the left endpoint, 2.5% to the right of the right endpoint.) The predata distribution of T_{200}, representing the TMRCA of the whole population, has a mean of 238,000 years, and an I95 of (113,000–504,000). In the 10,000 simulations, the observed fraction of times that the sample and the population had the same MRCA was 0.671, in good agreement with the theoretical value of 0.673 from (5.20).

The postdata distribution of T_5 has a mean of 108,000 years, and an I95 of (61,000–194,000) years. Note that the postdata distribution suggests a much shorter time for the TMRCA of the *sample*, and the postdata distribution is much more concentrated than the predata distribution. The estimated densities are plotted in Figure 5.3.

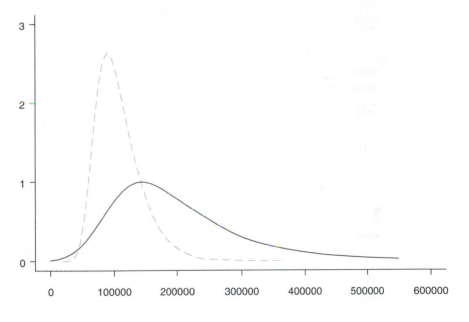

Time to MRCA (years)

Figure 5.3. Density functions for the pre-data (solid lines) and post-data distribution of T_5. *x*-axis is in years.

The proportion of the 10,000 runs that resulted in the population and the sample having a common MRCA was 0.233, markedly smaller than the predata fraction. On the basis of this, we anticipate that the additional time to the MRCA of the population will be much larger than the corresponding increment in the predata distribution. This is indeed the case; the former has a mean of 101,000 years, the latter 39,000 years. The mean time to the population MRCA is 209,000 years, with an I95 of (99,000–465,000) years. This interval is somewhat shorter than the I95 for the predata distribution. The estimated densities are plotted in Figure 5.4.

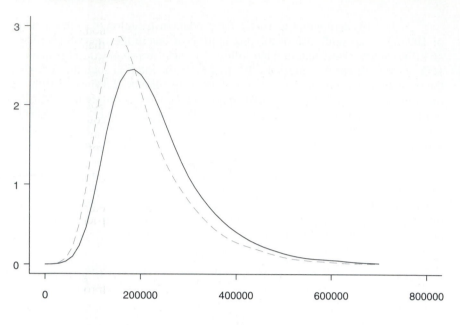

Time to MRCA (years)

Figure 5.4. Density functions for the predata (solid lines) and postdata distribution of T_{200}. x-axis is in years.

Two points from this analysis should be emphasized. First, the post-data TMRCA of the *sample* can be a serious underestimate of the corresponding time for the *population*. Second, the extra information contained in this small sample does rather little to refine our predata assessment of TMRCA for the population; as can be seen in Figure 5.4, the two densities are very similar.

Finally, what is the effect of the population expansion? For a constant-size model with $N = 4900$ and the same value of μ, the I95 of the postdata distribution of T_{200} is (59,000–410,000) years. The population expansion has shifted this interval to the right by about 54,000 years, essentially the time to the start of the constant phase. The intuition behind this is clear. By the beginning of the expansion phase at time V, a sample of moderate size m is likely to have almost m distinct ancestors. The time to the MRCA is therefore approximately V plus the corresponding time in a population of constant size.

5.8   DISCUSSION

This chapter gives a feel for one approach to an intriguing problem in the historical sciences: estimation of the time to the most recent common ancestor of a population of individuals given DNA-sequence data on a sample from the population. The approach described here, based on a stochastic model for the ancestral relationships between individuals, is intended to be illustrative of the field.

We based our inferences on a summary statistic (the number of segregating sites) of the sample of DNA sequences. Related methods for the full data (once more under the infinitely-many-sites mutation model) have also been developed. The theory of the reduced genealogical trees that represent the complete sequence information appears in Griffiths (1989) and Griffiths and Tavaré (1995). Computer-intensive inference methods are described in Griffiths and Tavaré (1994a).

There are many open problems that remain to be solved. Careful analysis of samples of molecular sequences should take account of the role of demography: nonrandom mating, population subdivision, and fluctuations in population size. With the availability of more data, more refined mutation models could also be exploited. The assumption of random sampling is implicit in most analyses, but nonrandom samples are more likely the rule. Finally, we have assumed that mutation rates and population-size fluctuations are known. Methods that allow for variability in these parameters, and further discussion of many related issues, appear in Donnelly et al. (1996) and Tavaré et al. (1996).

ACKNOWLEDGMENTS S.T. was supported in part by NSF grant BIR 95-04393. The article was completed while the author was Gastprofessor in Applied Mathematics at the University of Zürich.

REFERENCES

Ayala, F. J. 1995. Association Affairs: The myth of Eve: Molecular biology and human origins. *Science*, **270**, 1930–1936. {*81, 95*}

Cann, R., Stoneking, M., and Wilson, A. C. 1987. Mitochondrial DNA and human evolution. *Nature*, **325**, 31–36. {*81, 95*}

Donnelly, P. and Tavaré, S. 1995. Coalescents and genealogical structure under neutrality. *Annual Review of Genetics*, **29**, 401–421. {*86, 87, 95*}

Donnelly, P., Tavaré, S., Balding, D. J., and Griffiths, R. C. 1996. On the time since Adam. *Science*, **272**, 1357–1359. {*95*}

Dorit, R. L., Akashi, H., and Gilbert, W. 1995. Absence of polymorphism at the ZFY locus on the human Y chromosome. *Science*, **268**, 1183–1185. {*81, 95*}

Griffiths, R. C. 1979. Exact sampling distributions from the infinite neutral alleles model. *Advances in Applied Probability*, **11**, 326–354. {*85, 95*}

Griffiths, R. C. 1980. Lines of descent in the diffusion approximation of neutral Wright-Fisher models. *Theoretical Population Biology*, **17**, 37–50. {*81, 95*}

Griffiths, R. C. 1989. Genealogical-tree probabilities in the infinitely-many-sites model. *Journal of Mathematical Biology*, **27**, 667–68. {*95*}

Griffiths, R. C. and Tavaré, S. 1994a. Ancestral inference in population genetics. *Statistical Science*, **9**, 307–319. {*95*}

Griffiths, R. C. and Tavaré, S. 1994b. Sampling theory for neutral alleles in a varying environment. *Philosophical Transactions of the Royal Society of London Series B, Biological sciences*, **344**, 403–410. {*87, 95*}

Griffiths, R. C. and Tavaré, S. 1995. Unrooted genealogical tree probabilities in the infinitely-many-sites model. *Mathematical Biosciences*, **127**, 77–98. {*95*}

Hammer, M. F. 1995. A recent common ancestry for human Y chromosomes. *Nature*, **378**, 376–378. {*81, 92, 95*}

Hudson, R. R. 1983. Properties of a neutral allele model with intragenic recombination. *Theoretical Population Biology*, **23**, 183–201. {*81, 95*}

Hudson, R. R. 1991. Gene genealogies and the coalescent process. *Oxford Surveys in Evolutionary Biology*, **7**, 1–44. {*86, 95*}

Hudson, R. R. 1992. The how and why of generating gene genealogies. *Pages 23–36 of:* Takahata, Naoyuki and Clark, Andrew G. (eds), *Mechanisms of Molecular Evolution: Introduction to Molecular Paleopopulation Biology.* Sunderland, MA: Sinauer Associates, Inc. *{86, 96}*

Jobling, M. A. and Tyler-Smith, C. 1995. Fathers and sons: the Y chromosome and human evolution. *Trends in Genetics,* **11**, 449–456. *{81, 96}*

Kingman, J. F. C. 1982a. Exchangeability and the evolution of large populations. *Pages 97–112 of:* Koch, G. and Spizzichino, F. (eds), *Exchangeability in probability and statistics: proceedings of the International Conference on Exchangeability in Probability and Statistics, Rome, 6th–9th April, 1981, in honour of Professor Bruno de Finetti.* Amsterdam, The Netherlands: North-Holland Publishing Co. *{81, 84, 96}*

Kingman, J. F. C. 1982b. On the genealogy of large populations. *Journal of Applied Probability,* **19A**, 27–43. *{84, 87, 96}*

Ripley, Brian D. 1987. *Stochastic Simulation.* Wiley series in probability and mathematical statistics. New York, NY: John Wiley and Sons. ISBN 0-471-81884-4. Pages xi + 237. *{85, 91, 96}*

Saunders, I. W., Tavaré, S., and Watterson, G. A. 1984. On the genealogy of nested subsamples from a haploid population. *Advances in Applied Probability,* **16**, 471–491. *{96}*

Tajima, F. 1983. Evolutionary relationships of DNA sequences in finite population. *Genetics,* **105**, 437–460. *{81, 96}*

Tavaré, S., Balding, D. J., Griffiths, R. C., and Donnelly, P. 1996. Inferring coalescence times from DNA sequence data. *Genetics.* Submitted. *{92, 95, 96}*

Templeton, A. R. 1993. The "Eve" hypothesis: a genetic critique and reanalysis. *American Anthropologist,* **95**, 51–72. *{81, 96}*

Wallace, D. C. 1995. Mitochondrial DNA variation in human evolution, degenerative disease, and aging. *American Journal of Human Genetics,* **57**, 201–223. *{81, 96}*

Watterson, G. A. 1975. On the number of segregating sites in genetical models without recombination. *Theoretical Population Biology,* **7**, 256–276. *{90, 96}*

Watterson, G. A. 1982. Mutant substitutions at linked nucleotide sites. *Advances in Applied Probability,* **14**, 206–224. *{87, 96}*

Whitfield, L. S., Sulston, J. E., and Goodfellow, P. N. 1995. Sequence variation of the human Y chromosome. *Nature,* **378**, 379–380. *{81, 92, 96}*

Wills, C. 1995. When did Eve live? An evolutionary detective story. *Evolution,* **49**, 593–607. *{81, 96}*

PART II: CELL BIOLOGY

Mark A. Lewis

Organisms and their constituent cells have an immense number of tasks to perform, from information processing to growth and cell division to signaling to maintaining a constant internal environment in the face of external variation (homeostasis). While the disciplines of biochemistry, biophysics, genetics, and molecular biology are central to any understanding of how these tasks are performed, biological experience cannot always predict the consequences of complex physiological interactions. In this realm of unpredictability, mathematical of models can have a crucial role in our understanding of physiology. Mathematics provides a quantitative means with which to formulate precise hypotheses about physiology. The purpose of a model is not to "simulate" the biology, but to judiciously simplify the level of detail to where only the most important elements controlling the physiology remain. Analyses of the model then serve to show which hypotheses are consistent with the more complex experimental detail.

The chapters in this section discuss mechanisms governing physiological phenomena. Subjects include signal transduction via control of calcium and second-messenger dynamics (Chapter 6 by H. G. Othmer), control of cellular replication (Chapter 7 by J. J. Tyson, K. Chen and B. Novak), periodic diseases of the blood (Chapter 8 by M. C. Mackey), oscillatory responses of pupil nervous system to stimuli (Chapter 9 by J. Milton and J. Foss), and electrical bursting behaviors of cells (Chapter 10 by A. Sherman). Although each chapter deals with distinct phenomenon, mathematics allow us to look beneath the specific details to underlying themes that pertain to many of the chapters.

One theme common to each section is biological oscillations, whether in the intracellular calcium levels (H. G. Othmer), the control of cell cycling (J. J. Tyson, K. Chen, and B. Novak), blood counts in diseased individuals (M. C. Mackey), the diameter of the pupil responding to a light source (J. Milton and J. Foss) or the onset of active spiking interspersed with silent states in electrical behavior of cells (A. Sherman). Mechanisms governing each of the above oscillatory systems clearly depend upon specific biological milieu. The mathematics used in their study, however, have some common origins. For example, it may interest the reader to note that Mackey's discussion of periodic diseases of the blood (Chapter 8) and Milton and Foss' discussion of the pupil-light reflex (Chapter 9) use similar nonlinear delay differential equation (DDE) models. Although the processes differ in the biological specifics, oscillations arise in both cases from negative-feedback loops with delays, making DDEs suitable for their study. Both show that the results of long delays can be dramatic, leading not only to oscillations, but also to a variety of very complex temporal patterns. These patterns are not simply mathematical niceties; they are observed biologically either under disease conditions (Mackey) or under laboratory conditions (Milton and Foss).

The theme of excitable dynamics arises in Othmer's discussion of calcium dynamics (Chapter 6), Tyson et al.'s discussion of cellular control (Chapter 7) and Sherman's discussion of electrophysiology (Chapter 10). Models in each of these chapters possess a nonlinear threshold, so that small (subthreshold) stimuli have little effect, but larger (suprathreshold) stimuli cause a dramatic, large-scale response. The classic example of an excitable system arises in the firing of a nerve: the stimulus to a nerve must exceed a threshold before the nerve fires. Mathematical details of this behavior were pioneered in the early 1950s by Hodgkin and Huxley, with an ODE model for the nerve. This elegant work had a profound impact on electrophysiology that is still evident today (see Chapter 10 by A. Sherman) and earned Hodgkin and Huxley a Nobel prize in physiology. Many models in this section more closely resemble later simplifications of Hodgkin and Huxley's (1952) work by FitzHugh (1961) and Nagumo (1962), who showed how the essence of the nerve could be described by a pair of coupled ODEs with excitable dynamics. The themes of oscillations and excitability are connected. It turns out that small modifications of excitable dynamics can give rise to oscillatory dynamics. In the modified excitable system, for example, the nerve never returns completely to a rest state, but fires repeatedly. This kind of oscillation mechanism, exemplified in Chapters 6, 7 and 10, differs from the delayed-negative-feedback loop in Chapters 8 and 9.

Although this section examines a diverse body of work, readers may find other themes that unify. They are also encouraged to ask questions of each chapter:

- **How is the mathematics useful?** Are precise hypotheses formulated? Does analysis of the model yield specific predictions? If so, how do these compare with experiment? How would one proceed without mathematics?

- **Where do the equations come from?** Each chapter derives equations using specific modeling principles. How do the authors make the transition from scientific hypotheses to mathematical formulae?

- **Does the model work?** A key element of each chapter is the author's ability to tailor a mathematical model to focus on the specific question at hand. What is missing from each model? How do we know whether it is important or not?

REFERENCES

FitzHugh, R. 1961. Impulses and physiological states in theoretical models of nerve membrane. *Biophysical Journal*, **1**, 445–466. {*98, 299, 305*}

Hodgkin, A. L. and Huxley, A. F. 1952. A quantitative description of membrane current and its application to conduction and excitation in nerve. *Journal of Physiology*, **117**, 500–544. {*98*}

Nagumo, J. S., Arimoto, S., and Yoshizawa, S. 1962. An active pulse transmission line simulating nerve axon. *Proceedings of the Institute of Radio Engineers*, **50**, 2061–2071. {*98, 299, 307*}

6. SIGNAL TRANSDUCTION AND SECOND MESSENGER SYSTEMS

Hans G. Othmer

6.1 INTRODUCTION

A primary characteristic of living organisms, from the unicellular level to the largest plants and animals, is their active interaction with their environment. From the unicellular level upward organisms have developed sensitive signal- detection systems that enable them to find food and mates and to avoid harmful environments. However, most organisms also maintain a clear distinction between inside and outside, and as a result, the environmental signals rarely penetrate very far into an organism. Instead, there are mechanisms for transducing external signals into internal signals and, where appropriate, a response.

When the extracellular signal is a molecule such as a hormone, the detection step usually involves membrane-bound receptors that bind this molecule. Frequently the receptor-hormone complex has catalytic activity that can activate an intracellular *G protein* in the next step of the transduction process. These G proteins are heterotrimeric proteins that consist of α, β, and γ subunits. When activated, the G protein splits into the α subunit and the $\beta\gamma$ subunit. The α subunit may in turn activate intracellular enzymes such as adenylate cyclase, which produces cyclic 5' adenosine monophosphate (cAMP), or phospholipase C, an enzyme involved in the pathway that leads to mobilization of intracellular calcium (Ca^{2+}). Intracellular species such as cAMP or Ca^{2+} then serve as *second messengers* that can control a variety of components or processes, including the activity of protein kinases and other enzymes, cytoskeletal structure, and gene transcription (see Figure 6.1). Alternatively, G proteins can regulate the conductance of an ion channel in the membrane and thereby affect the transmembrane electrical potential difference. We shall discuss a specific transduction scheme further in the following section, and we refer the reader to Koshland (1988); Monk and Othmer (1989); Borsellino et al. (1990), and Tang and Othmer (1994a) for models of transduction in other systems.

Both the extracellular and intracellular signals can be characterized by their power spectrum, which gives the amplitude of the Fourier components of the signal as a function of the frequency. Broadly speaking, the problem in mathematical modeling of signal-transduction systems is to understand how the biochemical and biophysical steps in the transduction process determine the amplitude and temporal characteristics of the intracellular signal. In some

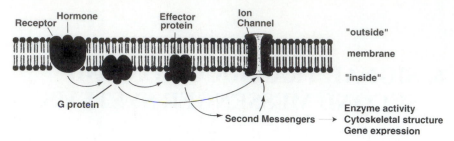

Figure 6.1. A schematic of a typical G-protein-based transduction scheme. The effector may be an enzyme such as adenylate cyclase or some other protein.

systems a sensitive detection system and a powerful amplifier are needed because extracellular signals are very weak. The gain in amplitude can be enormous: a signal consisting of a few molecules bound to receptors that lead to activation of an enzyme can be amplified into thousands of second-messenger molecules. In other systems transduction primarily involves conversion of a sufficiently large amplitude extracellular signal into a repetitively varying intracellular messenger. This type of amplitude-to-frequency conversion is often used in the nervous system, where the objective is to transmit information to some distant site with minimal loss of information. Frequency encoding of the amplitude of the external signal, coupled with the appropriate transmission system, is an efficient way to accomplish this. In this chapter we shall demonstrate that frequency encoding can arise in certain models, but space does not permit a theoretical analysis of the general characteristics of a model that produces frequency encoding. We refer the reader to Tang and Othmer (1995) for this aspect.

A further complication in understanding signal transduction and its intracellular effects arises from the fact that many sensory systems respond primarily to short-term *changes* in the signal intensity, rather than to the absolute intensity. More precisely, a step change in an external signal from one constant level to another elicits a transient change in some internal component of the state of the system, followed by a return to a basal level of activity. This is called desensitization, habituation, or adaptation, depending on the context. The visual system provides a common example: when one steps from a dark room into bright light, initially there is a rapid response, but the visual system quickly adapts to the higher light intensity. Adaptation should be distinguished from simple saturation of the sensory system, for it is also important to maintain sensitivity to further changes in the signal. These characteristics are shown schematically in Figure 6.2.

Signal-transduction/second-messenger mechanisms are often described by a finite number of state variables and an evolution equation that determines how the state changes under prescribed inputs or stimuli. Let us denote the state vector by $\mathbf{x}(\tau) \in \mathbf{R}^n$ and write the evolution equation in the form

$$\frac{d\mathbf{x}}{d\tau} = \mathbf{F}(\mathbf{x}, S), \tag{6.1}$$

where $S \in \mathbf{R}$ represents the stimulus or input to the system. Typical state variables are the concentration of a chemical second messenger, a transmembrane

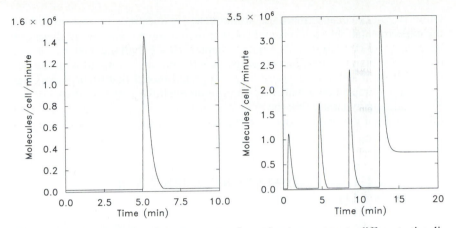

Figure 6.2. A schematic of the response of an adapting system to different stimuli. Shown here is the simulated cAMP relay response to extracellular cAMP stimuli in a model of the cellular slime mold *Dictyostelium discoideum*. In (a) a step change in extracellular cAMP elicits a single pulse of secreted cAMP. In (b) the system responds and adapts to a sequence of step increases in extracellular cAMP ranging from $10^{-4}\mu$M to 1μM (1μM = 1 micromolar), but at the highest stimulus the transduction system saturates. (After Tang and Othmer (1994a).)

potential difference, gating variables, etc. In the context of calcium dynamics as described in the next section, $\mathbf{x}$ is a vector comprised of the dimensionless concentrations of cytoplasmic calcium, sequestered calcium and channel states, and S represents the extracellular hormonal stimulus at the cell surface. In general a change in S leads to a change in both the transient and steady-state values of $\mathbf{x}$, but in systems that adapt some functional of $\mathbf{x}$ that we call the response should be independent of S when S is time-independent. Suppose that the response $\mathcal{R}$ of the system is given by

$$\mathcal{R}(\tau) = \mathcal{G}(\mathbf{x}(\tau)). \tag{6.2}$$

For instance, in the visual system $\mathcal{G}$ could represent the firing rate of a neuron. When the "basal dynamics" are time independent, which means that the system has an asymptotically-stable steady state in the presence of any constant stimulus, we say that the response $\mathcal{R}$ adapts to constant stimuli if the steady state response is independent of the magnitude of the stimulus S. By adapting to background levels of a signal (or equivalently, changing the sensitivity to the amplitude of signals) the sensory system can process a far greater range of amplitudes. For example, the visual system in certain amphibians can detect and respond to light stimuli whose amplitude ranges over five or more orders of magnitude (Koutalos and Yau 1996). A discussion of various models of adaptation, including some for bacterial chemotaxis, for signal relay in the cellular slime mold *Dictyostelium discoideum*, and for adenylate cyclase, can be found in Koshland (1988); Othmer et al. (1985); Tang and Othmer (1994a) and Spiro et al. (1996).

Our objective in this chapter is to illustrate how a model for a signal-transduction/second-messenger system can be derived, and thereby to illustrate the major steps in the process of developing a mathematical description of a physical or biological process. We have chosen to model intracellular calcium dynamics because there are some clear-cut questions that are raised by the experimental observations and some progress toward answering them can be made within the confines of a single chapter. As we shall see later, the model does not involve the transduction steps directly, but deals instead with the details of the intracellular response. An integrated model that incorporates the entire signal transduction pathway and the intracellular response does not exist for calcium dynamics, but the reader is referred to Tang and Othmer (1994a) for a model of the entire process in the cAMP pathway in *Dictyostelium discoideum*.

6.2   BACKGROUND MATERIAL ON CALCIUM DYNAMICS

6.2.1   The experimental observations

Many cell types use changes in intracellular calcium triggered by hormones, neurotransmitters, or growth factors as a second messenger to trigger a variety of intracellular responses, including contraction, secretion, growth and differentiation. One of the earliest experiments that revealed the complexity of the calcium response following hormonal stimulation is that due to Woods et al. (1986), who studied the response of hepatocytes (a type of liver cell) to vasopressin, a hormone that mobilizes intracellular calcium. They found that over the range of 200 nM to 1 μM (where μM stands for micromolar) in the hormone concentration, stimuli evoke repetitive spikes in the intracellular calcium concentration, rather than simply elevating the level of calcium. Moreover, they found that as the hormone concentration was raised, the frequency of spiking increased, but the amplitude remained nearly constant. Thus the analog extracellular hormone signal was converted into a frequency-encoded digital signal (the number of calcium spikes). Similar dynamic behavior has been found in a large number of cell types since then, and has led to the suggestion that calcium spiking and frequency encoding must have a physiological role. Since high calcium levels may be cytotoxic, it may be necessary to maintain a low average concentration of calcium, but since calcium frequently serves as a trigger for other processes, a transient elevation above a low mean concentration suffices for this purpose. Moreover, using a large transient increase as the trigger permits the use of a sharper threshold for response, and hence better noise discrimination. Other possible advantages are suggested by Meyer and Stryer (1991) and in the final discussion of the symposium volume edited by Bock and Ackrill (1995), but to date there is little hard evidence that the spiking plays an essential physiological role. As we shall see, the mathematical model developed later has the property that it is excitable, which means that small stimuli are damped but large enough stimuli give a large response (Alexander et al. 1990), and the observed dynamic behavior follows from this fact.

Perhaps of greater significance is the fact that the response in many cells is also spatially inhomogeneous. In a variety of cell types, waves of calcium release propagate across the cell in response to hormonal or other stimuli. For instance, in *Xenopus laevis* oocytes (frog eggs), penetration of a sperm into the egg triggers a localized increase in cytosolic calcium that propagates away from the point of entry at approximately 10 microns/sec, inducing cortical contraction, cell division, and structural rearrangement (Gilkey et al. 1978; Nuccitelli 1991). In addition, there is evidence that at least some of the hormone-sensitive calcium stores in *Xenopus* oocytes are localized at the animal pole (Lupu-Meiri et al. 1988; Berridge 1990), and thus in either fertilization or hormonal stimulation, propagation of the wave is essential for inducing the entire cell to respond to a localized stimulus. In hepatocytes the oscillations appear to originate from a single locus and propagate across the cell (Rooney et al. 1991). Moreover, the initiation site seems to be relatively constant in a given cell, even for different stimuli. These cells are polarized to a degree and a variety of receptors are known to be concentrated at the sinusoidal membrane, which may account for the origin of the wave. The speed of the waves in hepatocytes is typically 20–25 microns/sec.

In cardiac and smooth-muscle cells, calcium is sequestered primarily in a specialized organelle called the sarcoplasmic reticulum, whereas in cells that are not electrically excitable it is stored in the endoplasmic reticulum. In electrically nonexcitable cells at least part of the sequestered calcium can usually be released by binding of inositol 1,4,5-trisphosphate ($InsP_3$) to a receptor that controls the permeability of a calcium channel in the ER membrane. This channel is a tetramer which can open to four distinct conductance levels that are multiples of a unit step of 20 picoSiemans (Watras et al. 1991). $InsP_3$ binding is obligatory for opening of the channel (Harootunian et al. 1991), as is calcium (Watras et al. 1991). Low calcium levels promote opening of the channels, while high levels inhibit opening (Finch et al. 1991; Bezprozvanny et al. 1991). Some cells also exhibit calcium-induced calcium release via another class of channels whose permeability is controlled by calcium alone (Tang and Othmer 1994b).

A typical pathway for transduction of a hormonal signal into a variation in intracellular calcium is shown in Figure 6.3. The hormone binds to a plasma membrane receptor, and the complex catalyzes the GDP/GTP (guanosine diphosphate/triphosphate) exchange in a G protein. The G_α component in turn activates phospholipase C, the enzyme that catalyzes the hydrolysis of phosphatidylinositol 4,5-bisphosphate to $InsP_3$ and diacylglycerol. $InsP_3$ then binds to an $InsP_3$ receptor that is part of the calcium channel complex on the ER membrane, and if calcium is also bound at the calcium-activating site, the channel opens. Frequently sequestered calcium can also be released from either the $InsP_3$-sensitive store or from a distinct store through another type of calcium channel that is $InsP_3$ insensitive. For instance, Thomas et al. (1991) cite evidence that in hepatocytes only 30% to 50% of the nonmitochondrial calcium is released from the $InsP_3$-sensitive pool, even when the calcium pump is inhibited; the remainder is released by calcium ionophores, which are proteins that form pores in the membrane through which calcium can diffuse. In Figure 6.3 we indicate that these stores are distinct, but that is not known in general.

Under prolonged hormonal stimulation, or when $InsP_3$ is injected directly into the cell, calcium spiking occurs only in a range of hormonal or $InsP_3$ concentrations (DeLisle et al. 1990). An example is shown in Figure 6.4, where one sees that the frequency of oscillation depends on the hormone level. Rooney et al. (1989) found that there was a correlation between the latent pe-

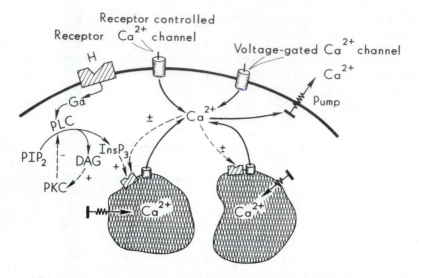

Figure 6.3. A schematic of the transduction system for hormonal stimuli, the calcium transport mechanisms, and the InsP$_3$-sensitive and InsP$_3$-insensitive stores. Key: G$_\alpha$: The G protein that activates phospholipase C (PLC); H: hormone; PIP$_2$: phosphatidylinositol 4,5-bisphosphate; InsP$_3$: inositol 1,4,5-trisphosphate; DAG: diacylglycerol; PKC: protein kinase C. Solid lines indicate a material flow, and dashed lines indicate a control influence. Not all cells have all the components shown.

riod (the time between application of the stimulus and the first response) and the subsequent oscillation period over a range of hormonal doses and oscillation frequencies. In some cells the shape of the oscillations depends on the hormone, but such details probably arise from differences in the earliest steps of the signal-transduction system.

6.2.2   An overview of models for calcium dynamics

Calcium oscillations will in general involve many intracellular components, but in order to understand which components are essential and which are not, it is helpful to identify major subsystems and determine how they are involved in the overall process. One can identify three major subsystems in Figure 6.3: (i) the transduction subsystem, whereby hormonal signals are transduced via G proteins into an InsP$_3$ signal, (ii) a system comprising the InsP$_3$-sensitive calcium store, and (iii), a system comprising an InsP$_3$-insensitive calcium store, which may include voltage- and receptor-gated membrane channels. Of course, not all cells will necessarily have all of these components.

Given this subdivision, one can identify a number of distinct ways in which calcium oscillations may be generated. First, the transduction system may generate an oscillatory InsP$_3$ signal in response to a constant hormonal stimulus. This could arise from a feedback loop in which DAG activates protein kinase C, which in turn phosphorylates the G protein that couples the receptor to phospholipase C. This would lead to a reduction in the production of InsP$_3$ and DAG, which in turn would reduce the inhibition of transduction,

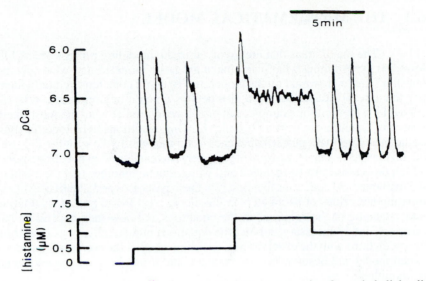

Figure 6.4. The experimentally observed calcium concentration for endothelial cells as a function of the histamine stimulus level. The solid piecewise-constant line shows the hormone level as a function of time. (From Jacob et al. (1988), with permission.)

and so on. Thus the transduction system produces an oscillatory input into the InsP$_3$-sensitive calcium subsystem, and in such systems the InsP$_3$ oscillations drive the calcium oscillations. The model proposed by Cobbold, et al. (1991) is of this type.

In a second class of models the oscillations are generated within the InsP$_3$-sensitive calcium-store subsystem. As we remarked earlier, it is known that in a number of systems the InsP$_3$-sensitive calcium channel is not activated unless calcium occupies an activating site on the receptor. Furthermore, the channel conductivity is reduced at high calcium concentrations, probably owing to the presence of a calcium-binding inhibitory site on the receptor. The models proposed be DeYoung and Keizer (1992) and by Othmer and Tang (1993) incorporate these observations. In both these models InsP$_3$ is treated as a parameter; oscillations in the cytosol calcium arise without any temporal variation in the level of InsP$_3$. As we show later, such a system is excitable in the sense used earlier, and thus there is a threshold level of the InsP$_3$ stimulus below which there is no significant response, but above which there is a large response. As frequently happens in such systems, tuning the system parameters appropriately also produces oscillatory behavior. In the model developed in the following section InsP$_3$ is both the stimulus that produces an excitable response, and the parameter that produces oscillatory behavior in a suitable range. This is in accord with the experimental observations, in that injection of InsP$_3$ can produce either single calcium transients or calcium oscillations, depending on the stimulus level (Dupont et al. 1991).

6.3   THE MATHEMATICAL MODEL

The major steps that are involved in the modeling process are as follows. (i) Determining what phenomena are to be modeled and what specific questions are to answered by the modeling. To do this one must not only identify the unresolved questions but also decide which if any of these can be resolved by a model. For example, can one hope to predict what the most likely source of calcium oscillations is under *in vitro* conditions? (ii) Identification of the most important physical and chemical processes in the system. This frequently involves a large measure of judgment, in that it is often not clear a priori which processes must be included and which can be ignored. (iii) Developing the mathematical equations that describe these processes and identifying or estimating the values of parameters in the model. (iv) Developing a qualitative understanding of the solutions of the equations and how they depend on parameters, and if necessary, solving the equations numerically. (v) Comparing the predictions with the observations, deciding whether the correspondence between model and observations is adequate, and if not, how the model should be modified.

One can identify two aspects to the experimental results in the preceding section: those such as the oscillations that can perhaps be understood by ignoring spatial variations in the calcium concentration within a cell, and those such as the traveling waves, in which the spatial variations in concentration are essential. We shall focus on the oscillations first and regard the system as spatially uniform, but at some point we must convince ourselves that spatial variations can be ignored, either in the intact system or in certain experimental contexts.

6.3.1   What are the important processes?

In constructing mathematical models of complicated systems such as that involved in calcium regulation, one should always try to simplify the model as much as possible, but not so much as to make it unrealistic. Thus a first step is to decide what processes are involved, which of them are essential and must be included in the model, and which are secondary and can be neglected. The objective is to develop a model that is complex enough to reproduce the major aspects of the experimental observations, yet simple enough to be analyzed qualitatively in order to understand how the component processes interact to produce the observed dynamics. As we indicated earlier, direct injection of $InsP_3$ produces the oscillatory behavior observed under hormonal stimulation, and therefore we can ignore the steps from hormonal stimulus to production of $InsP_3$ in a first model. Obviously the resulting model will not be complete, but it will lead to an understanding of how the $InsP_3$-Ca^{2+} system interacts at the level of the $InsP_3$ receptor.

A second issue concerns the role of extracellular Ca^{2+}. In small cells such as hepatocytes, extracellular calcium is needed to sustain the calcium oscillations: in its absence the frequency of oscillation is reduced and the oscillations eventually die out. In larger cells such as oocytes, the major features of the observed dynamics are not dependent on the presence of extracellular calcium or calcium-induced calcium release from $InsP_3$-insensitive stores. Consequently, we focus on a model in which there is no calcium transport between the interior and the exterior of the cell, and no calcium-induced calcium release from $InsP_3$-insensitive stores; the oscillations center on the dynamics of the $InsP_3$ receptor. Even though the model may not be strictly applicable to hepatocytes, some important insights can be gained even for this system.

Since the calcium is free to diffuse in the cytoplasm (except for buffering, which is the binding of calcium to fixed or mobile proteins, and which is neglected here), we must determine whether it is essential to incorporate diffusion in the model. Before we analyze this, we should recognize that the answer will depend on the type of cell under consideration. In very small cells we might expect that diffusion is fast enough to minimize spatial variations in the calcium concentration, but we also know or should suspect that diffusion is important in large cells such as *Xenopus* oocytes, because a variety of different types of waves are observed. How do we determine quantitatively whether or not we can neglect diffusion?

Consider a closed one-dimensional system of length L. If diffusion is the only process in the system, then the governing equation is the diffusion equation

$$\frac{\partial c}{\partial t} = D \, \frac{\partial^2 c}{\partial x^2} \tag{6.3}$$

for $x \in (0, L)$. In this equation c is the concentration of the diffusing substance, D is the diffusion coefficient, t denotes the time variable, and x denotes the space variable. This is a partial differential equation, so we need an initial function that gives the spatial distribution of c at $t = 0$, which we denote $c(x, 0) = \phi(x)$. In addition we have to specify conditions at the end of the interval, and since we have assumed that the system is closed the appropriate conditions are that the flux vanishes at the boundary, *i. e.*

$$- D \, \frac{\partial c}{\partial x} \, (0, t) = D \, \frac{\partial c}{\partial x} \, (L, t) = 0 \tag{6.4}$$

for $t > 0$. A (in fact, *the*) solution of (6.3) subject to the boundary conditions at (6.4) is given by the infinite series:

$$c(x, t) = \sum_{n=0}^{\infty} a_n e^{-(n\pi/L)^2 Dt} \cos\left(\frac{n\pi x}{L}\right). \tag{6.5}$$

If the diffusion equation is unfamiliar, the reader may simply verify by formal differentiation of the series and substitution that this is a solution. The coefficients a_n can be computed by constructing the Fourier cosine series of the initial function $\phi(x)$. In particular, $a_0 = L^{-1} \int_0^L \phi(x) \, dx$.

How does this solution help us to determine whether or not spatial variations in calcium can be neglected in an intact cell? In order to neglect them, any variations should decay rapidly compared to the characteristic time of other processes being considered. From (6.5) one sees that the characteristic time for decay of the lowest nonconstant component of the solution is

$$\tau_1 = \frac{L^2}{\pi^2 D}. \tag{6.6}$$

The diffusion coefficient of a small molecule in water is of order $5 \cdot 10^{-6}$ cm^2/sec, and for a small cell $L \sim \mathcal{O}(10)$ microns. From these we find that $\tau_1 \sim 2 \cdot 10^{-2}$ seconds, which is small compared, for example, to the period of oscillations of intracellular calcium (see Figure 6.4). Of course when certain components of the transduction system are spatially localized, as in some of the cells discussed in the earlier, diffusion plays an essential role. The reader can check that this remark also applies to large cells such as oocytes, for which $L \sim 1.0$ mm, even if the transduction system and the storage compartment are uniformly distributed.

6.3.2   Two states for the InsP$_3$ receptor do not suffice

As we stated earlier, we will not consider calcium transport across the plasma membrane at first. In this case calcium simply flows or is pumped back and forth between the storage compartment and the cytoplasm, and the total amount of intracellular calcium is conserved. Thus the dynamics can be described by a autonomous scalar equation, and such equations cannot give rise to periodic oscillations. Consequently, a two-variable model of a closed system in which the only dynamic variables are the concentrations of calcium in the cytosol and in the storage compartment cannot give rise to oscillations, whether or not InsP$_3$ plays a role in the release of calcium. This conclusion applies, in particular, to the model of Goldbeter and DuPont (1990), and to that of Somogyi and Stucki (1991). It is conceivable, however, that a model in which the InsP$_3$ receptor has two distinct states may give rise to oscillation, since there are now at least two independent dynamic variables. However, it can be shown that oscillations cannot arise in response to hormonal stimulation in such a model under very mild restrictions on the functional form of the receptor response to InsP$_3$ and calcium (Othmer and Tang 1993).

DeYoung and Keizer (1992) first proposed a kinetic scheme for the InsP$_3$ receptor which assumes that there is an activating site at which InsP$_3$ binds and both activating and inhibiting sites at which Ca^{2+} can bind. They assume that binding can occur in any order, which leads to eight possible states of the receptor. Their work is based on experimental results of Bezprozvanny et al. (1991), discussed later, and they showed that their model can reproduce many of the observed phenomena. However, it is difficult to obtain a qualitative understanding of such a complex model, and difficult to identify all of the rate constants involved in the transitions.

This leads us to a less complex four-state model proposed by Othmer and Tang (1993), which we describe next and analyze in the following section. The major assumptions of the model are that calcium binds to the activating site on the channel only after InsP$_3$ has bound to the receptor, and that the binding of calcium to the inhibitory site occurs only after calcium is bound to the activating site. This assumption of sequential binding leads to a model with only four states, which lends itself to qualitative analysis of the effect of various parameters. As we shall see later, the oscillations arise from the biphasic response of the InsP$_3$-sensitive calcium channel to calcium.

6.3.3 A four-state single-pool model

We suppose that the transitions between the states occur according to the following scheme:

$$I + R \quad \underset{k_{-1}}{\overset{k_1}{\rightleftarrows}} \quad RI,$$

$$RI + C \quad \underset{k_{-2}}{\overset{\bar{k}_2}{\rightleftarrows}} \quad RIC^+, \tag{6.7}$$

$$RIC^+ + C \quad \underset{k_{-3}}{\overset{\bar{k}_3}{\rightleftarrows}} \quad RIC^+C^-.$$

Here R denotes the bare receptor, I denotes InsP$_3$, and C denotes the cytosol calcium concentration. Further, RI denotes the receptor-InsP$_3$ complex, and RIC^+ (respectively, RIC^+C^-) denotes RI with calcium bound at the activating site (respectively, the activating and inhibitory sites). Each of these steps is analogous to a chemical reaction, and we assume that each obeys mass-action kinetics (*i. e.*, the rate is proportional to the products of the amounts of the reacting species). This means that the forward rate of the first step is $k_1 I \cdot R$, whereas the rate of the reverse step is $k_{-1} RI$. We assume that the only open (conducting) state in the above diagram is RIC^+

In addition, there is calcium exchange between the ER (the calcium store) and the cytoplasm, and a calcium pump between the cytosol and the ER. At present I is treated as a parameter for reasons stated earlier. Let $x_i, i = 2, \dots, 5$, denote the fractions in states R, RI, RIC^+, and RIC^+C^-, respectively. Then the governing equations are

$$\begin{aligned}
\frac{dC}{dt} &= v_r(\gamma_0 + \gamma_1 f(x_4))(C_s - C) - \bar{g}(C), \\
\frac{dx_2}{dt} &= -k_1 I x_2 + k_{-1} x_3, \\
\frac{dx_3}{dt} &= -(k_{-1} + \bar{k}_2 C) x_3 + k_1 I x_2 + k_{-2} x_4, \\
\frac{dx_4}{dt} &= -(k_{-2} + \bar{k}_3 C) x_4 + \bar{k}_2 C x_3 + k_{-3} x_5, \\
\frac{dx_5}{dt} &= \bar{k}_3 C x_4 - k_{-3} x_5.
\end{aligned} \tag{6.8}$$

Here v_r is the ratio of the ER volume to the cytoplasmic volume, γ_0 is the basal permeability of the ER membrane in the absence of InsP$_3$ and γ_1 the density of InsP$_3$-sensitive channels, both per unit volume of the ER, and C_s is the calcium concentration in the store. In the first equation we have neglected the small loss term due to Ca^{2+} binding to the receptor. The function $f(x_4)$ represents the dependence of the channel conductivity on the fraction of channels in the conducting state RIC^+, and we assume that $f(x_4) = x_4$, i.e., that the total channel conductivity is a linear function of the fraction of channels open. This choice is somewhat arbitrary, and the correct functional form is not known at present. In addition, we have the conservation condition $\sum_{k=2}^{5} x_k = 1$, which expresses the fact that the receptor must be in one of the four states.

The calcium pump between the cytoplasm and the ER is known to be a tetramer with four calcium binding sites. Thus we assume that

$$\bar{g}(C) = \frac{\bar{p}_1 C^4}{C^4 + \bar{p}_2{}^4}. \tag{6.9}$$

However it is not firmly established what the exponent should be, and some authors use 2 instead of 4. The equations can be simplified by defining the volume-average intracellular calcium concentration, C_0, as

$$C_0 = \frac{C + v_r C_s}{(1 + v_r)}. \tag{6.10}$$

We then define $x_1 = C/C_0$, and we can write the governing equations in the form

$$
\begin{aligned}
\frac{dx_1}{dt} &= \lambda(\gamma_0 + \gamma_1 x_4)(1 - x_1) - \frac{p_1 x_1^4}{p_2^4 + x_1^4}, \\
\frac{dx_2}{dt} &= -k_1 I x_2 + k_{-1} x_3, \\
\frac{dx_3}{dt} &= -(k_{-1} + k_2 x_1) x_3 + k_1 I x_2 + k_{-2} x_4, \\
\frac{dx_4}{dt} &= k_2 x_1 x_3 + k_{-3} x_5 - (k_{-2} + k_3 x_1) x_4, \\
\frac{dx_5}{dt} &= k_3 x_1 x_4 - k_{-3} x_5,
\end{aligned}
\tag{6.11}
$$

where $\lambda \equiv 1 + v_r$, $k_2 = \bar{k}_2 C_0$, $k_3 = \bar{k}_3 C_0$, $p_1 = \bar{p}_1/C_0$, and $p_2 = \bar{p}_2/C_0$. In this scaling all x_i range between 0 and 1, and it can be shown that the set $\{x_i \mid 0 \leq x_i \leq 1\}$ is invariant under the flow of (6.11). As we show later, this scheme is sufficiently robust to reproduce many of the observations. Since the sum of the channel fractions is one we could eliminate one fraction, and as we shall see later, the one most conveniently eliminated depends on the context.

Frequently a major problem in modeling biological phenomena is the parameter-estimation step, because rarely are all the parameters measured in the same system. We shall not confront this problem in detail here, but rather simply list the parameters that have been measured and refer the reader to Othmer and Tang (1993) for a discussion as to how the remainder were chosen. The ratio of the ER volume to cytoplasmic volume v_r is taken to be 0.185, after Alberts et al. (1989). The leakage rate between the calcium store and cytoplasm is known to be small. We use the value $\gamma_0 = 0.1 \ s^{-1}$, but the value of this parameter does not significantly influence the dynamics as long as it is small. The average calcium concentration is taken as $C_0 = 1.56 \ \mu M$. This leads to a maximal calcium concentration in the ER (with zero calcium concentration in the cytoplasm) of $10 \ \mu M$, which is in the physiological range (De Young and Keizer 1992).

The parameter values are listed in Table 6.1. The parameters needed for the nondimensional versions of the equations can easily be calculated from these values. We remark that the table originally given in Othmer and Tang (1993) contains a number of misprints that are corrected here.

Table 6.1. Parameter values for the four-state model.

Parameter	Value	Parameter	Value
v_r	0.185	k_1	12.0 $(\mu M \cdot s)^{-1}$
γ_0	0.1 s^{-1}	$\bar{k}_2$	15.0 $(\mu M \cdot s)^{-1}$
γ_1	20.5 s^{-1}	$\bar{k}_3$	1.8 $(\mu M \cdot s)^{-1}$
$\bar{p}_1$	8.5 $\mu M \cdot s^{-1}$	k_{-1}	8.0 s^{-1}
$\bar{p}_2$	0.065 μM	k_{-2}	1.65 s^{-1}
C_0	1.56μM	k_{-3}	0.21 s^{-1}

6.4 ANALYSIS OF THE LOCAL DYNAMICS

6.4.1 Steady-state analysis

The first step is to determine how the steady-state (*i. e.* time-independent) values of the fractions in the different channel states depend on the parameters, and whether the results agree with experimental observations. If we set the time derivatives to zero in (6.11) and solve the last four equations, we find that

$$x_5 = \left(\frac{k_3 x_1}{k_{-3}} \right) x_4 \equiv K_3 x_1 x_4,$$

$$x_4 = \left(\frac{k_2 x_1}{k_{-2}} \right) x_3 \equiv K_2 x_1 x_3,$$

$$x_3 = \left(\frac{k_1 I}{k_{-1}} \right) x_2 \equiv K_1 I x_2,$$

$$x_2 = \frac{1}{1 + K_1 I + K_1 K_2 I x_1 + K_1 K_2 K_3 I x_1^2} \equiv \frac{1}{K(I, x_1)}.$$

(6.12)

Therefore the fractions in the states RI, RIC^+, and RIC^+C^- are given by

$$x_3 = \frac{K_1 I}{K(I, x_1)},$$

$$x_4 = \frac{K_1 K_2 I x_1}{K(I, x_1)} \equiv X_4(I, x_1),$$

(6.13)

$$x_5 = \frac{K_1 K_2 K_3 I x_1^2}{K(I, x_1)}.$$

From these one sees that $x_3, \cdots, x_5 \to 0$, as the InsP$_3$ concentration goes to zero, and that x_4 and $x_5 \to 0$ as the calcium concentration goes to zero.

Since the fractions in the various receptor states at steady state can be expressed in terms of the dimensionless calcium concentration, these fractions can be computed, once we know the solution of the scalar equation

$$\lambda(\gamma_0 + \gamma_1 x_4)(1 - x_1) = \frac{p_1 x_1^4}{p_2^4 + x_1^4},\tag{6.14}$$

where x_4 is given by (6.13). Before we analyze this equation, however, we shall consider the case in which calcium is held at a fixed concentration, as in the experiments done by Bezprozvanny et al. (1991).

It follows from (6.13) and the definition of $K(I, x_1)$ that the fraction of channels open at a fixed calcium concentration increases with the calcium concentration at low concentrations and decreases at high concentrations. One finds that the maximum open fraction occurs at the dimensionless calcium concentration

$$x_1^* = \sqrt{\frac{1 + K_1 I}{K_1 K_2 K_3 I}},\tag{6.15}$$

and that the fraction open at this value is

$$x_4^* = \frac{\sqrt{K_1 K_2 I}}{\sqrt{K_1 K_2 I} + 2\sqrt{K_3(1 + K_1 I)}}.\tag{6.16}$$

It follows that the maximum increases as $\sqrt{I}$ for small I. The width of the graph of $X_4(x_1)$ at half the maximal open fraction is determined by the roots of the equation

$$x_4^* = 2X_4(x_1),\tag{6.17}$$

and the reader can show that this width depends very weakly on the InsP$_3$ concentration. The graph of $X_4(x_1)$ is shown in Figure 6.5(a), and the experimentally derived curve is shown in Figure 6.5(b). Certainly the model parameters could be tuned to reproduce the experimental results as well as desired, but because there is no complete set of data available for the system from which Figure 6.5 is obtained, we have not done this.

The results shown in Figure 6.5 suggest a mechanism for generating oscillations when calcium is allowed to vary in time. Suppose that calcium is at a low level, but increases slowly owing to the leak between the ER and the cytoplasm. Since the pump response is sublinear at low calcium levels, the calcium may rise to a level that opens a significant number of channels. If the channel density is sufficiently large, this in turn releases a large amount of calcium. However, at high calcium the channels close, because calcium binds at the inhibitory site. This terminates the rise of calcium, and since the decay of channel inactivation (i. e., the transition from RIC^+C^- to RIC^+) is slow, the pump can reduce the calcium level to the resting level, and the cycle may begin anew. This qualitative explanation is essentially correct, but as we shall see later, the oscillations occur only if the parameters are suitably tuned.

Now consider the steady-state equation (6.14), which reads

$$\lambda\left(\gamma_0 + \frac{\gamma_1 K_1 K_2 I x_1}{K(I, x_1)}\right)(1 - x_1) = \frac{p_1 x_1^4}{p_2^4 + x_1^4}.\tag{6.18}$$

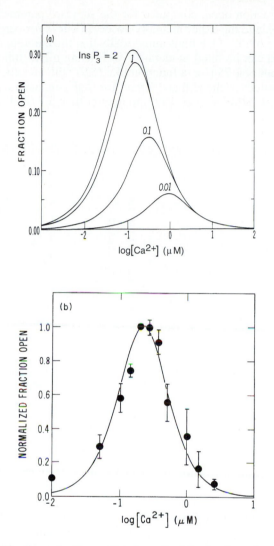

Figure 6.5. (a) The theoretically-predicted graph of the fraction of activated channels as a function of cytosol calcium. (b) The experimentally-measured curve for cerebellar cells with $\mathsf{InsP_3} = 2.0 \ \mu M$ (From Bezprozvanny et al. (1991).)

The lefthand side of (6.18) has the value $\lambda \gamma_0$ at $x_1 = 0$, it vanishes at $x_1 = 1$, and it has a unique maximum in $(0,1)$. On the other hand, the right-hand side vanishes at $x_1 = 0$, is monotone-increasing in x_1, and attains its half-maximal value at $x_1 = p_4$. Depending on the choice of parameters, there may be one or three steady states at a fixed $\mathsf{InsP_3}$ concentration. However, if p_4 is chosen sufficiently large there is a unique positive steady state for all values of I, and in this case one can show that the steady-state calcium concentration is a monotone-increasing function of I. It then follows from (6.13) that the fraction of channels open at the steady state is monotone increasing in I, provided that the steady-state level of calcium is such that $x_1 < x_1^*$ for all $I \in (0, \infty)$.

A plot of the fraction open, computed for the standard parameters, is shown in Figure 6.6(a), and the experimentally observed curve for cerebellar cells is shown in Figure 6.6(b). One finds numerically that the theoretical curve given in Figure 6.6(a) can be fitted moderately well using a Hill function in $InsP_3$ with an exponent of 1.75 (*i. e.* a function like (6.9) with an exponent of 1.75). Experimental values of the Hill coefficient for $InsP_3$-induced calcium release range from 1.0–1.3 (Volpe et al. 1990) to greater than 3.7 (Delisle 1991).

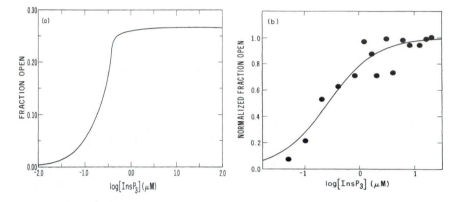

Figure 6.6. (a) The theoretically-predicted fraction of channels open as a function of the $InsP_3$ concentration. In this figure the calcium is not clamped, but rather is obtained from the solution of the full set of steady state equations. (b) The experimentally-measured curve for cerebellar cells. 100% corresponds to 14% open. (From Watras et al. (1991).)

The maximum fraction of channels open in Figure 6.6(a) is almost twice that observed experimentally. The reason may be that the parameters used here have not been measured for the cerebellar cells used for 6.6(b), but it may also be that the sequential binding model is too simple. Bezprozvanny and Ehrlich (1994) propose a modification of the sequential scheme that adds one additional state, and with this the maximum fraction of channels open is reduced to approximately the experimentally observed value. However, we have not attempted to tune parameters in the simpler model to achieve this, and it may be possible to do it without additional states.

6.4.2   Excitability

As we remarked earlier, the experimentally observed dynamics of the $InsP_3$-triggered calcium release have the hallmarks of an excitable system as the phrase is used, for example, in the context of nerve conduction equations (Othmer 1991). Here this means that small changes of $InsP_3$ or Ca^{2+} are damped, but sufficiently large stimuli produce a large response. To understand the origin of excitability in this system it is helpful to represent the kinetic mechanism schematically as follows:

$$R \underset{8}{\overset{12}{\rightleftarrows}} RI \underset{1.65}{\overset{15}{\rightleftarrows}} RIC^+ \underset{0.21}{\overset{1.8}{\rightleftarrows}} RIC^+C^-. \tag{6.19}$$

The first forward reaction involves binding of $InsP_3$, while the last two forward reactions involve binding of calcium. The numerical values of the rate constants are indicated, and from these one sees that binding of calcium to the activating site is an order of magnitude faster than binding of calcium to the inhibitory site at all Ca^{2+} concentrations. $InsP_3$ binding occurs on the same time scale as Ca^{2+} binding as long as the $InsP_3$ concentration is comparable to that of Ca^{2+}. Of course, any of these steps is slow when the concentration of the binding species is very low.

Two aspects of excitability are of interest here: (i) the response to pulses of $InsP_3$ and (ii) the response to a pulse of calcium in a system with fixed $InsP_3$. The former case corresponds to what occurs in response to a hormonal stimulus, and we consider this case first. In the absence of $InsP_3$, $x_2 = 1$ (*i. e.* all receptors are in the "bare" state) and the calcium concentration is low. If a step change in $InsP_3$ is made, $InsP_3$ binds rapidly with R, and calcium rapidly binds with the RI complex at the activating site. Since RIC^+ opens the calcium channels, the cytosol calcium increases, which in turn produces more activated receptors. Because the rate at which channels open is proportional to $x_1 x_3$, and the closing of channels by the reverse step is linear in x_4, the latter will dominate at low calcium, and we can anticipate that a small $InsP_3$ stimulus will not produce a large calcium response. To gain further insight into this, we note that, because binding at the inactivating site is much slower, the initial rapid response only involves $x_1, \cdots, x_4$. Thus if we neglect x_5 and eliminate x_2 by using the conservation condition, we arrive at the system

$$\frac{dx_1}{dt} = \lambda(\gamma_0 + \gamma_1 x_4)(1 - x_1) - \frac{p_1 x_1^4}{p_2^4 + x_1^4}, \tag{6.20}$$

$$\frac{dx_3}{dt} = -(k_{-1} + k_2 x_1 + k_1 I)x_3 + (k_{-2} + k_1 I)x_4 + k_1 I, \tag{6.21}$$

$$\frac{dx_4}{dt} = k_2 x_1 x_3 - (k_{-2} + k_3 x_1)x_4. \tag{6.22}$$

On the surfaces along which $\dot{x}_i = 0$, which are called null or balance surfaces, the different components that comprise the rate of change of species i sum to zero, and this imposes a relationship between the species. For instance, on the surface $\dot{x}_3 = 0$, where the rate of change of RI vanishes, one has

$$x_3 = \frac{k_1 I + (k_{-2} - k_1 I)x_4}{k_{-1} + k_2 x_1 + k_1 I}. \tag{6.23}$$

Therefore, on the intersection of $\dot{x}_4 = 0$ with this surface one has that

$$x_4 = \frac{k_1 k_2 x_1 I}{k_{-1}k_{-2} + k_3 x_1 (k_2 + k_{-1}) + k_1 I (k_{-2} + k_3 + k_2 x_1)}. \tag{6.24}$$

Similarly, on $\dot{x}_1 = 0$ one has

$$x_4 = \left(\frac{g(x_1)}{\lambda(1 - x_1)} - \gamma_0 \right) / \gamma_1. \tag{6.25}$$

In Figure 6.7(a) we show the intersections of the surfaces $\dot{x}_1 = 0$, and $\dot{x}_4 = 0$ with the null surface $\dot{x}_3 = 0$ (which are curves in three-space), projected into the $x_1 - x_4$ plane. One sees from (6.22) that on $\dot{x}_4 = 0$, x_4 is an increasing function of x_1, while on $\dot{x}_1 = 0$, x_4 is independent of x_3. It is easy to see that there may be one or three steady states in the fast dynamics, depending on the value of I. For instance, when $I = 0.36$ there are three, and the significance of this particular value of I will become clear later. In Figure 6.7(a) we also show the x_1 and x_4 components of the full system (6.11), starting at the steady state for $I = 0$ and applying a square-wave stimulus of I of different durations. One sees there that a 3-second stimulus is subthreshold, whereas a 4-second stimulus is superthreshold. It is clear in that figure that the unstable manifold of the intermediate steady state of the fast system, which is a saddle point, serves as a threshold surface. Stimuli which carry the state above this manifold produce a significant amplification of cytosol calcium, and stimuli that leave the state below this manifold can be termed subthreshold. The time course of the calcium component of the response for the foregoing stimuli is shown in Figure 6.7(b), and these confirm that this system is excitable in the usual sense (Othmer 1991).

The response for an InsP$_3$ stimulus of fixed duration and varying amplitude is shown in Figure 6.8. This figure shows that the latency in the response is inversely related to the amplitude of the stimulus, as is observed experimentally. The figure also shows that a sufficiently large stimulus generates a secondary response that coincides time wise with the response generated to a stimulus of smaller amplitude, again in agreement with experimental observations. For a stimulus of this duration the threshold lies between 0.34 and 0.35. One could certainly compute the threshold stimulation for a variety of amplitude-duration pairs and thereby construct an amplitude-duration curve for this excitable system, but this is not done here.

A similar analysis can be done to understand excitability with respect to calcium perturbations at fixed InsP$_3$. In this case we freeze x_5 at its steady-state value for the given value of InsP$_3$ and analyze the fast dynamics as before. In Figure 6.9(a) we show the time course of calcium and in Figure 6.9(b) the time course of the fraction of open channels for two initial perturbations in calcium. It is clear from these figures that a sufficiently large pulse triggers an excitable response, and in view of this, one expects that these dynamics can generate propagating calcium waves when InsP$_3$ is spatially-uniform and a spatially-localized perturbation of calcium is introduced. It should be noted in Figure 6.9(a) that the initial pulse triggers a secondary response, which is what is observed experimentally when certain cells that are treated with vasopressin are exposed to a calcium pulse (Harootunian et al. 1991). In the experimental context the secondary response is broad (see Figure 2(b) in Harootunian et al. (1991)), but this may be the result of cell-to-cell differences in the population.

6.4.3   Periodic oscillations

It is very common to find in excitable systems that they can also be spontaneously oscillatory if the parameters are tuned properly. In the simplest scenario this means that for suitable parameter values the steady state solu-

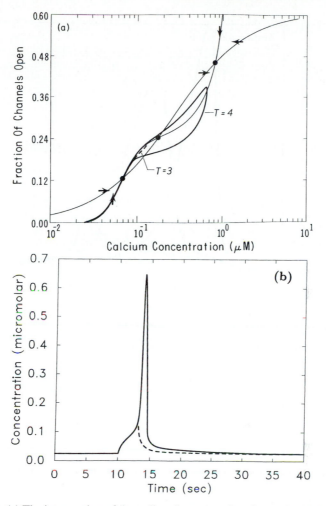

Figure 6.7. (a) The intersection of the null surfaces $\dot{x}_1 = 0$ and $\dot{x}_4 = 0$ with the null surface $\dot{x}_3 = 0$, projected into the $x_1 - x_4$ plane. The dashed curve represents the response of a system at steady state with $I = 0$ to a square-wave impulse in $InsP_3$ of amplitude $I = 0.36$ and of 3 seconds duration, and the heavy solid curve is the response to a 4-second stimulus of the same duration. (b) The time course of the calcium component for these stimuli.

tion is unstable and there is a stable periodic solution that coexists with the unstable steady state. In the Appendix we discuss how this occurs in a general one-parameter family of ordinary differential equations. In the present case, the connection between excitability and oscillations suggests that there may also be periodic solutions, in addition to excitability, if I is adjusted properly in (6.20)-(6.22). In fact we find that as I is increased oscillations set in at $I \sim 0.366$, where the steady state becomes unstable, and persist until $I \sim 2.57$, where the steady state regains stability. We show the amplitude of the calcium component of the periodic solution in this range in Figure 6.10. Note that the steady state-level of calcium is essentially constant over most of the range of I

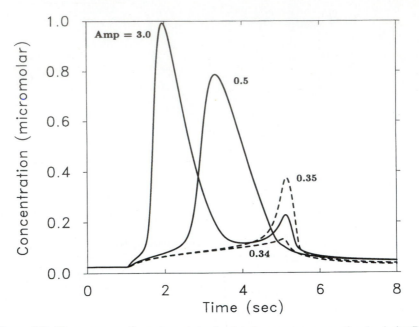

Figure 6.8. The response of a system at steady state to a square-wave stimulus in InsP$_3$ of 4 seconds duration and of the amplitude indicated on the figure.

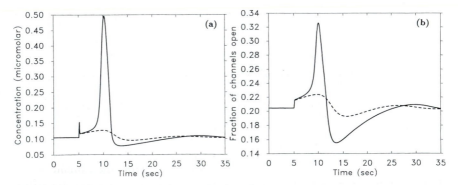

Figure 6.9. The time course of calcium (a) and the fraction of open channels (b). The system is in a steady state for fixed $I = 0.36$ when $T \in (0, 5)$, and a square wave of calcium of amplitude 1.2 (solid lines) or amplitude 1.15 (dashed lines) is imposed for 0.1 second at $T = 5$.

in which oscillations exist. It should also be noted that the maximum amplitude of the oscillations decreases somewhat over the oscillatory range of InsP$_3$, but the period changes by an order of magnitude. Thus the system exhibits the experimentally observed frequency coding, a point that we shall return to shortly. In Figure 6.11 we show the calcium concentration and the fraction of conducting channels at $I = 0.4$, which is in the oscillatory regime. Note that there are two phases in the early part of each cycle, a phase in which calcium rises slowly and the fraction of channels open begins to rise, followed by a phase in which the channels open very rapidly and calcium rises rapidly.

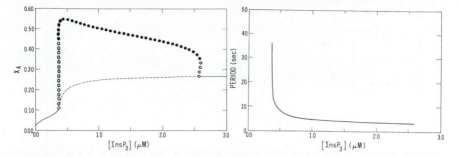

Figure 6.10. (a) The amplitude at steady state (solid and dashed lines) and the maximum amplitude of a periodic solution (open and filled circles) of calcium as a function of the InsP$_3$ concentration. Solid lines and circles indicate stable solutions; dashed lines and open circles indicate unstable solutions. (b) The period in seconds of the periodic solutions in (a). These results were computed using the software package AUTO (Doedel 1986).

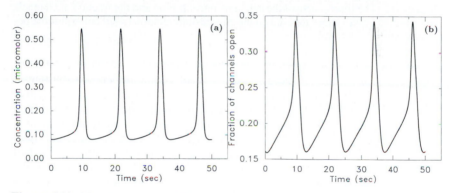

Figure 6.11. The time course of the calcium component (a) and the fraction of conducting channels (b) for the periodic solution at $I = 0.4$.

Earlier we gave a qualitative explanation of how oscillations could arise, and we now want to see whether it is correct. In Figure 6.12 we show the calcium concentration and the conducting channels superimposed on the equilibrium relation for $I = 0.4$. One sees that qualitatively the explanation is correct, in that the cycle can be followed from a low calcium level and a low fraction of channels open, through the rising phase in which the channels open and calcium increases, and into the declining phase in which the channels become inactivate and the calcium level falls. However, there is one aspect of the behavior that could not have been predicted readily, and that is that the fraction of channels open rises far above the maximum equilibrium value. Thus there is a very significant disequilibrium between calcium and the fraction of open channels during the oscillation; the two are not equilibrated except at two instants in the cycle.

In the preceding figures the concentration of InsP$_3$ has been held fixed. However, under normal circumstances InsP$_3$ is degraded, and thus the foregoing results strictly apply only when a nonhydrolyzable analog of InsP$_3$ is used. However, it is clear that if InsP$_3$ is degraded, and if the initial concentration is larger than the upper limit of the oscillatory range of InsP$_3$, then

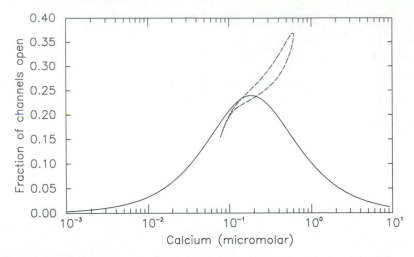

Figure 6.12. The time course of the calcium concentration and the fraction of channels open (dashed curve) superimposed on the equilibrium relation (solid curve) for the periodic solution at $I = 0.4$.

as $InsP_3$ decreases the system sweeps through the entire range of oscillatory dynamics. The details of the dynamics will of course depend on the rate at which $InsP_3$ is degraded. By controlling the $InsP_3$ level experimentally one can control the passage through various dynamical regimes. This is illustrated in Figure 6.13, where we show the calcium concentration as a function of time when the time course of $InsP_3$ is as given in the figure caption. The experimentally observed behavior for endothelial cells under the same stimulus protocol (albeit on a different time scale) is shown in Figure 6.4. The disparity in the time scales stems from the fact that the parameters used here come from a variety of cells, and by changing them one can obtain closer agreement between theory and experiment (Tang et al. 1996).

6.5   WAVES

A striking aspect of calcium dynamics in oocytes is the wide variety of wave patterns that are observed (Lechleiter et al. 1991; Lechleiter and Clapham 1992). A model based on calcium-induced calcium release that can qualitatively reproduce these observations is proposed in Girad et al. (1992). This model uses the model of Goldbeter et al. (1990) for the local dynamics. However, as we noted earlier, calcium transport across the plasma membrane plays an essential role in this model, yet the major features of the calcium dynamics in oocytes do not depend on this, at least on a short time scale. Furthermore, it is known that there are no ryanodine receptors, and therefore there is no $InsP_3$-insensitive calcium pool, in *Xenopus* oocytes. In hamster eggs the $InsP_3$-sensitive pool plays an essential role in the calcium waves, for when the eggs are treated with antibodies to the $InsP_3$ receptor, the waves are blocked (Miyazaki et al. 1992).

In this section we briefly describe one of the wave patterns that is predicted by the model studied in earlier sections. Further details can be found in

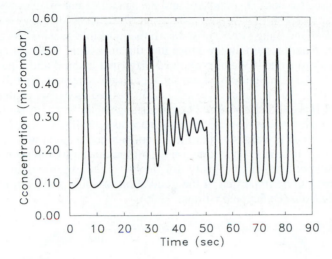

Figure 6.13. The theoretically predicted calcium concentration for varying InsP$_3$. The InsP$_3$ concentration is held at 0.5 for $T \in (3, 30)$, then is increased to InsP$_3$ = 5 for 20 seconds and held at InsP$_3$ = 1 thereafter.

Othmer and Tang (1993), and a review of models for calcium waves is given in Sneyd et al. (1995). We include diffusion of both calcium and InsP$_3$, we neglect calcium buffering as before, and we use the conservation condition to eliminate x_5. Consequently the governing equations in a spatially distributed system are as follows:

$$\frac{\partial x_1}{\partial t} = D_{CA}\Delta x_1 + \lambda(\gamma_0 + \gamma_1 x_4)(1 - x_1) - \frac{p_1 x_1^4}{p_2^4 + x_1^4},$$

$$\frac{\partial y}{\partial t} = D_I \Delta y + H - y,$$

$$\frac{dx_2}{dt} = -k_1 I x_2 + k_{-1} x_3 \tag{6.26}$$

$$\frac{dx_3}{dt} = -(k_{-1} + k_2 x_1)x_3 + k_1 I x_2 + k_{-2}x_4,$$

$$\frac{dx_4}{dt} = k_{-3} - k_{-3}x_2 + (k_2 x_1 - k_{-3})x_3 - (k_{-2} + k_3 x_1 + k_{-3})x_4.$$

Here y is the dimensionless concentration of InsP$_3$, Δ denotes the Laplace operator, D_{CA} and D_I are the diffusion coefficients of calcium and InsP$_3$, respectively, and H is a functional of the hormone concentration determined by the transduction step. Note that we have augmented the previous equations (6.11) to incorporate source terms of InsP$_3$ and first-order decay. In a spatially uniform system InsP$_3$ relaxes exponentially to the value H. We can thereby mimic experiments in which there are spatially distributed sources of InsP$_3$ by specifying the spatial variation of H. We impose homogeneous Neumann (*i. e.*, zero-flux) boundary conditions on the boundary of the domain. The numerical procedure used is based on a fully implicit time-stepping algorithm and finite-difference approximations to the spatial derivatives. We use the value $5. \times 10^{-4}$ mm^2/sec for both D_{CA} and D_I, and the spatial extent of all systems is 1 millimeter.

In Othmer and Tang (1993) it is shown that multiple pacemakers can coexist in a distributed system, in the sense that the fastest one does not necessarily entrain slower ones. We also showed there that a spiral and a pacemaker can also coexist. In Figure 6.14 we show two time snapshots that illustrate this coexistence. In this figure $H = 0.5$ in a disk of radius 0.0707 centered at $(0.5, 0.5)$, and $H = 0.35$ elsewhere. The coexistence of these waves is more sensitive to parameter values than the multiple pacemakers. If one slightly alters the combination of H-values, either the pacemaker or the spiral will dominate the asymptotic behavior, depending on the alteration. However, such solutions are more than a passing curiosity, for they are seen experimentally in oocytes, and prove to be important in the generation of spiral waves in *Dictyostelium discoideum* (Dallon and Othmer 1996).

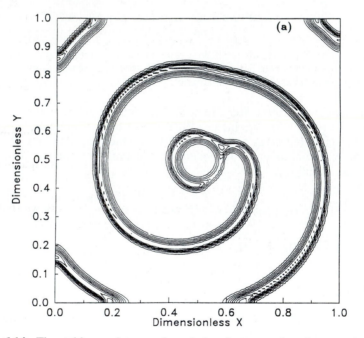

Figure 6.14. The stable coexistence of a spiral and a pacemaker shown at time T = 87.5. (From Othmer and Tang (1993)).

6.6   DISCUSSION

We have described a model for calcium dynamics in certain types of cells and have shown that predictions made by the model agree with many of the experimental observations. The local dynamics in this model can be excitable or self-oscillatory, depending on the level of InsP$_3$. The different types

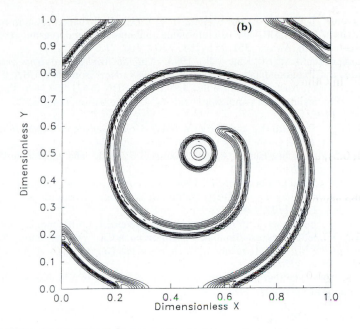

Figure 6.14. (cont'd) The spiral and pacemaker shown in the previous figure at a later time (T = 100).

of responses cover essentially all the local aspects of calcium dynamics observed to date. We have not discussed the spatial aspect in detail here, but the distributed version of the model shows traveling waves in one spatial dimension, and target patterns and spiral waves in two spatial dimensions (Othmer and Tang 1993). Thus this model seems to be sufficiently robust to reproduce a variety of experimentally observed phenomena. In future work we shall apply this scheme to the fertilization waves in the *Xenopus* oocytes and in sea urchin eggs (Swann and Whitaker 1986).

REFERENCES

Alberts, B., Bray, D., Lewis, J., Raff, M., Roberts, K., and Watson, J. D. 1989. *Molecular biology of the cell*. Second edn. New York, NY, USA: Garland Publishing, Inc. ISBN 0-8240-3695-6, 0-8240-3696-4 (paperback). Pages xliii + 1294 + 67. {*110, 123*}

Alexander, J. C., Doedel, E. J., and Othmer, H. G. 1990. On the resonance structure in a forced excitable system. *SIAM Journal on Applied Mathematics*, **50**(5), 1373–1418. {*102, 123*}

Berridge, M. J. 1990. Inositol 1,4,5-trisphosphate-induced calcium mobilization is localized in *Xenopus* oocytes. *Proceedings of the Royal Society of London. Series B. Biological sciences*, **238**, 235–343. {*103, 123*}

Bezprozvanny, I. and Ehrlich, B. 1994. Inositol (1,4,5)-trisphosphate (InsP$_3$)-gated Ca channels from cerebellum: Conduction properties for divalent cations and regulation by intraluminal calcium. *Journal of General Physiology*, **104**, 821–856. {*114, 123*}

Bezprozvanny, I., Watras, J., and Ehrlich, B. E. 1991. Bell-shaped calcium-response curves of Ins(1,4,5)P$_3$- and calcium-gated channels from endoplasmic recticulum of cerebellum. *Nature*, **351**, 751–754. *{103, 108, 112, 113, 123}*

Bock, Gregory and Ackrill, Kate (eds). 1995. *Calcium Waves, Gradients and Oscillations.* Ciba Foundation symposium, vol. 188. New York, NY, USA; London, UK; Sydney, Australia: John Wiley and Sons. ISBN 0-471-95234-6. Pages ix + 291. Proceedings of a symposium held at Ciba Foundation, London, 26–28 April 1994. *{102, 124}*

Borsellino, A., Cervetto, L., and Torre, V. (eds). 1990. *Sensory Transduction.* New York, NY, USA; London, UK: Plenum Press. *{99, 124}*

Cobbold, P. H., Sanchez-Bueno, A., and Dixon, C. J. 1991. The hepatocyte Calcium Oscillator. *Cell calcium*, **12**, 87–95. *{105, 124}*

Dallon, John. C. and Othmer, Hans G. 1996. *A discrete cell model with adaptive signalling for aggregation of* Dictyostelium discoideum. To appear in Phil. Trans. Roy. Soc. Lon. *{122, 124}*

De Young, G. and Keizer, J. 1992. A single-pool inositol 1,4,5-trisphophate-receptor-based model for agonist-stimulated oscillations in Ca^{+2} concentration. *Proceedings of the National Academy of Sciences of the United States of America*, **89**, 9895–9899. *{110, 124}*

Delisle, S. 1991. The four dimensions of calcium signalling in *Xenopus* oocytes. *Cell calcium*, **12**, 217–227. *{114, 124}*

DeLisle, S., Krause, K. H., Denning, G., Potter, B. V. L., and Welsh, M. J. 1990. Effect of Inositol Trisphosphate and Calcium on Oscillating Elevations of Intracellular Calcium in *Xenopus* oocytes. *The Journal of biological chemistry*, **265**, 11726–11730. *{103, 124}*

Doedel, E. 1986. *AUTO: Software for continuation and bifurcation problems in ordinary differential equations.* Tech. rept. California Institute of Technology. *{119, 124}*

Dupont, G., Berridge, M. J., and Goldbeter, A. 1991. Signal-induced Ca^{2+} oscillations: Properties of a model based on Ca^{2+}-induced Ca^{2+} release. *Cell calcium*, **12**, 73–85. *{105, 124}*

Finch, E. A., Turner, T. J., and Goldin, S. M. 1991. Calcium as a coagonist of inositol 1,4,5-trisphosphate-induced calcium release. *Science*, **252**, 443–446. *{103, 124}*

Gilkey, J. C., Jaffe, L. F., Ridgeway, E. B., and Reynolds, G. T. 1978. A free calcium wave traverses the activating egg of the medaka, *Oryzias latipes. Journal of Cell Biology*, **76**, 448–466. *{103, 124}*

Girard, S., Luckhoff, A., Lechleiter, J., Sneyd, J., and Clapham, D. 1992. Two-dimensional model of calcium waves reproduces the patterns observed *Xenopus* oocytes. *Biophysical Journal*, **61**, 509–517. *{120, 124}*

Goldbeter, A., Dupont, G., and Berridge, M. J. 1990. Minimal model for signal-induced Ca^{2+} oscillations and for their frequency encoding through protein phosphorylation. *Proceedings of the National Academy of Sciences of the United States of America*, **87**, 1461–1465. *{108, 120, 124}*

Harootunian, A. T., Kao, J. P. Y., Paranjape, S., and Tsien, R. Y. 1991. Generation of Calcium Oscillations in Fibroblasts by positive feedback between calcium and IP3. *Science*, **251**, 75–78. *{103, 116, 124}*

Jacob, R., Merritt, J. E., Hallam, T. J., and Rink, T. J. 1988. Repetitive spikes in cytoplasmic calcium evoked by histamine in human endothelial cells. *Nature*, **335**(Sept.), 40–45. *{105, 124}*

Keizer, J. and De Young, G. W. 1992. Two roles for Ca^{2+} in agonist stimulated Ca^{2+} oscillations. *Biophysical Journal*, **61**, 649–660. *{105, 108, 124}*

Koshland, D. E. 1988. Chemotaxis as a model second-messenger system. *Biochemistry*, **27**(16), 5829–5834. *{99, 101, 124}*

Koutalos, Y. and Yau, K. Y. 1996. Regulation of sensitivity in vertebrate rod photoreceptors by calcium. *Trends in neurosciences*, **19**(2), 73–81. *{101, 124}*

Lechleiter, J. and Clapham, D. 1992. Molecular mechanisms of intracellular calcium excitability in *X. laevis* oocytes. *Cell*, **69**, 283–294. {*120, 125*}

Lechleiter, J., Girard, S., Peralta, E., and Clapham, D. 1991. Spiral calcium wave propagation and annihilation in *Xenopus laevis* oocytes. *Science*, **252**, 123–126. {*120, 125*}

Lupu-Meiri, M., Shapira, H., and Oron, Y. 1988. Hemispheric asymmetry of rapid chloride responses to inositol trisphosphate and calcium in *Xenopus* oocytes. *FEBS Letters*, **240**, 83–87. {*103, 125*}

Meyer, Tobias and Stryer, Lubert. 1991. Calcium spiking. *Annual review of biophysics and biophysical chemistry*, **20**, 153–174. {*102, 125*}

Miyazaki, S., Yuzaki, M., Nakada, K., Shirakawa, H., Nakanishi, S., Nakade, S., and Mikoshiba, K. 1992. Block of Ca^{2+} wave and Ca^{2+} oscillation by antibody to the Inositol 1,4,5-Trisphosphate receptor in fertilized hamster eggs. *Science*, **257**, 251–255. {*120, 125*}

Monk, P. B. and Othmer, H. G. 1989. Cyclic AMP oscillations in suspensions of *Dictyostelium discoideum*. *Philosophical transactions of the Royal Society of London*, **323**(1215), 185–224. {*99, 125*}

Nuccitelli, R. 1991. How do sperm activate eggs? *Current topics in developmental biology*, **25**, 1–16. {*103, 125*}

Othmer, H. G. 1991. The dynamics of forced excitable systems. *Pages 213–232 of:* Holden, A. V., Markus, M., and Othmer, Hans G. (eds), *Nonlinear Wave Processes in Excitable Media*. NATO ASI series. Series B, Physics, vol. 244. New York, NY, USA; London, UK: Plenum Press, for NATO. {*114, 116, 125*}

Othmer, H. G. and Tang, Y. 1993. Oscillations and waves in a model of $InsP_3$-controlled calcium dynamics. *Pages 277–313 of:* Othmer, H. G., Maini, P. K., and Murray, J. D. (eds), *Experimental and Theoretical Advances in Biological Pattern Formation*. NATO ASI series. Series A, Life sciences, vol. 259. New York, NY, USA; London, UK: Plenum Press. {*105, 108, 110, 121–123, 125*}

Othmer, H. G., Monk, P. B., and Rapp, P. E. 1985. A model for signal relay and adaptation in *Dictyostelium discoideum*. Part II. Analytical and numerical results. *Mathematical Biosciences*, **77**, 77–139. {*101, 125*}

Rooney, T. A., Renard, D. C., Sass, E. J., and Thomas, A. P. 1991. Oscillatory cytosolic calcium waves independent of stimulated inositol 1,4,5-trisphosphate formation in hepatocytes. *The Journal of biological chemistry*, **266**(19), 12272–12282. {*103, 125*}

Rooney, Thomas A., Sass, Ellen J., and Thomas, Andrew P. 1989. Characterization of cytosolic Calcium Oscillations Induced by Phenylephrine and Vasopressin in Single Fura-2-loaded Hepatocytes. *The Journal of biological chemistry*, **264**(29), 17131–17141. {*103, 125*}

Sneyd, J., Keizer, J., and Sanderson, M. J. 1995. Mechanisms of calcium oscillations and waves: A quantitative analysis. *The FASEB journal: official publication of the Federation of American Societies for Experimental Biology*, **9**, 1463–1472. {*121, 125*}

Somogyi, R. and Stucki, J. W. 1991. Hormone-induced calcium oscillations in liver cells can be explained by a simple one pool model. *The Journal of biological chemistry*, **266**(17), 11068–11077. {*108, 125*}

Spiro, Peter, Parkinson, John S., and Othmer, H. G. 1996. *A model of excitation and adaptation in bacterial chemotaxis*. Submitted to PNAS. {*101, 125*}

Swann, K. and Whitaker, M. 1986. The part played by inositol trisphosphate and calcium in the propagation of the fertilization wave in sea urchin eggs. *Journal of Cell Biology*, **103**(6 (part 1)), 2333–2342. {*123, 125*}

Tang, Y., Stephenson, J., and Othmer, H. G. 1996. Simplification and analysis of models of calcium dynamics based on $InsP_3$-sensitive calcium channel dynamics. *Biophysical Journal*, **70**, 246–263. {*120, 125*}

Tang, Yuanhua and Othmer, Hans G. 1994a. A G-protein-based model of adaptation in *Dictyostelium discoideum. Mathematical Biosciences*, **120**, 25–76. {*99, 101, 102, 125*}

Tang, Yuanhua and Othmer, Hans G. 1994b. A model of calcium dynamics in cardiac myocytes based on the kinetics of ryanodine-sensitive calcium channels. *Biophysical Journal*, **67**, 2223–2235. {*103, 126*}

Tang, Yuanhua and Othmer, Hans G. 1995. Frequency encoding in forced excitable systems with applications to calcium oscillations. *Proceedings of the National Academy of Sciences of the United States of America*, **92**, 7869–7873. {*100, 126*}

Thomas, A. P., Renard, D. C., and Rooney, T. A. 1991. Spatial and temporal organization of calcium signalling in hepatocytes. *Cell calcium*, **12**, 111–126. {*103, 126*}

Volpe, P., Alderson-Lang, B. H., and Nickols, G. A. 1990. Regulation of inositol 1,4,5-trisphosphate-induced Ca^{2+} release. I. Effect of Mg^{2+}. *American journal of physiology*, **258**(6), C1077–C1085. {*114, 126*}

Watras, J., Bezprozvanny, I., and Ehrlich, B. E. 1991. Inositol 1,4,5-trisphosphate-gated channels in cerebellum: Presence of multiple conductance states. *Journal of Neuroscience*, **11**(10), 3239–3245. {*103, 114, 126*}

Woods, N. M., Cuthberson, K. S. R., and Cobbold, P. H. 1986. Repetitive transient rises in cytoplasmic free calcium in hormone-stimulated hepatocytes. *Nature*, **319**, 600–602. {*102, 126*}

7. THE EUKARYOTIC CELL CYCLE: MOLECULES, MECHANISMS, AND MATHEMATICAL MODELS

John J. Tyson, Kathy Chen and Bela Novak

7.1 INTRODUCTION

In this chapter we describe what is known about the machinery of the cell-division cycle and the control signals that enforce a constant cell size and strict alternation of DNA synthesis and cell-division. The machinery, as one might expect, is quite complex, and it is impossible to comprehend how all the pieces interrelate by casual verbal arguments. We show how the mathematics of dynamical systems can be used to construct simple models of the major checkpoints in the cell cycle. Although mathematical modeling will always play second fiddle to molecular biology in the quest to understand the cell cycle, it is our opinion that, without the guidance of comprehensive mathematical models, molecular biologists will soon be overwhelmed by their own success.

7.2 PHYSIOLOGY OF THE CELL CYCLE

The cell cycle is the sequence of events that mark the passage of a growing cell from birth to division (Mitchison 1971). By this process of growth and division all cells come into existence and pass their genetic material on to future generations. To understand the molecular machinery and regulatory signals of the cell cycle would be a great advance in basic biology and would revolutionize the treatment of human maladies attributable to too much cell division (cancer) or too little (nonregenerative tissues).

Prokaryotic cells (bacteria and blue-green algae) differ in many ways from eukaryotic cells (which have a membrane-bound nucleus containing DNA — the genetic information of the cell — organized into chromosomes), including the machinery of cell replication. In this chapter we shall discuss the eukaryotic cell cycle only. The most dramatic events of the eukaryotic cell cycle are DNA replication, when one double-stranded DNA molecule (chromatid) is copied into two, and mitosis, when the two sister chromatids are physically separated to the incipient daughter cells. These two events are temporally separated: the period of DNA synthesis is called "S phase" and the later stage of mitosis is called "M phase." S and M phases are usually separated by two gaps (G1 and G2) when the cell is presumably preparing for DNA synthesis and mitosis, respectively. As we shall see, specific molecular signals must be generated to trigger the start of S phase and M phase.

The logic of the eukaryotic cell cycle requires strong inhibitory signals to enforce a strict alternation of S phase and M phase. Cells in G1 or S must be prohibited from entering M phase until DNA replication is completed, and cells in G2 or M must not begin a new round of DNA replication until sister chromatids are successfully separated. Bypassing M and reinitiating S is not so bad (the cell becomes polyploid; i.e., it will contain twice the original complement of DNA), but bypassing S phase and trying to divide is lethal (daughter cells receive only bits and pieces of the genome and soon die).

The interdependence of S and M phases was demonstrated in a classic set of experiments by Rao and Johnson (1970). By fusing mammalian cells at different stages in the cell cycle to form binucleate cells, Rao and Johnson were able to show that S-phase cytoplasm contains an inducer of DNA synthesis that can drive G1 nuclei prematurely into S phase, but G2 and M nuclei are somehow blocked from replicating their DNA even when exposed to the S-phase inducer. When G1 and G2 cells are fused, the G2 nucleus is delayed from entering mitosis (relative to unfused cells), whereas the G1 nucleus is accelerated into mitosis. The fact that both nuclei enter mitosis synchronously suggests that there is a cytoplasmic signal for mitosis. In fact, M-phase cells contain a powerful cytoplasmic inducer of mitosis which is able to force G1- and S-phase nuclei into premature chromosome condensation; i.e., these nuclei enter mitosis, even though their DNA is not fully replicated, with catastrophic consequences.

In addition to the nuclear cycle of chromosome replication and separation, there is also a cytoplasmic cycle of synthesis and distribution of all other cellular components: enzymes, ribosomes, membranes, and organelles. Whereas the nuclear cycle is noted for its precision (DNA is accurately copied and carefully distributed to daughter cells), the cytoplasmic cycle is more sloppy because it is unnecessary that each daughter inherit precisely half of all the other components. Nonetheless, in the long run, it is necessary that the processes of growth and division be coordinated.

Single-celled microorganisms show clear signs of maintaining cell size at a constant level (cell-size homeostasis). Many years ago Hartmann (1928) showed that cell division in *Amoeba proteus* could be delayed indefinitely by regular removal of cytoplasmic material, and later Frazier (1973) showed that cell division in the ciliate, *Stentor coeruleus*, could be advanced by excising DNA from the macronucleus. These experiments and others suggest that cells can measure the ratio of nuclear material (DNA) to cytoplasmic material (volume, protein, content, etc.), and that they trigger cell division and a new round of DNA synthesis when the nucleocytoplasmic ratio drops below some critical value.

A simple idea that accounts for these experiments is the nuclear-sites-titration model (Sachsenmaier et al. 1972; Tyson 1984). An unstable activator, synthesized at a rate proportional to cytoplasmic volume, enters the nucleus and binds to sites on the DNA. When all the sites are occupied, the free activator can accumulate in the cytoplasm and trigger mitosis. In this model, the cell-cycle is driven by growth (expansion of cytoplasmic volume). Stopping growth will stop division.

In some circumstances, however, growth and division proceed independently. During egg development, the immature egg grows very large without compensatory DNA replication. Following fertilization, the fertilized egg then undergoes a series of rapid mitotic cycles without growth. In fact, it is even possible to remove the nucleus from a fertilized egg and still observe certain periodic events of the cell cycle (Harvey 1940). Clearly the nucleocytoplasmic ratio and binding of nuclear sites has nothing to do with the timing of cell-cycle events in an enucleated egg! Rather, there must be a cytoplasmic biochemical oscillator which keeps track of time in these eggs independently of growth and/or DNA synthesis.

7.3 GENETICS OF THE CELL CYCLE

Yeast cells, both budding yeast (*Saccharomyces cerevisiae*) and fission yeast (*Schizosaccharomyces pombe*), are ideal organisms for studying the genetics of the eukaryotic cell cycle. Hartwell led the way with the isolation of *cdc* (*c*ell *d*ivision *c*ycle) mutants of budding yeast. A *cdc* mutant cell continues to grow but is blocked at a specific phase in the cell cycle and therefore cannot divide (these characteristics defined the mutant *cdc* "phenotype"). All *cdc* mutants are isolated as temperature-sensitive lethals: they grow and divide at 25°C, but are unable to reproduce at the "restrictive" temperature (typically 35°C). Hartwell isolated many different genes which showed the same set of externally expressed characteristics (phenotype) as *cdc* (Hartwell et al. 1974) (for review, see Pringle and Hartwell (1981)) and used these genes to define the major events and regulatory points of the budding-yeast division cycle.

While looking for *cdc* mutants in fission yeast, Nurse (1975; 1981) discovered an interesting mutant with a different phenotype: unlike *cdc* mutants, these mutant cells continue to grow and divide, but they divide at an abnormally small size, about one-half the size of wild-type cells. Nurse, working in Edinburgh at the time, called this phenotype *wee*, the Scottish word for small. Whereas the *cdc* phenotype will be exhibited by a cell lacking one of the enzymes necessary to complete the tasks of chromosome replication and segregation, the *wee* phenotype indicates that the mutation has struck the regulatory components of the cycle. The "nuts and bolts" of cell division are all in place, but the "thermostat" for cell division has been reset.

Nurse isolated 52 different strains showing the *wee* phenotype and, by genetic crosses, showed that 50 strains were different alleles of a single gene called *wee*1. Each allele is a slightly different nucleotide sequence at the same location on the DNA (a genetic "locus"). Each nucleotide sequence codes for a slightly different protein with altered properties. Usually the mutant proteins are non-functional compared to the "wild-type" (normal) protein. The two other strains with *wee* phenotype mapped to a second genetic locus already identified as *cdc*2. (Loss-of-function *cdc*2 mutants block in late G2, just before cells enter M phase.) These facts indicate that the normal allele for *wee*1 (*wee*1$^+$) codes for an inhibitor of cell division but the many different mutant versions of the protein have all lost their ability to inhibit division, hence the cells divide at an abnormally small size. The normal allele for *cdc*2 (*cdc*2$^+$) codes for an activator of cell division, with most mutations destroying the activating role of the protein, so progress through the division cycle is blocked, but with two rare mutations producing "superactivators" which drive cells into division at small size.[1]

Fantes (1979) soon showed that another gene, *cdc25*, interacts closely with *wee1* because the single mutant *cdc25⁻* is inviable (blocked in the cell cycle) but the double mutant *cdc25⁻ wee1⁻* is viable. So *cdc25* presumably codes for an activator of division which is not needed if the inhibitor gene *wee1* is missing. Russell and Nurse (1986; 1987) showed that size at division depends on gene dosage of *wee1* and *cdc25*: overproduction of *wee1* gene product (*wee1^{OP}*) makes cells larger at division and *cdc25^{OP}* makes cells smaller. What happens when one combines the size-reducing tendencies of *wee1⁻* and *cdc25^{OP}*? The double mutant, *wee1⁻ cdc25^{OP}*, is lethal (Russell and Nurse 1986): the cell enters mitosis at a very small size before it can finish DNA synthesis. Daughter cells do not receive a full genome and subsequently die.

These genetic studies led to a simple picture of mitotic control in fission yeast (Figure 7.1). The Cdc2 protein is present in the cell throughout the

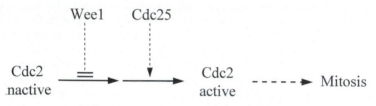

Figure 7.1. Mitotic control in fission yeast.

cell cycle. Its concentration does not fluctuate (Simanis and Nurse 1986), but it is kept in an inactive state by the action of Wee1. When the cell is ready to divide, Cdc25 somehow activates Cdc2. In this view, the story of cell cycle control must involve the regulation of the enzymes Wee1 and Cdc25.

7.4 BIOCHEMISTRY OF THE CELL CYCLE

Because eggs are large cells that proceed rapidly and synchronously through the cell cycle, they are ideal for studying the biochemistry of cell division control. The frog, *Xenopus laevis*, has been a favorite organism for such studies (Murray and Hunt 1993). Frog eggs develop in several stages (Figure 7.2). The egg grows to about 1mm in diameter without dividing and arrests in G2 phase just before meiosis I. This stage is called the immature oocyte. In response to the hormone progesterone, immature oocytes are induced to com-

---

[1]A word about notation. Yeast genes are always denoted by a combination of three italicized letters (indicative of the mutant phenotype) and numbers to distinguish different genes exhibiting the same phenotype. Different mutant alleles of the same gene are distinguished by adding a hyphenated extender to the gene name, e.g., *wee1*-50 or *cdc2*-3w. To distinguish the "wild-type" (i.e., normal) allele from the mutant alleles, a superscript "+" is often used, say *cdc25⁺*. Other superscripts indicate types of mutant alleles; e.g., *wee1⁻* would denote any one of the many "wee" alleles at the *wee1* genetic locus, *wee1^{ts}* refers to a "temperature-sensitive" allele of *wee1* that is normal at 25°C but small at 35°C, *wee1^{OP}* is a mutant with extra copies of the wild-type (*wee1⁺*) gene. The messenger RNA transcribed from a gene is called, for example, *cdc25* mRNA, and the protein it codes for will be denoted Cdc25 (note, not italicized, first letter capitalized). This notation for protein was proposed by Murray and Hunt (1993), but, unfortunately, it is not universally adopted by molecular biologists.

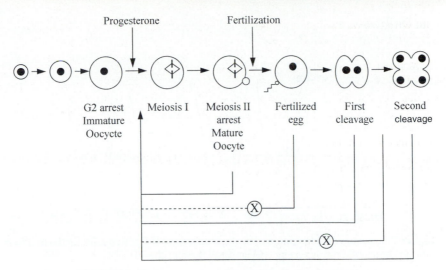

Figure 7.2. Frog-egg maturation and early development.

plete the first meiotic division and stop at metaphase of meiosis II (mature oocyte). Fertilization triggers the third stage: exit from meiosis II, fusion of the maternal and paternal nuclei, and then a series of twelve rapid synchronous mitotic divisions to form a hollow ball of 4096 cells (the blastula). For a detailed description of frog-egg development see Murray et al. (1993).

In 1971 Masui and Markert (1971) discovered that progesterone stimulation could be bypassed by injecting into immature oocytes a cytoplasmic extract prepared from mature oocytes (Figure 7.2). That is, mature oocytes contain some cytoplasmic factor that induces oocyte maturation; hence it was called "maturation promoting factor" (MPF). Gerhardt, Wu and Kirschner (1984) soon showed that MPF activity was present also in M phase but not in S phase of the mitotic cycles of fertilized eggs (Figure 7.2).

Working at about the same time on sea urchin embryos, Evans *et al.* (1983) uncovered the existence of a protein (called "cyclin") that accumulates steadily during the early synchronous mitotic cycles and is destroyed abruptly at mitosis (Figure 7.3).

In an elegant series of experiments, Murray and coworkers (1989a; 1989b) showed that cyclin synthesis and degradation correlate closely with MPF activation and inactivation. In this view, the regulation of entry into and exit from mitosis is the story of cyclin synthesis and degradation. It would seem that yeast (Figure 7.1) and frog-egg (Figure 7.4) mitotic cycles are driven by different mechanisms.

7.5 MOLECULAR BIOLOGY OF THE CELL CYCLE

With the advance of genetic engineering techniques in the 1980s, many cell-cycle genes were cloned, sequenced, and characterized. The sequence of *cdc*2 predicts that its protein product is a protein kinase, i.e., an enzyme that phosphorylates (adds a phosphate group to) other proteins:

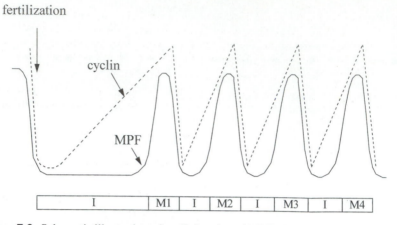

Figure 7.3. Schematic illustration of cyclin levels and MPF activity during the first four mitotic cycles of a fertilized egg. (M=mitosis, I=interphase).

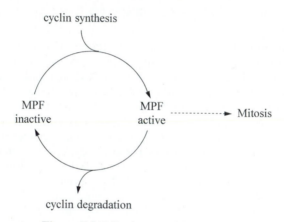

Figure 7.4. Mitotic control in frog eggs.

$$\text{Protein} + \text{ATP} \xrightarrow{\text{Cdc2}} \text{Protein-P} + \text{ADP}$$

Other important fission-yeast genes are *cdc*13 (which codes for a B-type cyclin), *wee*1 (another protein kinase), and *cdc*25 (a protein phosphatase):

$$\text{Protein-P} + \text{H}_2\text{0} \xrightarrow{\text{Cdc25}} \text{Protein} + \text{P}_i$$

where P_i stands for "inorganic" phosphate ions.

The two simple pictures of mitotic control (Figure 7.1 and Figure 7.4) came together in 1988, when MPF was finally purified and identified as a dimer comprised of Cdc2 and cyclin-B (Lohka et al. 1988). Cdc2 has no kinase activity unless it is associated with cyclins. In addition, the kinase activity of dimers can be controlled by phosphorylation of the Cdc2 subunit at a tyrosine residue (Y15; "Y" stands for tyrosine and the number "15" indicates its position in the protein sequence (Nurse 1990)). Y15 phosphorylation inhibits the kinase activity of Cdc2. The phosphorylation state of Y15 is controlled by Wee1 and Cdc25. From this diagram it is clear why *wee*1 codes for an inhibitor and *cdc*25 for an activator of mitosis. Somehow cell size must be encoded in the ratio of Wee1 activity to Cdc25 activity.

Wee1 and Cdc25 are also phosphoproteins: Wee1-P is less active and Cdc25-P is more active than the unphosphorylated forms. Cdc2 can phosphorylate these proteins (Murray 1993), so there are two positive-feedback loops involving active MPF (Figure 7.5).

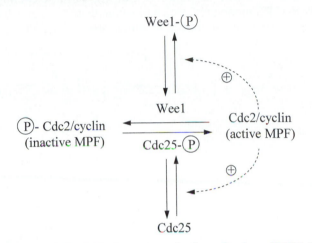

Figure 7.5. The two positive-feedback loops in the activation of MPF. The activity of MPF (Cdc2/cyclin dimer) is controlled by phosphorylation at an inhibitory tyrosine residue. The tyrosine kinase Wee1 is inhibited by MPF, and the tyrosine phosphatase Cdc25 is activated by MPF.

To exit from mitosis, the cyclin subunit of MPF must be destroyed by the ubiquitin pathway of enzymatic protein breakdown (proteolysis) (Murray 1995). First a chain of ubiquitin molecules is attached to cyclin (ubiquitin is a small protein) by the specific ubiquitin-ligating enzymes that recognize cyclin molecules and attach the "eat me" label. Ubiquitin-labeled proteins are then recognized by proteasomes and broken down. The activities of cyclin-ubiquitin ligating enzymes are apparently regulated during the cell cycle, because cyclin degradation is turned on and off at specific stages in the cycle. The control signals on cyclin degradation are still poorly understood, but it is known that, in frog-egg extracts, cyclin degradation can be induced by active MPF, after a significant time lag. We can diagram this effect as

$$\begin{array}{cc} \text{MPF} \xrightarrow{\hspace{3cm}} & \text{Cdc2} + \text{cyclin fragment} \\ \cdots\cdots\text{?}\cdots\blacktriangleright \;\; \text{Ub ligase} & \end{array}$$

where '?' represents unknown intermediate(s) that introduce a time delay into this control signal.

The material we have described so far is covered in more detail in Chapter 17 of Alberts *et al.* (1994) and Chapter 25 of Lodish *et al.* (1995).

7.6   MATHEMATICAL MODEL OF THE M-PHASE TRIGGER

We now have enough information to build a mathematical model of the biochemical processes that activate MPF and degrade cyclin, driving cells into and out of M phase, respectively. This model will be sufficient to understand many features of the frog egg cell cycle (Tyson et al. 1995). The simplest realistic model is shown in Figure 7.6 (for simplicity, we consider only one positive feedback loop of MPF activation of Cdc25). Because cyclin monomers bind rapidly to Cdc2 monomers, we can assume that k_3C is much larger than the other rate constants in the model, in which case $L \cong 0$ and

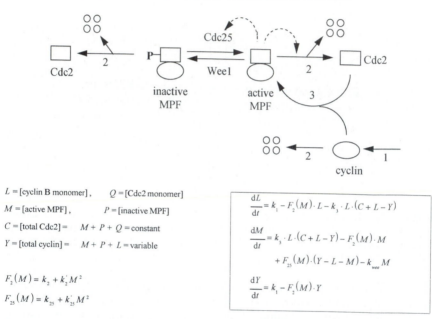

$L = $ [cyclin B monomer], $\quad Q = $ [Cdc2 monomer]

$M = $ [active MPF], $\quad P = $ [inactive MPF]

$C = $ [total Cdc2] $= M + P + Q = $ constant

$Y = $ [total cyclin] $= M + P + L = $ variable

$F_2(M) = k_2 + k_2' M^2$

$F_{25}(M) = k_{25} + k_{25}' M^2$

$$\frac{dL}{dt} = k_1 - F_2(M) \cdot L - k_3 \cdot L \cdot (C + L - Y)$$

$$\frac{dM}{dt} = k_3 \cdot L \cdot (C + L - Y) - F_2(M) \cdot M + F_{25}(M) \cdot (Y - L - M) - k_{wee} M$$

$$\frac{dY}{dt} = k_1 - F_2(M) \cdot Y$$

Figure 7.6. G2 checkpoint control. The biochemical network at the top can be converted into the differential equations at the bottom, using the Law of Mass Action. The functions $F_2(M)$ and $F_{25}(M)$ describe the feedback of MPF activity on the rates of cyclin degradation and tyrosine phosphorylation, respectively. The k_i's are rate constants. From Tyson *et al.* (1995), used by permission.

$$\frac{dM}{dt} = k_1 - (k_2 + k_2' M^2)M + (k_{25} + k_{25}' M^2)(Y - M) - k_{wee}M, \quad (7.1)$$

$$\frac{dY}{dt} = k_1 - (k_2 + k_2' M^2)Y. \quad (7.2)$$

This system of equations can be analyzed by phase-plane techniques (Appendix B). The Y nullcline is easy, a sigmoidal curve in the $M - Y$ plane:

$$\frac{dY}{dt} = 0 \Leftrightarrow Y = \frac{k_1}{k_2 + k_2' M^2}. \quad (7.3)$$

The M nullcline, given by

$$\frac{dM}{dt} = 0 \Leftrightarrow Y = M\left(1 + \frac{k_{wee} + k_2 + k_2'M^2}{k_{25} + k_{25}'M^2}\right) - \frac{k_1}{k_{25} + k_{25}'M^2}, \quad (7.4)$$

is an N-shaped curve. As we adjust the rate constants of the model (k_1 for cyclin synthesis, k_2 and k_2' for cyclin degradation, k_{wee} for Wee1 activity, and k_{25}, k_{25}' for Cdc25 activity), the nullclines move around in the phase plane, and the qualitative features of the control system change. In Figure 7.7 we illustrate several qualitatively different phase portraits that can be expected from this model.

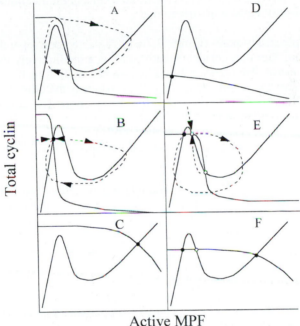

Figure 7.7. Typical phase portraits for the M-phase trigger. • = stable steady state, ○ = unstable steady state. Dashed curves are trajectories. Solid curves are nullclines. The N-shaped curve is the MPF nullcline ($\dot{M} = 0$), and the sigmoidal curve is the cyclin ($\dot{Y} = 0$). Refer to text for detailed description.

- Autonomous oscillations (Figure 7.7(a)). An unstable steady state on the middle branch of the N-shaped nullcline is surrounded by a stable limit cycle. Like the autonomous cytoplasmic oscillator during the early mitotic cycles of a fertilized frog egg, the limit cycle runs independently of growth and DNA synthesis (requiring neither).

- G2 arrest (Figure 7.7(b)). A stable steady state on the lefthand branch of the N-shaped nullcline corresponds to a cell filled with inactive MPF, e.g., an immature oocyte (Figure 7.2). This steady state, though globally asymptotically stable, is "excitable." That is, small perturbations return directly to the steady state, but a suitably large injection of active MPF will stimulate the activation of the store of inactive MPF, reminiscent of the effect of "maturation promoting factor" on immature oocytes (Masui and Markert 1971).

- M arrest (Figure 7.7(c)). A stable steady state on the righthand branch of the N-shaped nullcline corresponds to a cell filled with active MPF, e.g., a mature oocyte (Figure 7.2). Because cyclin is not being effectively degraded, the cell cannot exit mitosis.

- G1 arrest (Figure 7.7(d)). If B-type cyclin synthesis is slow and degradation fast, the steady state will correspond to a cell lacking mitotic cyclins. This is the typical state of differentiated cells in multicellular organisms, and of yeast cells preparing to mate.

- Multiple steady states (Figure 7.7(e)). A stable node separated by a saddle point from an unstable steady state is a qualitatively different kind of G2 arrest. This arrest is lifted by a saddle-node-loop bifurcation (Figure 7.7(e) converted to Figure 7.7(a)), whereas the other kind of G2 arrest (Figure 7.7(b)) is lifted by a Hopf bifurcation. The saddle-node-loop route to oscillation predicts that the newly born limit cycle has a very long temporal period, whereas the Hopf route predicts that the incipient limit cycle has a "normal" temporal period. This difference should be detectable experimentally.

- Bistability (Figure 7.7(f)). Two stable steady states separated by a saddle point would correspond to a cell that could be arrested either in G2 or in M phase. Large perturbations could kick the mitotic control system back and forth between these two states.

Other more complicated portraits are possible even for this simple model of the mitotic control system. Interested readers are encouraged to use a standard software package like AUTO or LOCBIF to investigate the qualitative behavior of this system in dependence on parameter values.

Taking off from this point, Novak and Tyson (1993; 1995) have constructed much more elaborate versions of the mitotic control system, comparing the model in great detail to experiments in frog-egg extracts, intact frog embryos, and fission yeast cells. From these papers, we choose two examples (Sections 7.7 and 7.8) of the kind of insight mathematical models can provide to molecular biologists.

7.7   CYCLIN THRESHOLD FOR MPF ACTIVATION

A clever experiment was designed and executed by Solomon *et al.* (1990) to investigate the role of cyclin in activating MPF. By standard methods, they prepared cell-free frog-egg extracts that contained Cdc2, Wee1, Cdc25, and all the other proteins involved in the mitotic control system except cyclin. The extract was prevented from synthesizing cyclin from its own stores of cyclin mRNA. In place of missing endogenous cyclin, the experimenters

provided exogenously synthesized cyclin, produced by bacteria carrying a eukaryotic cyclin-B gene. The cyclin made by the bacteria was genetically en-
gineered so that it could not be degraded by the frog-egg extract (presumably
it is not recognized by the cyclin-ubiquitin ligating enzyme). Because the extract could neither synthesize its own cyclin nor degrade added cyclin, the experimentalists could precisely control the amount of cyclin in the extract. For
different amounts of added cyclin, they measured the amount of MPF activity
by monitoring the MPF-catalyzed reaction

$$\text{H1-histone} + \text{ATP} \xrightarrow{\text{active MPF}} \text{H1-histone-P} + \text{ADP}.$$

They found that exogenously introduced cyclin could activate the endogenous
pool of Cdc2 protein, but only if they added enough cyclin to exceed a threshold. Such a threshold is exactly to be expected if the MPF nullcline is N-shaped
(Figure 7.8) (Tyson et al. 1996). The threshold represents the total cyclin concentration at the local maximum (C_{max}) of the nullcline.

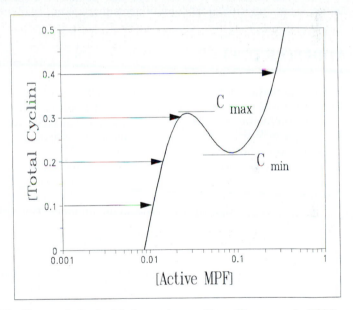

Figure 7.8. Hysteresis in the M-phase trigger. (From Tyson *et al.* (1996), used by
permission.)

If this interpretation is correct, it leads to two clear predictions that
have yet to be tested experimentally. First, there should be a second threshold (C_{min} in Figure 7.8), for MPF inactivation, if the extract is prepared in the
active state. Total cyclin must be reduced below this threshold before MPF
is massively inactivated by tyrosine phosphorylation. The cyclin threshold for
MPF inactivation, the local minimum of the N-shaped nullcline, should be significantly smaller than the cyclin threshold for MPF activation.

Second, for cyclin levels just marginally above the threshold, there
should be a dramatic "slowing-down" of the rate of MPF activation as the solution trajectory passes by the MPF nullcline, where the rate of MPF dephospho-

rylation is only slightly greater than the rate of MPF phosphorylation. Therefore the "lag" time, from first introduction of exogenous cyclin to the appearance of a significant level of MPF activity, will get longer and longer as the exogenous cyclin level approaches the local maximum of the MPF nullcline from above. This prediction is in direct contradiction to the conclusion of Solomon *et al.* that the lag time is independent of added cyclin. Presumably the researchers came to this conclusion because they looked only at cyclin levels considerably larger than the threshold level. We predict that, if they were to repeat the experiment with cyclin levels approaching the threshold from above, they would see this "slowing-down" effect.

The N-shaped nullcline for MPF is sometimes referred to as a hysteresis loop. Hysteresis is a common feature of switch-like processes of human design and of natural occurrence. Hysteresis assures that the switch is not easily reversed: the switch must be pushed hard in one direction (cyclin synthesis) before it turns on, and then it must be pushed hard in the other direction (cyclin degradation) before it turns off. The hysteresis loop in the MPF activation system, caused by the positive feedback loops, assures that both entry into and exit from mitosis are physiological commitments that are not easily revoked.

7.8   ADDITIVE EFFECTS OF *WEE*1$^-$ AND *CDC*25^{OP}

As mentioned earlier, *wee*1$^-$ rescues the lethality of *cdc*25$^-$, which suggests that *wee*1 and *cdc*25 gene products act in opposition, because Cdc25 is necessary for completing the cell cycle in the presence of Wee1 but unnecessary in the absence of Wee1. This is perfectly consistent with the idea of Wee1 and Cdc25 as a kinase-phosphatase pair. In addition, *cdc*25^{OP} *wee*1$^+$ is "wee" but *cdc*25^{OP} *wee*1$^-$ is a mitotic catastrophe (cells try to divide before they have finished replicating their DNA). Geneticists would say that *wee*1$^-$ and *cdc*25^{OP} have additive effects (Russell and Nurse 1986), i.e., the double mutant has a more extreme phenotype than either of the single mutants. But this is hard to explain on the basis of a kinase-phosphatase pair: if one knocks out the kinase (*wee*1$^-$), what difference could it make to overproduce the phosphatase (*cdc*25^{OP})?

Part of the explanation is a second tyrosine kinase, Mik1, that phosphorylates tyrosine-15 during S phase, even in *wee*1$^-$ cells. In *wee*1$^-$ *cdc*25^{OP} cells, the excess Cdc25 phosphatase activity apparently overwhelms Mik1 so that cells accumulate MPF in the active form even while DNA synthesis is ongoing. There is a race to see if M phase is triggered before S phase can be completed. To describe this race we need a quantitative model. Novak and Tyson (1995) have calculated the delay until mitosis in idealized cells overexpressing *cdc*25 in combination with either *wee*1$^+$ or *wee*1$^-$ (Figure 7.9). In a *wee*1$^+$ background, *cdc*25^{OP} leads to a dramatic shortening of G2, because cells activate MPF faster. Cells overexpressing Cdc25 moderately (10-fold) are small but otherwise normal; however, a higher level of overexpression (20-fold) leads to a "conditional mitotic catastrophe." That is, if such cells are transferred to media containing hydroxyurea (HU) (a potent inhibitor of DNA synthesis), the nuclei will enter mitosis even though the chromosomes are not replicated.

In a *wee*1$^-$ background, the whole pattern is shifted to lower levels

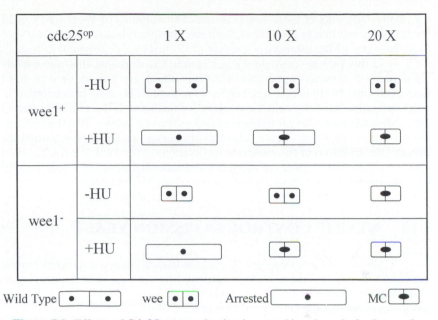

Figure 7.9. Effects of Cdc25 overproduction in a *wee*1⁺ and *wee*1⁻ background.

of overexpression of *cdc*25. At 20-fold overexpression, where *wee*1⁺ cells show conditional mitotic catastrophe, *wee*1⁻ cells show unconditional mitotic catastrophe. The mathematical model clearly shows additivity of the effects of *wee*1⁻ and $cdc25^{OP}$, even though Wee1 and Cdc25 catalyze opposing reactions in the molecular mechanism.

7.9 SIZE CONTROL

Despite the major role that cell size plays in the physiology of the division cycle in yeasts and other single-celled organisms, we still know very little about the molecular mechanism by which the cell monitors its nucleocytoplasmic ratio and triggers events of the cell cycle when it reaches a critical size. This cannot be accomplished by mechanisms involving cytoplasmic concentrations only, because concentration is independent of size. We need at least one variable indicative of overall cell size (e.g., volume, surface area, protein content, number of ribosomes, etc.) and one variable indicative of DNA content (e.g., number of origins of replication, number of copies of a specific gene, number of kinetochores, histone content, etc.). By comparing one variable to the other, the cell could sense its nucleocytoplasmic ratio.

Curiously, in the flood of new molecular information about the cell cycle, the old problem of size control seems to have been forgotten. We are hardly any closer to a molecular mechanism of this essential piece of the puzzle than we were in 1980. The lack of any molecular details about size control is a great hindrance to any comprehensive theory of the cell cycle, mathematical or not. In fission yeast, there are good reasons to suspect that the nucleocytoplasmic ratio is sensed by Wee1 and Cdc25, because size at division is very sensitive to the relative amounts of these two proteins. The amounts of Wee1 and Cdc25 seem to be compared to each other, with the phosphorylation of Cdc2 serving as an indicator of the titration.

In this picture, how do we get around the dilemma that our models are expressed in terms of the concentrations which are independent of size? One likely possibility is that some of the components become concentrated in a subcompartment of the cell, the nucleus for instance. If the effective size of the subcompartment stays roughly constant as the cell grows, then the concentration of the components in the subcompartment can increase even though the overall concentration of the components (averaged over total cell size) remain constant. This is the picture we adopt in our mathematical models, at least until we have evidence for a more realistic mechanism of size control.

7.10   "START" CONTROL IN FISSION YEAST

Control of the wild-type cell cycle in fission yeast is organized around the G2/M transition, where MPF is activated by tyrosine dephosphorylation when DNA replication is complete and cells are large enough. There also exists a size control at the G1/S transition point ("Start"), but wild-type cells are already large enough at birth to satisfy the G1/S size requirement (Fantes and Nurse 1978). Cells proceed into S phase almost immediately after mitosis. The *wee1⁻* mutation disables tyrosine phosphorylation in late G2 phase, so cells enter mitosis at an abnormally small size, and immediately after division these *wee1⁻* cells are too small to execute Start. For this reason, *wee1⁻* cells have short G2 and long G1 periods. They must grow to a minimal size before they can begin DNA synthesis.

The key factor at Start in fission yeast is a gene called *rum1* (replication *u*ncoupled from *m*itosis), first identified by Moreno and Nurse (1994) by its unusual phenotype: when the protein is overexpressed from a multicopy plasmid containing the wild-type gene (*rum1⁺*), cells cease to divide but replicate their DNA over and over again. They become very large, with 8, 16, 32, ... copies of nuclear DNA. Genetic and biochemical analysis showed that Rum1 is a potent inhibitor of Cdc13/Cdc2-kinase (i.e., MPF): Rum1 binds strongly to Cdc13/Cdc2 dimers and completely inhibits the dimer's kinase activity (Correa-Bordes and Nurse 1995). Cells that overexpress Rum1, therefore, cannot enter mitosis. But Rum1 apparently allows enough Cdc2-kinase activity, associated with other cyclins (Cig1 and Cig2), for cells to enter S phase repeatedly. This is not the place for a thorough analysis of G1/S control in fission yeast (Novak and Tyson, in preparation), but we can describe the basic hysteresis loop that we believe underlies the G1/S transition (Start) in *S. pombe* (see Figure 7.10). We build the model around the fact that Rum1 and Cdc13 are antagonistic proteins. In G1 phase, Rum1 level is high and Cdc13 level is low, whereas the reverse situation prevails in G2. We propose that this antagonism is produced by mutual acceleration of the degradation of the other component (Figure 7.10(a)). Rum1 is known to have many potential Cdc2 phosphorylation sites, and the unphosphorylated form is possibly more stable (Figure 7.10(b)). Although Cdc13/Cdc2-kinase does not phosphorylate Rum1 when the purified components are mixed in a test tube, we suppose that it does phosphorylate Rum1 in the living cell (perhaps indirectly). Similarly,

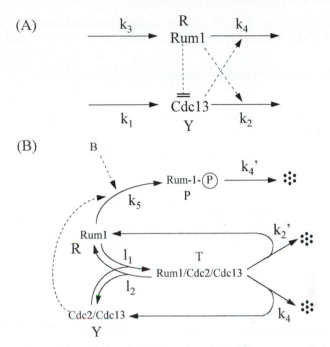

Figure 7.10. G1 checkpoint control. (A) Rum1 and Cdc13 are antagonistic proteins. (B) more details.

although there is no direct biochemical evidence, we assume that Cdc13, when bound to Rum1, is more sensitive to degradation (Figure 7.10(b)). If these interactions or similar cross-coupling inhibitions take place *in vivo*, then Rum1 and Cdc13 will be mutually exclusive factors.

A simple mathematical model of these interactions (Figure 7.10) is provided by the following set of ODEs for Rum1 monomers (R), Cdc13/Cdc2 dimers (Y), phosphorylated Rum1 (P), and trimeric complexes of Cdc13/Cdc2/Rum1 (T):

$$\frac{dY}{dt} = k_1 - k_2 Y - l_1 RY + l_2 T + k_4 T, \tag{7.5}$$

$$\frac{dR}{dt} = k_3 - k_4 R - l_1 RY + l_2 T + k_2' T - \frac{k_5 R(B + Y)}{J_5 + R}, \tag{7.6}$$

$$\frac{dP}{dt} = \frac{k_5 R(B + Y)}{J_5 + R} - k_4' P, \tag{7.7}$$

$$\frac{dT}{dt} = l_1 RY - l_2 T - k_2' T - k_4 T, \tag{7.8}$$

where B is a parameter representing the activity of a "Starter" kinase, and the other rate constants are labeled according to the figure. Our assumptions that phosphorylation destabilizes Rum1 and that trimer formation destabilizes Cdc13 translate into the conditions that $k_2' \gg k_2$ and $k_4' \gg k_4$.

We can simplify these equations, first of all, by noticing that P is not involved in any other differential equation but its own. Phosphorylation removes Rum1 from the picture, so P can be ignored in calculating the dynamics of the other three variables. Second, we assume that the trimer T is in equilibrium with R and Y, because the binding reaction is known to be fast and tight:

$$T = LRY, \text{ where } L = \frac{l_1}{l_2 + k_2' + k_4}. \tag{7.9}$$

We are left with two ODEs, which are conveniently written in terms of total Rum1 ($\widehat{R} = R + T$) and total Cdc13 ($\widehat{Y} = Y + T$):

$$\frac{d\widehat{Y}}{dt} = k_1 - k_2(\widehat{Y} - T) - k_2'T, \tag{7.10}$$

$$\frac{d\widehat{R}}{dt} = k_3 - k_4\widehat{R} - \frac{k_5(B + \widehat{Y} - T)(\widehat{R} - T)}{J_5 + \widehat{R} - T}, \tag{7.11}$$

where

$$T(\widehat{R}, \widehat{Y}) = \frac{2L\widehat{R}\widehat{Y}}{1 + L\widehat{R} + L\widehat{Y} + \sqrt{(1 + L\widehat{R} + L\widehat{Y})^2 - 4L^2\widehat{R}\widehat{Y}}}. \tag{7.12}$$

The phase plane for this pair of ODEs is illustrated in Figure 7.11. The system

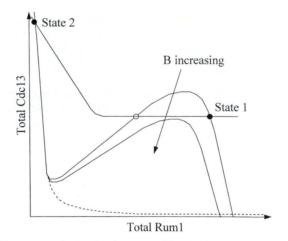

Figure 7.11. Phase portrait for the Start transition in fission yeast. The solid curves are nullclines: the Rum 1 nullcline is N-shaped, the Cdc13 nullcline is hyperbolic. The dashed curve represents the fraction of Cdc13 bound to Cdc2 but not to Rum1 (i.e., active MPF).

can persist in two alternative states: (1) with high levels of Rum1 and very low Cdc13/Cdc2-kinase activity, and (2) with very little Rum1 and lots of active Cdc13/Cdc2-kinase. The Start transition (State 1 → State 2) can be accomplished by a starter kinase (B) that increases in activity as the cell grows. As a result, the Rum1 nullcline is pulled down (Figure 7.11) and State 1 disappears by a saddle-node bifurcation.

7.11   METAPHASE CHECKPOINT

Cell biologists have identified three checkpoints in the division cycle (Alberts et al. 1994). Cells can arrest in G1, before DNA replication, to check that the previous mitosis has been completed properly, that cell size is large enough, that appropriate proliferation factors are present in the environment, that signals for mating or differentiation are absent, etc. In G2, before committing to nuclear division, cells can stop to check that DNA replication is complete and that cell size is sufficient. And at metaphase, cells can arrest to make sure that all chromosomes are properly arranged on the mitotic spindle. We have described some of the mechanisms used for G1 and G2 arrest, and now we turn briefly to the metaphase checkpoint.

Not much detail can be given because little is known about the molecular machinery of the metaphase-to-anaphase transition. We know that it involves the activation of a ubiquitin-mediated proteolytic pathway. After the pathway is activated, the first step seems to be dissolution of a protein that binds sister chromatids together at the kinetochore. As the "glue" dissolves, motor proteins drag sister chromatids to opposite poles of the spindle. This event is sometimes called anaphase I. During anaphase II, the incipient daughter nuclei at the spindle poles are pushed apart in preparation for cell division. This latter step requires proteolysis of mitotic cyclins (nondegradable B-type cyclins block cells in late anaphase) (Holloway et al. 1993). The dramatic drop in MPF activity in anaphase also seems to be necessary to reprime replication origins, so that DNA synthesis can recommence in the next cell cycle.

In budding yeast the ubiquitin-mediated pathway of mitotic cyclin degradation seems to have two stable states: it is "on" in G1 phase and "off" in the rest of the cycle (Amon et al. 1994). The pathway is turned off at Start by the rising activity of SPF ("S-phase promoting factor"), and it stays off during G2 and M, even though SPF activity drops and MPF activity rises. At the metaphase checkpoint it is turned on, perhaps by high levels of MPF activity, but it stays on throughout G1, even though both SPF and MPF activities seem to be low. This behavior has all the earmarks of hysteresis, but not enough is known about the component parts to identify with confidence the positive-feedback loop responsible for the hysteretic response. Nonetheless, we take the liberty to speculate how the metaphase checkpoint might work. Suppose cells synthesize an inhibitor (I) of the ubiquitination pathway (U) that is itself degraded by the ubiquitination (Figure 7.12). We might have a dynamical system like this:

$$\frac{dI}{dt} = k_1 + k_1' F - (k_2 + k_2' U)I, \tag{7.13}$$

$$\frac{dU}{dt} = \frac{k_3(M+G)(\widehat{U}-U)}{J_3 + \widehat{U} - U} - \frac{k_4(I+X)U}{J_4 + U}. \tag{7.14}$$

The interpretation of the rate constants should be obvious by now. $\widehat{U} = U + U^*$ = total concentration of the ubiquitin-conjugating enzyme = constant. F represents a transcription factor which is turned on by SPF and/or unreplicated DNA, and G represents a background kinase which cooperates with MPF (M) to keep the ubiquitination pathway active. X is a strong signal from misaligned

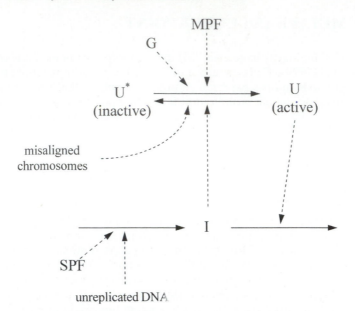

Figure 7.12. A speculative model of the metaphase checkpoint.

chromosomes that keeps U inactive. The phase portraits in Figure 7.13 tell the story. We choose parameters so that the system has two alternative steady states (U "on" or U "off") and suppose that at birth the cell is trapped in the "on" state (labeled G1 in Figure 7.13(a)). At Start the transcription factor is activated ($F = 1$), the "on" state disappears (by a saddle-node bifurcation), and the ubiquitination pathway turns off. Later, when F switches off ($F = 0$), the cyclin degradation system finds itself trapped in the U "off" state (labeled G2 in Figure 7.13(a)). As MPF activity rises and the cell enters mitosis ($M = 1$), the "off" state is temporarily lost, and the ubiquitination pathway turns on (labeled An, for anaphase, in Figure 7.13(b)). The resulting degradation of MPF brings us back to G1 phase ($M = 0$, $F = 0$) with U "on" (Figure 7.13(b)).

If for any reason the metaphase spindle has not assembled properly, a strong signal ($X = 1$) keeps U "off" even for high levels of MPF (labeled Me, for metaphase arrest).

There is no experimental evidence for an inhibitor of the type we have postulated. Maybe it does not exist. However, we expect that some hysteresis loop, like the one in Figure 7.13, is operative at the anaphase checkpoint, as at the G1 and G2 checkpoints.

7.12   CONCLUSION

In this chapter we have reviewed the experimental basis on which to build realistic mathematical models of the molecular mechanisms controlling key events of the cell cycle, and we have described some simple two-component models and phase-plane portraits of the G1, G2, and M-phase checkpoints. By emphasizing the physiological, genetic, and biochemical evidence, we want to stress the importance of model building driven by exper-

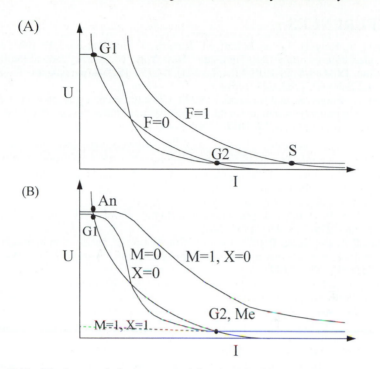

Figure 7.13. The hysteresis loop proposed for the ubiquitination pathway of cyclin degradation. The solid curves are nullclines: the U-nullcline is sigmoidal and the I-nullcline hyperbolic. (A) At Start the ubiquitin-cyclin ligating enzyme (*U*) switches from its active form (at G1) to its inactive form (at S). (B) As cells enter mitosis with properly aligned chromosomes (*X* = 0), *U* is switched back to its active form (at An). If the chromosomes are misaligned (*X* = 1), the proteolytic machinery remains inactive (at Me).

imental facts rather than mathematical convenience or elegance. We suspect that readers of this chapter may need more coaching in the art of modeling (getting the right equations to start with) than in the science of analysis (bifurcation theory, singular perturbation theory, numerical analysis, etc.). We have limited our model building, here, to the first step: identifying the crucial reactions involved in isolated stages of the cycle (tyrosine phosphorylation of Cdc2, synthesis and degradation of stoichiometric inhibitors, and activation of the ubiquitination pathway) and constructing simple models of each stage. Elsewhere (Novak and Tyson 1993; Novak and Tyson 1995) we show how to do the next steps, putting the pieces together to get working models of frog egg extracts or yeast-cell cultures and comparing the model in quantitative detail with physiological and genetic observations.

7.13   ACKNOWLEDGMENTS

We gratefully acknowledge the National Science Foundation (MCB-9207160 and INT-9212471) and the Howard Hughes Medical Institute (75195-512302) for supporting the research described in this chapter.

REFERENCES

Alberts, B., Bray, D., Lewis, J., Raff, M., Roberts, K., and Watson, J. D. 1994. *Molecular Biology of the Cell*. Third edn. New York, NY, USA: Garland Publishing, Inc. ISBN 0-8153-1619-4 (hard cover), 0-8153-1620-8 (paperback). Pages xliii + 1294 + 67. {*133, 143, 146*}

Amon, A., Irniger, S., and Nasmyth, K. 1994. Closing the cell cycle circle in yeast: G2 cyclin proteolysis initiated at mitosis persists until the activation of G1 cyclins in the next cycle. *Cell*, **77**, 1037–1050. {*143, 146*}

Correa-Bordes, J. and Nurse, P. 1995. p25^{rum1} orders S-phase and mitosis by acting as an inhibitor of the p34^{cdc2} mitotic kinase. *Cell*, **83**, 1001–1009. {*140, 146*}

Evans, T., Rosenthal, E. T., Youngbloom, J., Distel, D., and Hunt, T. 1983. Cyclin: a protein specified by maternal mRNA in sea urchin eggs that is destroyed at each cleavage division. *Cell*, **33**, 389–396. {*131, 146*}

Fantes, P. A. 1979. Epistatic gene interactions in the control of division in fission yeast. *Nature*, **279**, 428–430. {*130, 146*}

Fantes, P. A. and Nurse, P. 1978. Control of the timing of cell division in fission yeast: cell size mutants reveal a second control pathway. *Experimental Cell Research*, **115**, 317–329. {*140, 146*}

Frazier, E. A. J. 1973. DNA synthesis following gross alterations of the nucleocytoplasmic ratio in the ciliate *Stentor coeruleus*. *Developmental Biology*, **34**, 77–92. {*128, 146*}

Gerhart, J., Wu, M., and Kirschner, M. W. 1984. Cell cycle dynamics of an M-phase-specific cytoplasmic factor in *Xenopus laevis* oocytes and eggs. *Journal of Cell Biology*, **98**, 1247–1255. {*131, 146*}

Hartmann, M. 1928. Über experimentelle unsterblichkeit von protozoen-Individuen. Ersatz der fotpflanzung von amoeba proteus durch fortgesetzte regenerationen. *Zoologisches Jahrbuch*, **45**, 973–987. {*128, 146*}

Hartwell, L. H., Culotti, J., Pringle, J. R., and Reid, B. J. 1974. Genetic control of the cell division cycle in yeast. *Science*, **183**, 46–51. {*129, 146*}

Harvey, E. B. 1940. A comparison of the development of nucleate and non-nucleate eggs of *Arbacia punctulata*. *Biological Bulletin*, **79**, 166–187. {*129, 146*}

Holloway, S. L., Glotzer, M., King, R. W., and Murray, A. W. 1993. Anaphase is initiated by proteolysis rather than by the inactivation of maturation-promoting factor. *Cell*, **73**, 1393–1402. {*143, 146*}

Lodish, Harvey, Baltimore, D., Berk, A., Zipursky, S. L., Matsudaira, P., and Darnell, James E. 1995. *Molecular Cell Biology*. Third edn. New York, NY, USA: W. H. Freeman. ISBN 0-7167-2380-8. Page various. {*133, 146*}

Lohka, M. J., Hayes, M. K., and Maller, J. L. 1988. Purification of maturation-promoting factor, an intracellular regulator of early mitotic events. *Proceedings of the National Academy of Sciences of the United States of America*, **85**, 3009–3013. {*132, 146*}

Masui, Y. and Markert, C. L. 1971. Cytoplasmic control of nuclear behavior during meiotic maturation of frog oocytes. *Journal of Experimental Zoology*, **177**, 129–146. {*131, 136, 146*}

Mitchison, J. M. 1971. *The Biology of the Cell Cycle*. Cambridge, UK: Cambridge University Press. ISBN 0-521-08251-X. 0-521-09671-5 (paperback). Pages v + 313. {*127, 146*}

Moreno, S. and Nurse, P. 1994. Regulation of progression through the G1 phase of the cell cycle by the *rum1$^+$* gene. *Nature*, **367**(6460), 236–242. {*140, 146*}

Murray, A. W. 1993. Turning on mitosis. *Current Biology*, **3**, 291–293. {*133, 146*}

Murray, A. W. 1995. Cyclin ubiquitination: the destructive end of mitosis. *Cell*, **81**, 149–152. {*133, 146*}

Murray, A. W. and Kirschner, M. W. 1989a. Cyclin synthesis drives the early embryonic cell cycle. *Nature*, **339**(6222), 275–280. {*131, 146*}

Murray, A. W., Solomon, M. J., and Kirschner, M. W. 1989b. The role of cyclin synthesis and degradation in the control of maturation promoting factor activity. *Nature*, **339**(6222), 280–286. {*131, 147*}

Murray, Andrew and Hunt, Tim. 1993. *The Cell Cycle: an Introduction.* New York, NY, USA: W. H. Freeman. ISBN 0-7167-7044-X (hard), 0-7167-7046-6 (soft), 0-19-509529-4. Pages xii + 251. {*130, 131, 147*}

Novak, B. and Tyson, J. J. 1993. Numerical analysis of a comprehensive model of M-phase control in *Xenopus* oocyte extracts and intact embryos. *Journal of Cell Science*, **106**, 1153–1168. {*136, 145, 147*}

Novak, B. and Tyson, J. J. 1995. Quantitative analysis of a molecular model of mitotic control in fission yeast. *Journal of Theoretical Biology*, **173**, 283–305. {*136, 138, 145, 147*}

Nurse, P. 1975. Genetic control of cell size at division in yeast. *Nature*, **256**, 547–551. {*129, 147*}

Nurse, P. 1981. Genetics analysis of the cell cycle. *Pages 291–315 of:* Glover, S. W. and Hopwood, D. A. (eds), *Genetics as a tool in microbiology.* Cambridge, UK: Cambridge University Press. {*129, 147*}

Nurse, P. 1990. Universal control mechanism regulating onset of M-phase. *Nature*, **344**, 503–508. {*133, 147*}

Pringle, J. R. and Hartwell, L. H. 1981. The *Saccharomyces cerevisiae* cell cycle. *Pages 97–142 of:* Strathern, Jeffrey N., Jones, Elizabeth W., and Broach, James R. (eds), *The Molecular Biology of the Yeast Saccharomyces.* Cold Spring Harbor, NY, USA: Cold Spring Harbor Laboratory. {*129, 147*}

Rao, P. N. and Johnson, R. T. 1970. Mammalian cell fusion: studies on the regulation of DNA synthesis and mitosis. *Nature*, **225**, 159–164. {*128, 147*}

Russell, P. and Nurse, P. 1986. $cdc25^+$ functions as an inducer in the mitotic control of fission yeast. *Cell*, **45**, 145–153. {*130, 138, 147*}

Russell, P. and Nurse, P. 1987. Negative regulation of mitosis by $wee1^+$, a gene encoding a protein kinase homolog. *Cell*, **49**, 559–567. {*130, 147*}

Sachsenmaier, W., Remy, U., and Plattner-Schobel, R. 1972. Initiation of synchronous mitosis in *Physarum polycephalum. Experimental Cell Research*, **73**, 41–48. {*128, 147*}

Simanis, V. and Nurse, P. 1986. The cell cycle control gene *cdc2* of fission yeast encodes a protein kinase potentially regulated by phosphorylation. *Cell*, **45**, 261–268. {*130, 147*}

Solomon, M. J., Glotzer, M., Lee, T. H., Philippe, M., and Kirschner, M. W. 1990. Cyclin activation of p34^{cdc2}. *Cell*, **63**, 1013–1024. {*136, 147*}

Tyson, J. J. 1984. The control of nuclear division in *Physarum polycephalum. Pages 175–190 of:* Nurse, Paul and Streiblova, Eva (eds), *The Microbial Cell Cycle.* 2000 Corporate Blvd., Boca Raton, FL 33431, USA: CRC Press. {*128, 147*}

Tyson, J. J., Novak, B., Chen, K. C., and Val, J. 1995. Checkpoints in the cell cycle from a modeler's perspective. *Progress in Cell Cycle Research*, **1**, 1–8. {*134, 147*}

Tyson, J. J., Novak, B., Odell, G. M., Chen, K., and Thron, C. D. 1996. Chemical kinetic theory as a tool for understanding the regulation of M-phase promoting factor in the cell cycle. *Trends in Biochemical Sciences*, **21**, 89–96. {*137, 147*}

8. MATHEMATICAL MODELS OF HEMATOPOIETIC CELL REPLICATION AND CONTROL

Michael C. Mackey

8.1 INTRODUCTION

The biological sciences offer an abundance of examples in which the dynamics are dependent on the past history of the system being studied. In the attempt to construct mathematical models for these processes, the dynamic equations are often framed either as functional equations or, more specifically, as differential delay equations. Often by studying the solution behavior of these model systems one can begin to understand the normal biological responses noted in the laboratory.

Even more interesting are those clinical situations in which the normal dynamics of a system are replaced, in disease situations, by dynamics with different characteristics. Sometimes these characteristics involve the destabilization of a steady state in favor of periodic or aperiodic behavior, or the replacement by a new periodic regime of the normal periodic one. Such diseases are called "periodic diseases" (Glass and Mackey 1988; Mackey and Glass 1977; Reimann 1963) and are quite interesting in that they can often give clues about the nature of the underlying disease expressed as a shift in parameters giving rise to a bifurcation.

Some of the most fascinating of these periodic diseases are the periodic diseases of the blood (periodic hematological diseases). It has long been suspected that periodic hematological diseases arise because of abnormalities in the feedback mechanisms which regulate blood-cell number (Dunn 1983; Kirk et al. 1968; Mackey 1978; Mackey 1979b; Mackey 1979a; Morley 1979; Wichmann and Loeffler 1988). Indeed this observation has provided a major impetus for mathematicians to determine the conditions for oscillation onset in these mechanisms. There have been two surprising predictions of these studies (Glass and Mackey 1979; Mackey and Glass 1977): (1) qualitative changes can occur in blood-cell dynamics as quantitative changes are made in feedback control; and (2) under appropriate conditions, these feedback mechanisms can produce aperiodic, irregular fluctuations ("chaotic" in the current vernacular) which could easily be mistaken for noise and/or experimental error (Bai-Lin 1984; Degn et al. 1987; Glass and Mackey 1988). The clinical significance is that it may be possible to develop new diagnostic and therapeutic strategies based on manipulation of feedback (Glass and Mackey 1979; Glass and Mackey 1988; Mackey and Glass 1977; Mackey and an der Heiden 1982; Mackey and Milton 1987). In this chapter I examine some of these theoretical developments and discuss their clinical implications, using several different disease entities as case studies.

8.2   CONTROL OF BLOOD-CELL PRODUCTION

The organization of normal hematopoiesis (the production and maintenance of circulating blood cell numbers) is shown in Figure 8.1. It is generally believed that there exists a self-maintaining population of undifferentiated cells (pluripotential stem cells) capable of producing committed stem cells specialized for the red blood (erythroid), white blood (myeloid) or blood-clotting (thromboid) cell lines (Quesenberry 1990). The dynamics of pluripotential stem cells and the committed stem cells are regulated by two types of feedback mechanisms: (1) long-range humoral mechanisms (Cebon and Layton 1994; Sachs 1993; Sachs and Lotem 1994), e.g., renal erythropoietin for the erythrocytes, the colony-stimulating factors (G-colony-stimulating factor and GM-colony-stimulating factor) for the leukocytes (white blood cells), and thrombopoietin for the platelets; and (2) local environmental mechanisms, such as stem-cell factor (SCF) which are as yet poorly characterized (labeled as LR in Figure 8.1). An intrinsic property of these feedback mechanisms is the presence of time delays which arise because of finite nonzero cell-maturation times and cell-replication times. Thus many investigators have studied models of delayed feedback in order to investigate the periodic hematological diseases (Dunn 1983; Kazarinoff and van den Driessche 1979; Kirk et al. 1968; Glass and Mackey 1979; MacDonald 1978; Mackey and Glass 1977; Mackey 1978; Mackey 1979b; Mackey 1979a; Morley 1979; Nazarenko 1976; von Schulthess and Mazer 1982; Wheldon et al. 1974; Wheldon 1975).

In order to appreciate how oscillations develop in blood-cell number and their properties such as period and morphology, three steps are necessary: (1) development of a simple, but physiologically realistic, model for the relevant control mechanism; (2) investigation of the properties of the model, typically by the use of stability analysis and computer simulations; and (3) comparison of the model's predictions to experimental and/or clinical observations.

Current analytic and numerical work determine the time-dependent changes in blood-cell number as certain quantities, referred to as control parameters, are varied. Control parameters are those quantities which in comparison to blood-cell number either do not change with time, or change very little and hence are regarded by the investigator to be constant. Examples of control parameters in the regulation of hematopoiesis are the maturation times and the peripheral destruction rate(s).

8.3   RANDOM CELL LOSS FROM A COMPARTMENT

Let's consider the simplest possible situation, in which we have cells in a compartment that die randomly as time proceeds (Figure 8.2). We will let $X(t)$ be the number of cells (usually measured as a density, e.g., numbers

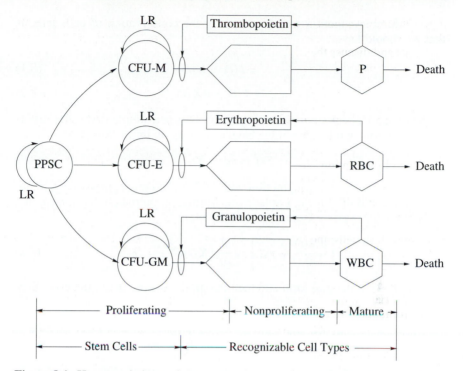

Figure 8.1. Hematopoietic regulation architecture. A schematic representation of the control of platelet (P), erythrocyte (RBC), and white blood cell (WBC) production (adapted from Quesenberry (1990)), showing loops mediated by the various poietins, as well as local regulatory (LR) loops within the various stem-cell compartments. CFU refers to the various colony-forming units (M = megakaryocytic, E = erythroid, GM = granulocyte/macrophage) which are the *in vitro* analogs of the *in vivo* committed stem-cell (CSC) populations, all of which arise from the pluripotential stem cells (PPSCs).

per liter, numbers per unit body weight, etc.) at a given time t, and assume that cells are arriving in the compartment (for example, the peripheral circulation) with an input flux of $I(t)$ and are being lost at a random rate γ, so the efflux to death is $D(t) = \gamma X(t)$. In hematology, a typical example of a process whose dynamics are described by this situation is the kinetics of circulating white blood cells.

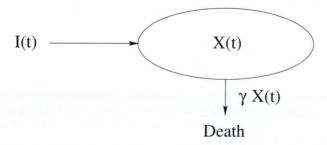

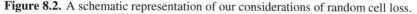

Figure 8.2. A schematic representation of our considerations of random cell loss.

Simple common sense concerning the conservation of cells tells us that we should have

$$\frac{dX}{dt} = I(t) - D(t) \tag{8.1}$$

$$= I(t) - \gamma X(t) \tag{8.2}$$

as the dynamical equation governing $X(t)$. This is a simple first order differential equation, and to solve it we have to specify an initial condition

$$X(t = t_0) = X_0. \tag{8.3}$$

The solution to (8.2) in conjunction with (8.3) is, of course, pretty easy to obtain. One could do it using Laplace transforms, or alternately by multiplying (8.2) by the integrating factor $\exp(\gamma t)$ to obtain

$$e^{\gamma t}\left[\frac{dX(t)}{dt} + \gamma X\right] = \frac{d}{dt}\left(X(t)e^{\gamma t}\right) = I(t)e^{\gamma t} \tag{8.4}$$

Writing (8.4) in integral form and then integrating, using the initial condition, (8.3) and rearranging, we have

$$X(t) = X_0 e^{-\gamma(t-t_0)} + e^{-\gamma t}\int_{t_0}^{t} I(z)e^{\gamma z}\,dx. \tag{8.5}$$

If the input of cells I is a constant, then the solution (8.4) of the original problem (8.2)-(8.3) is even simpler and can be written as

$$X(t) = X_0 e^{-\gamma(t-t_0)} + \frac{I}{\gamma}\left[1 - e^{-\gamma(t-t_0)}\right]. \tag{8.6}$$

8.3.1   Example

To illustrate how hematologists can make use of such a formulation, consider the real situation in which a radioactive tracer is given in what is known as a *flash-labeling* mode, such that all of the circulating white blood cells pick up some of the label, but none of their precursors has any. Then, although unlabeled white blood cells will still be entering the circulation through the flux I, their absence of label will distinguish them from their labeled older sisters. If we denote the population of labeled white blood cells by X^*, then their dynamics are going to be given by

$$X^*(t) = X_0^* e^{-\gamma t}. \tag{8.7}$$

Now we can actually use (8.7) to determine the rate of peripheral cell death γ in the following straightforward way. If we let the fraction of labeled cells at a time t after the administration of the label be $F(t)$, then we can rewrite (8.7) in the more useful form

$$F(t) = \frac{X^*(t)}{X_0^*} = e^{-\gamma t}. \tag{8.8}$$

Equation (8.8) tells us that the fraction of labeled cells decreases exponentially with time, and if we can measure what F is at a given time, then we can find out what γ is. What is typically done is to find the time (denoted by $t_{1/2}$) at which the fraction F is $\frac{1}{2}$, i.e., $F(t_{1/2}) = \frac{1}{2}$. Using this notation in conjunction with (8.8), we have after a bit of algebra that

$$\gamma = \frac{\ln 2}{t_{1/2}}. \tag{8.9}$$

When this experiment is done with white blood cells from humans, the typical result is that the semilogarithmic plot of $F(t)$ versus time is a straight line, so it really does decay exponentially, and $t_{1/2}$ is approximately 6.9 hours, so that for humans

$$\gamma = \frac{\ln 2}{6.9}\,\mathrm{hr}^{-1} = 0.1\mathrm{hr}^{-1}. \tag{8.10}$$

8.4 DYNAMICS OF CELLS THAT AGE

As a next step in building our toolbox of techniques for dealing with cell-dynamics problems, let us consider a situation slightly more complicated that the random-loss one of the previous section.

Now, we consider the case in which cells are in a compartment and are lost by one of two means (Figure 8.3). First, as before, these cells are lost at a random rate (again denoted by γ), and second, they are lost because they just become too old. In addition, cells arrive into the population with a certain influx $I(t)$.

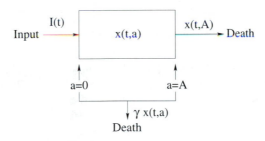

Figure 8.3. The situation in which we have cells living to a maximal age A, but also dying at a random rate γ.

This sounds somewhat like the situation facing humans or any other species — the birth rate is I, there are random deaths in the population as we age, and we eventually die when we get too old if something doesn't knock us off randomly. Clearly, there is a certain arbitrary nature to the distinction between death due to random events versus death due to senescence, but that won't bother us here.

We are going to let $x(t, a)$ be the number of cells at a given time t and age a. Further, we assume that cells can live to a maximum age A and that cells that immigrate with the influx $I(t)$ do so at an age $a = 0$. From a dynamic point of view, the equation that governs the evolution of $x(t, a)$ is the first-order partial differential equation (or conservation equation)

$$\frac{\partial x(t, a)}{\partial t} + \frac{\partial x(t, a)}{\partial a} = -\gamma x(t, a). \tag{8.11}$$

To complete the specification of the problem, we will have to supplement (8.11) with the initial condition

$$x(t = 0, a) = f(a) \tag{8.12}$$

as well as the boundary condition

$$x(t, a = 0) = I(t). \tag{8.13}$$

How to deal with the system (8.11) through (8.13)? Well, the first thing to realize is that we can write the total number of cells of all ages between the minimum age $a = 0$ and the maximum age $a = A$ as

$$X(t) = \int_0^A x(t, a) \, da. \tag{8.14}$$

Thus, integrating (8.11) over the entire range of ages and using the definition (8.14), we have

$$\frac{dX(t)}{dt} + x(t, A) - x(t, 0) = -\gamma X(t), \tag{8.15}$$

or, using the boundary condition (8.13), this becomes

$$\frac{dX(t)}{dt} = I(t) - \gamma X(t) - x(t, A). \tag{8.16}$$

Its easy to understand what each term in (8.16) means, for it just says that the total rate of change of all of the cells in the compartment is a balance between the input $I(t)$, the random loss $-\gamma X(t)$, and the loss to death of those individuals who made it to age A.

However, we can go even further by noting that, using the method of characteristics, the general solution of (8.11) is given by

$$x(t, a) = \begin{cases} x(0, a - t)e^{-\gamma t}, & 0 \le t < a, \\ x(t - a, 0)e^{-\gamma a}, & a \le t. \end{cases} \tag{8.17}$$

Further, making use of the initial condition (8.12) and the boundary condition (8.13), we can write the general solution (8.17) in the more explicit form

$$x(t, a) = \begin{cases} f(a - t)e^{-\gamma t}, & 0 \le t < a \\ I(t - a)e^{-\gamma a}, & a \le t. \end{cases} \tag{8.18}$$

Now (8.16) contains the term $x(t, A)$, and we clearly have, from (8.18), that

$$x(t, A) = \begin{cases} f(A - t)e^{-\gamma t}, & 0 \le t < A, \\ I(t - A)e^{-\gamma A}, & A \le t. \end{cases} \tag{8.19}$$

Thus we can finally write (8.16) in the final and useful form

$$\frac{dX(t)}{dt} = I(t) - \gamma X(t) - \begin{cases} f(A - t)e^{-\gamma t}, & 0 \le t < A, \\ I(t - A)e^{-\gamma A}, & A \le t. \end{cases} \tag{8.20}$$

8.5 PERIODIC AUTOIMMUNE HEMOLYTIC ANEMIA: CONTROL OF RED BLOOD CELL PRODUCTION AND DELAYED NEGATIVE FEEDBACK

Periodic autoimmune hemolytic anemia is a rare form of hemolytic anemia in humans (Gordon and Varadi 1962; Ranlov and Videbaek 1963), but it has been induced in rabbits (Figure 8.4) by using red blood cell auto-antibodies (Orr et al. 1968). At certain levels of administration, these antibodies oscillate in red blood cell (erythrocyte) precursors (reticulocytes) with a period of 16 to 17 days. Rabbit autoimmune hemolytic anemia is one of the best understood periodic hematological diseases and arises from increases in the destruction rate of circulating erythrocytes.

The concept of delayed negative feedback can be introduced by considering the control of erythrocyte production as represented schematically in Figures 8.1 and 8.5.

To formulate this sequence of physiological processes in a mathematical model, we will make the following definitions (Mackey 1979b). Let:

$E(t)$ (cells/kg) = circulating density of red blood cell as a function of time

F (cells/kg-day) = cell influx from erythroid colony forming units, under erythropoietin control

τ (days) = time required to pass through recognizable precursors

γ (days^{-1}) = loss rate of red blood cells in the circulation

Using this notation, we can write a balance equation stating that the rate of change of erythrocyte numbers is a balance between their production and their destruction:

$$\frac{dE(t)}{dt} = \text{production} - \text{destruction}$$
$$= F(E(t - \tau)) - \gamma E(t). \tag{8.21}$$

It is important to remember that once a pluripotential stem cell is committed to the erythroid series, it undergoes a series of nuclear divisions and enters a

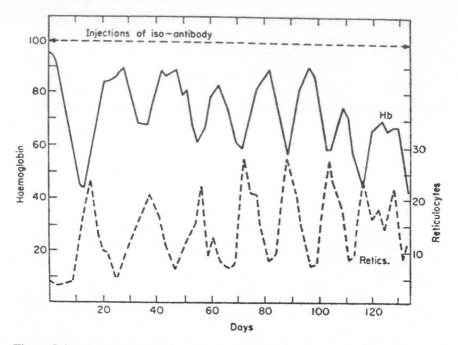

Figure 8.4. Laboratory-induced autoimmune hemolytic anemia. Oscillations in circulating hemoglobin and reticulocyte counts in a rabbit during constant application of red blood cell iso-antibody. (Redrawn from Kirk et al. (1968).)

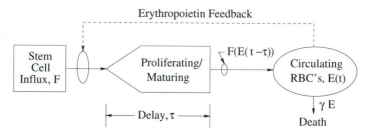

Figure 8.5. A fall in circulating erythrocyte numbers leads to a decrease in hemoglobin levels and thus in arterial oxygen tension. This decrease in turn triggers the production of renal erythropoietin, which increases the cellular production within the early committed erythrocyte series cells and thus the cellular efflux from the erythroid colony forming units into the identifiable proliferating and nonproliferating erythroid precursors, and ultimately augments circulating erythrocyte numbers (i.e., negative feedback) after a delay τ.

maturational phase for a period of time (τ approximately 5.7 days) before release into circulation. The argument in the production function is $E(t - \tau)$, and not $E(t)$, because a change in the peripheral red blood cell numbers can only augment or decrease the influx into the circulation after a period of time τ has elapsed. Thus, changes that occur at time t were actually initiated at a time $t - \tau$ in the past. To avoid the cumbersome notation $E(t - \tau)$, I will adopt here the usual convention of $E_\tau(t) = E(t - \tau)$, and I will not explicitly denote the time unless necessary. Thus, we can write our simple model (8.21) for red blood cell dynamics in the alternate form

$$\frac{dE}{dt} = F(E_\tau) - \gamma E. \tag{8.22}$$

The next step in our model construction is to define some appropriate form for the production function F. *In vivo* measurements of erythrocyte production rates F in rats (Hodgson and Eskuche 1966) and other mammals including humans indicate that the feedback function saturates at low erythrocyte numbers and is a decreasing function of increasing red blood cell levels (i.e., negative feedback). A convenient function that captures this behavior, and which has sufficient flexibility to be able to fit the data, as well as easily handled analytic properties, is given by

$$F(E_\tau) = F_0 \frac{\theta^n}{E_\tau^n + \theta^n}, \tag{8.23}$$

where F_0 (units of cells/kg-day) is the maximal red blood cell production rate that the body can approach at very low circulating red blood cell numbers, n is a positive exponent, and θ (units of cells/kg) is a shape parameter. These three parameters have to be determined from experimental data related to red blood cell production rates.

Combining (8.22) and (8.23), we have the final form for our model of red blood cell control given in the form

$$\frac{dE}{dt} = F_0 \frac{\theta^n}{E_\tau^n + \theta^n} - \gamma E. \tag{8.24}$$

Equation (8.24) is a differential delay equation (DDE). In contrast to ordinary differential equations, for which we need only to specify an initial condition as a particular point in the phase space for a given time, for DDEs we have to specify an initial condition in the form of a function defined for a period of time equal to the duration of the time delay. Thus we will select

$$E(t') = \phi(t'), \qquad -\tau \leq t' \leq 0. \tag{8.25}$$

Usually here we will only consider initial functions that are constant, but it must be noted that differential delay equations (like their ordinary cousins) can display multistable behavior in which two or more coexisting locally stable solutions result, depending on the initial function (an der Heiden et al. 1981; an der Heiden and Mackey 1982; an der Heiden and Mackey 1987; Campbell et al. 1995; Foss et al. 1996; Mackey and an der Heiden 1984; Losson et al. 1993).

8.5.1 Oscillations in differential delay equations

Based on what you have seen so far, it might seem somewhat strange to claim that the apparently first-order differential delay equation (8.24) can have oscillatory solutions. It is well established that under appropriate circumstances, delayed negative-feedback mechanisms as in (8.24) can produce oscillations, and one of the reasons is that (8.24) isn't really a first-order equation at all. Rather, it is an infinite-dimensional system; the reason is that the initial *function* is infinite dimensional in the sense that it takes an infinite number of points to specify it.

To illustrate the fact that differential delay equations can in fact oscillate, we continue our analysis of (8.24) by using an approximation system for which we can actually derive the analytic solution.

To do this, we imagine that in the nonlinear Hill function (8.23) we let $n \to \infty$, so the nonlinearity becomes progressively closer to the step-function nonlinearity illustrated in Figure 8.6. Under this circumstance, (8.24) becomes

$$\frac{dE(t)}{dt} = -\gamma E(t) + \begin{cases} F_0, & 0 \le E_\tau < \theta \\ 0, & \theta \le E_\tau. \end{cases} \tag{8.26}$$

In reality, the nonlinear differential delay equation (8.26) can be alternately viewed as a pair of ordinary differential delay equations, and which one we solve at any given time will depend on the value of the retarded variable E_τ with respect to the parameter θ. This method of solution, which we are going to carry out, is usually called the *method of steps*.

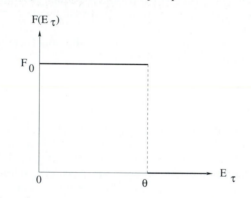

Figure 8.6. A piecewise constant nonlinearity approximating the negative-feedback function of (8.23) that is obtained in the limit as $n \to \infty$.

In preparation, we must first specify an initial function for (8.26) of the type in (8.25), and in point of fact it is almost immaterial what type of initial function we pick. Here, for concreteness, we will pick one that satisfies $\phi(t') > \theta$ for $-\tau \le t' \le 0$ and specify that $\phi(0) \equiv E_0$, a constant.

With this assumption (you could carry out the same analysis as here with a different initial function, and should do so to convince yourself that the ultimate conclusion we reach is independent of the initial function), we must first solve the equation

$$\frac{dE}{dt} = -\gamma E, \quad \theta < E_\tau, \quad E(t = 0) \equiv E_0. \tag{8.27}$$

The solution is this equation is, of course, almost trivial to write down and is given by

$$E(t) = E_0 e^{-\gamma t} \tag{8.28}$$

and this solution will be valid until a time t_1 determined by the condition $\theta = E(t_1 - \tau)$ or, more specifically,

$$E(t_1 - \tau) = \theta \equiv E_0 e^{-\gamma(t_1 - \tau)} \tag{8.29}$$

from which we immediately deduce that

$$t_1 = \frac{1}{\gamma} \ln \left\{ \frac{E_0 e^{\gamma\tau}}{\theta} \right\} \tag{8.30}$$

Having this value of t_1, it is a simple matter to then show that the value of E at $t = t_1$ is given by

$$E(t = t_1) \equiv E_1 = \theta e^{-\gamma\tau}. \tag{8.31}$$

Thus, for the particular form of the initial condition that we have chosen we conclude that the solution is a decaying exponential given by (8.29), and that this solution is valid until a time t_1 given by (8.30), at which point the solution has the value E_1 given by (8.31).

Now to proceed for times longer than t_1 we must solve the other differential equation as given in (8.26), namely

$$\frac{dE}{dt} = -\gamma E + F_0, \quad E_\tau \le \theta, \quad E(t_1) = E_1. \tag{8.32}$$

This is almost as easy as the first case, and the solution is given by the slightly more complicated relation

$$E(t) = E_1 e^{-\gamma(t - t_1)} + \frac{F_0}{\gamma} \left[1 - e^{-\gamma(t - t_1)} \right]. \tag{8.33}$$

According to our assumptions, the solution (8.33) will be valid until a time t_2 defined by $\theta = E(t_2 - \tau)$, or

$$E(t_2 - \tau) \equiv \theta = E_1 e^{-\gamma(t_2 - t_1)} + \frac{F_0}{\gamma} \left[1 - e^{-\gamma(t_2 - t_1)} \right], \tag{8.34}$$

from which we have, with a bit of algebra, that

$$t_2 = \frac{1}{\gamma} \ln \left\{ \left(\frac{E_0}{\theta} \right) \left[\frac{E_1 - (F_0/\gamma)}{\theta - (F_0/\gamma)} \right] e^{2\gamma\tau} \right\}. \tag{8.35}$$

We may then calculate the value of the solution at time t_2 $[E(t = t_2) \equiv E_2]$ to obtain

$$E_2 = \frac{F_0}{\gamma} + \left(\theta - \frac{F_0}{\gamma} \right) e^{-\gamma\tau}. \tag{8.36}$$

Therefore, to summarize, for the period of time $t \in [t_1, t_2]$ the solution to (8.26) is an exponentially increasing function that terminates at a time t_2 given by (8.35) with a value of E_2 given in (8.36). These values then constitute the initial conditions for our determination of the third portion of the solution and the method of steps is sufficiently clear that we need not write out things quite so explicitly.

In the computation of this third portion we must once again solve (8.27) subject to the end-point conditions just determined in the last computation. This then yields

$$E(t) = E_2 e^{-\gamma(t-t_2)}, \tag{8.37}$$

and from $\theta = E(t_3 - \tau)$ we find

$$t_3 = \frac{1}{\gamma} \ln \left\{ \left(\frac{E_0 E_2}{\theta^2} \right) \left[\frac{E_1 - (F_0/\gamma)}{\theta - (F_0/\gamma)} \right] e^{3\gamma\tau} \right\} \tag{8.38}$$

so $E(t_3) \equiv E_3$ is given by

$$E_3 = \theta e^{-\gamma\tau}. \tag{8.39}$$

Now comparison of (8.31) and (8.39) reveals that in point of fact $E_1 \equiv E_3$, and this in turn means that we have shown that (8.26) has a periodic solution! The argument that leads to this conclusion is as follows. Starting from two different initial values, namely E_0 and E_2 we have found that we arrive at precisely the same level $E_1 \equiv E_3$, and thus we know that proceeding further from E_3 at time t_3 will lead to a value of $E_4 \equiv E_2$ at a time $t_4 \equiv t_3 + (t_2 - t_1)$, and that this repeated cycling of the solution between a minimum value of E_1 and E_2 will continue indefinitely. Furthermore, since E_0 and E_2 both lead to the same minimum value of the exponentially decreasing portion of the solution, we could simply pick $E_0 \equiv E_2$ without any loss of generality, and thus t_3 takes the slightly simpler form

$$t_3 = \frac{1}{\gamma} \ln \left\{ \left(\frac{E_2^2}{\theta^2} \right) \left[\frac{E_1 - (F_0/\gamma)}{\theta - (F_0/\gamma)} \right] e^{3\gamma\tau} \right\}. \tag{8.40}$$

We can calculate the period of this periodic solution that we have derived by using $t_3 - t_1$, but this should just be equivalent to t_2 when we pick the special initial condition $E_0 \equiv E_2$. Denoting the period by T we obtain

$$T = \frac{1}{\gamma} \ln \left\{ \left(\frac{E_2}{\theta} \right) \left[\frac{E_1 - (F_0/\gamma)}{\theta - (F_0/\gamma)} \right] e^{2\gamma\tau} \right\}, \tag{8.41}$$

or after substituting the explicit value for E_2 from (8.36) we have

$$T = 2\tau + \frac{1}{\gamma} \ln \left\{ \left[\frac{F_0/(\gamma\theta)}{F_0/(\gamma\theta) - 1} - e^{-\gamma\tau} \right] \left[\frac{F_0}{(\gamma\theta)} - e^{-\gamma\tau} \right] \right\} \tag{8.42}$$

8.5.2   Steady states

A *steady state* (Appendix B) for the model (8.24) is defined by the requirement that the red blood cell number is not changing with time. This, in turn, can be translated to mean that

$$E(t) = E(t - \tau) = E_\tau(t) = \text{a constant, the steady state} = E^*, \tag{8.43}$$

and

$$\frac{dE}{dt} = 0 \qquad \text{so} \qquad F(E^*) = F_0 \frac{\theta^n}{E^* + \theta^n} = \gamma E^*. \qquad (8.44)$$

Now generally we can't solve (8.44) to get an analytic form for E^*, but a simple graphical argument shows that only one value of E^* will satisfy (8.44). This value of the steady state occurs at the intersection of the graph of γE^* with the graph of $F(E^*)$, as indicated in Figure 8.7.

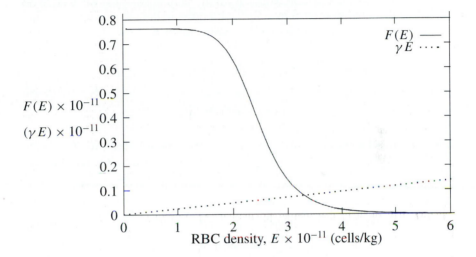

Figure 8.7. Determination of steady-state red blood cell numbers in rabbits. The unique steady state E_* of the model defined by (8.24) is determined by the intersection of the steady-state production curve $F(E)$ and the loss curve γE. Values of the parameters are selected to conform with the estimates of the normal parameter values as detailed in Section 8.5.4 below.

8.5.3 Stability of the steady state

Knowing the steady state of a model like (8.25), and how it depends on the various parameters of the problem, is certainly useful but is not of much help in understanding the dynamics of the oscillations seen in autoimmune hemolytic anemia as illustrated in Figure 8.4. To further our investigation, we must examine the *stability* of the steady-state E^* that we determined above.

What does stability mean? In words, it just means that if the body had a steady-state value E^* of red blood cell numbers and we perturbed this number (by, for example, blood donation or transfusion) to a value E less than or greater than E^*, then over time we would find that

$$\lim_{t \to \infty} E(t) = E^*. \qquad (8.45)$$

We would like to know what conditions on the parameters of our model (destruction rate, maximal production rate, etc.) are required to make sure that the stability condition (8.45) holds, and even further we would like to know what happens if the stability condition (8.45) is violated.

Unfortunately, because of the nature of model (8.24) that describes this physiological process we cannot answer these questions in total generality. Rather, we must be content with understanding what happens when we make a **small** perturbation of E away from E^*. Our assumption that the perturbation is small allows us to carry out a linear stability analysis of the steady state E^*, which we now proceed to do.

The nonlinearity of (8.24) comes from the term involving the red blood cell production function, which is highly nonlinear. What we want to do is replace this nonlinear term by a linear function in the vicinity of the steady state E^*. This involves writing out the expansion of F in the vicinity of E^*:

$$F(E_\tau) \simeq F(E^*) + (E_\tau - E^*)F'(E_\tau = E^*)$$

$$+\frac{1}{2}(E_\tau - E^*)^2 F''(E_\tau = E^*) + \cdots \tag{8.46}$$

remembering that we are assuming that $E_\tau - E^*$ is pretty small, so $(E_\tau - E^*)^2$ is even smaller and therefore negligible, and finally writing the approximate version of (8.24) in the vicinity of E^* (with the notation that $F'(E^*) \equiv S$, for slope) as

$$\frac{dE}{dt} \simeq F(E^*) + (E_\tau - E^*)S - \gamma E. \tag{8.47}$$

Remember from (8.44) that the defining equation for the steady-state is $F(E^*) = \gamma E^*$. Using this in (8.47), we can rewrite it as

$$\frac{dE}{dt} \simeq (E_\tau - E^*)S - \gamma(E - E^*). \tag{8.48}$$

If we further set $z(t) = E(t) - E^*$, so z is the deviation of the red blood cell numbers from their steady-state value, then $z_\tau(t) = E_\tau(t) - E^*$ and $dE/dt = dz/dt$, so (8.48) can be rewritten as the linear differential delay equation

$$\frac{dz}{dt} = Sz_\tau - \gamma z. \tag{8.49}$$

Now linear equations, with or without delays, are a lot easier to work with than nonlinear ones. The usual procedure, loosely speaking, to find a solution is to assume that the solution has the form $z(t) \simeq e^{\lambda t}$ and to find out the requirements on the parameters of the equation such that there is an eigenvalue λ allowing z to be written in this form. Usually the eigenvalue λ is a complex number, $\lambda = \mu + i\omega$, so the solution can actually be written in the alternative form

$$z(t) \simeq e^{\lambda t} = e^{\mu t}\{\cos(\omega t) + i\sin(\omega t)\}. \tag{8.50}$$

If $\mu = \text{Re }\lambda < 0$, then the solution is a decaying oscillating function of time that approaches zero, so we have a stable situation. If $\mu = \text{Re }\lambda > 0$ on the other hand, then the solution diverges to infinity in an oscillatory fashion and the solution is unstable. The boundary between these two situations, where $\mu = \text{Re } \lambda = 0$, defines a *Hopf bifurcation*, which is characterized by an eigenvalue pair crossing from the left-hand to the right-hand complex plane (Appendix B).

After that wordy digression, let's assume that $z(t) \simeq e^{\lambda t}$ in (8.49). If we make this substitution, and carry out the algebra, then we are left with the equation

$$\lambda = Se^{-\lambda \tau} - \gamma \tag{8.51}$$

that λ must satisfy. In general, determining the criteria such that $\text{Re } \lambda = \mu < 0$ involves a lot of messy algebra. However, determining the relation between the parameters such that $\text{Re}\lambda = \mu = 0$ is a lot easier, and so lets look at it.

Under this assumption, $\lambda = i\omega$, and substituting this into (8.51) gives

$$i\omega = S[\cos (\omega\tau) - i \sin (\omega\tau)] - \gamma, \tag{8.52}$$

or after separating real and imaginary parts

$$\omega = -S \sin (\omega\tau) \tag{8.53}$$
$$\gamma = S \cos (\omega\tau) \tag{8.54}$$

If $|\gamma/S| < 1$ then (8.54) can be solved for $\omega\tau$ to give

$$\omega\tau = \cos^{-1} \left(\frac{\gamma}{S}\right). \tag{8.55}$$

Further, squaring and adding the two equations (8.53 and 8.54) gives

$$\omega = \sqrt{S^2 - \gamma^2}. \tag{8.56}$$

Combining (8.55) and (8.56), we find that the relation connecting τ, S, and γ that must be satisfied in order for the eigenvalues to have real part identical to zero is given by

$$\tau = \frac{\cos^{-1}(\gamma/S)}{\sqrt{S^2 - \gamma^2}}, \qquad |\gamma/S| < 1. \tag{8.57}$$

In general (Hayes 1950), the real parts of λ will be negative, and thus the linear equation (8.49) will have a locally stable steady state, if and only if $|\gamma/S| > 1$ *or* if $|\gamma/S| < 1$ *and*

$$\tau < \frac{\cos^{-1}(\gamma/S)}{\sqrt{S^2 - \gamma^2}}. \tag{8.58}$$

When the parameters satisfy (8.57), then we say that there has been a Hopf bifurcation. The period of the periodic solution that is guaranteed when (8.57) is satisfied can be easily derived by noting that $\omega = 2\pi f = 2\pi/T$, where

f and T are the frequency and period of the solution, respectively, and thus from (8.55) through (8.57)

$$T = \frac{2\pi \tau}{\cos^{-1}(\gamma/S)}.$$

(8.59)

Since the inverse cosine ranges from 0 to π, from (8.59) we know that at the Hopf bifurcation the period of the periodic solution must satisfy

$$2\tau \leq T.$$

(8.60)

In general, the period of an oscillation produced by a delayed negative-feedback mechanism is at least twice the delay (Hayes 1950; Mackey 1978). Moreover, for our model of erythrocyte production it can be shown that the period of the oscillation should be no greater than four times the delay (Mackey 1979b), i.e.

$$2\tau \leq T \leq 4\tau.$$

(8.61)

Since the maturational delay for erythrocyte production τ is approximately 6 days, we would expect to see oscillations in erythrocyte numbers with periods ranging from 12 to 24 days. This is in agreement with the observed periods of 16 to 17 days in rabbit autoimmune hemolytic anemia (Orr et al. 1968). What is surprising is the fact that these oscillations are so rarely observed. This paradox is illuminated in the following sections.

8.5.4   Parameter estimation

Having extracted about as much information as is possible from the linear analysis of our model for red blood cell production, we now need to turn to numerical simulations to see the full behavior. This, by necessity, requires that we have some estimation of the parameters in (8.24). Through a variety of data, we finally conclude that in the normal situation (i.e., not autoimmune hemolytic anemia):

$$
\begin{aligned}
\gamma &= 2.31 \times 10^{-2} \text{ day}^{-1}, \\
F_0 &= 7.62 \times 10^{10} \text{ cells/kg-day}, \\
n &= 7.6, \\
\theta &= 2.47 \times 10^{11} \text{ cells/kg}, \\
\tau &= 5.7 \text{ days}.
\end{aligned}
$$

These parameters correspond to a steady-state circulating red blood cell mass of $E^* = 3.3 \times 10^{11}$ cells/kg, and from the linear analysis of the previous section it is predicted that this steady state is stable.

8.5.5   Explaining laboratory-induced autoimmune hemolytic anemia

The one fact we know about the induced autoimmune hemolytic anemia shown in Figure 8.4 is that the red blood cell destruction rate γ is increased through the action of cell damage (lysis) by the injected iso-antibody. The linear analysis presented above predicts that the steady-state E^* will be stable in the face of increased γ until $\gamma \simeq 5.12 \times 10^{-2}$ day^{-1}, and when it becomes unstable at this point there will be an oscillation about the steady state with a period of $T \simeq 20.6$ days, as given by (8.59). Our linear analysis tell us nothing about what happens in the full nonlinear equation (8.24) after this stability is lost, but it does predict that when γ is further elevated to $\gamma \simeq 2.70 \times 10^{-1}$, then another (reverse) Hopf bifurcation should occur and that the period of the periodic solution will be $T \simeq 16.6$ days. Further increases in γ such that $\gamma > 2.70 \times 10^{-1}$ are predicted to result in a stabilization of E^* about a low value.

8.5.6 Numerical simulations

Numerical simulations of (8.24) with the parameters given above show that the linear analysis results quoted above give a very accurate picture of the full nonlinear behavior, including the values of γ at which stability of E^* is lost and regained, and the period of the solutions at these stability boundaries (Figure 8.8).

Figure 8.8 shows a computer simulation of the model as a function of the peripheral destruction rate (γ). As can be seen, when γ is low, as normally occurs, oscillations in erythrocyte numbers do not occur. As γ increases, regular oscillations occur whose period increases as γ increases. However, for high γ, no oscillation occurs. Interestingly, depending on the severity of the hemolytic anemia induced in the rabbit model, reticulocyte levels were observed to be either depressed at constant levels or to oscillate (Orr et al. 1968). A much more comprehensive model for the control of red blood cell production, and mathematically much more complicated, has properties (Bélair et al. 1995) similar to those of the simple model presented here. The observations in Figure 8.8 indicate that whether or not a proposed mechanism for periodic autoimmune hemolytic anemia produces an oscillation critically depends on whether the value of the control parameter, i.e., the peripheral destruction rate (γ), lies in some crucial range. This may explain why oscillations in erythrocyte number are so rarely seen in patients with autoimmune hemolytic anemia.

It should be noted that the morphology of the oscillations shown in Figure 8.8 are quite simple; i.e., there is only one maximum per period. All studies to date of first-order delayed negative-feedback mechanisms have indicated that only oscillations with this simple morphology can be produced (an der Heiden and Mackey 1982; Glass and Mackey 1979; Longtin and Milton 1988; Mackey and Glass 1977), though second-order systems with delayed negative-feedback can have more complicated solution behavior as well as displaying multistability (an der Heiden et al. 1990; Campbell et al. 1995). More complex waveforms (i.e., more than one maximum per period) are, however, possible in first-order systems with multiple delayed negative-feedback loops (Glass et al. 1988) and with multiple time delays (Bélair and Mackey 1989).

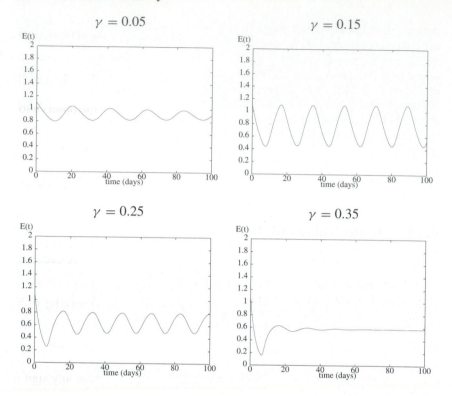

Figure 8.8. Computer simulations of the model (8.24) for erythrocyte production for four different peripheral destruction rates, γ, as indicated below each panel, all other parameters being held at the estimated normal values as given in the text. In every case, the plotted erythrocyte numbers are normalized to the normal steady state of $E^* = 3.3 \times 10^{11}$ cells/kg.

8.6   PERIODIC HEMATOPOIESIS AND PLURIPOTENTIAL STEM-CELL POPULATION STABILITY

The most common periodic hematological disease is periodic hemato-poiesis. In humans, periodic hematopoiesis is a disease characterized by 17- to 28-day periodic oscillations in circulating white blood-cell numbers from ap-proximately normal values to barely detectable numbers (Dale and Hammond 1988; Hoffman et al. 1974; Lange 1983; Wright et al. 1981). In addition to white blood cells, oscillations are seen for all the formed elements of blood with the same period. The oscillations for each of the blood-cell lines are out of phase, and the phase differences between the cell lines are consistent with the differences in the maturation times (Hoffman et al. 1974). These neutropenic (low circulating neutrophil numbers) episodes place the patients at increased risk for infective processes (e.g., abscesses, pneumonia, septicemia).

Experimental study of periodic hematopoiesis has been facilitated by the availability of suitable animal models. All gray collies have periodic hema-topoiesis (Dale et al. 1972a; Dale et al. 1972b; Dale and Hammond 1988; Jones et al. 1975; Jones and Lange 1983), and the only demonstrable quantitative difference between the human and canine form of periodic hematopoiesis is the period — in dogs the period ranges from about 10 to 17 days.

An abnormality in the regulation of the pluripotential stem cells in periodic hematopoiesis is suggested by the observation that the disorder can be transferred by bone-marrow transplantation (Jones et al. 1975; Krance et al. 1982; Quesenberry 1983). This is view is further supported by the observation (Abkowitz et al. 1988) that in the grey collie 12- to 13-day-period oscillations are found in marrow erythroid burst-forming units and that these are shifted in phase by two days from the oscillations in the erythroid colony-forming units. Furthermore, in the same study it was found that there were oscillations in the granulocyte-macrophage colony-forming cells and that these have phase differences of about five days preceding the neutrophils. These findings indicate that the defect in periodic hematopoiesis is resident in a cell population more primitive than these identifiable stem-cell populations.

Consequently most investigators have looked to abnormalities in the regulation of the pluripotential stem cells and delayed negative-feedback mechanisms as an explanation for periodic hematopoiesis (Dunn 1983; Mackey 1978; Mackey 1979a; Morley 1979; Wheldon et al. 1974; Wheldon 1975; Nazarenko 1976; MacDonald 1978; Kazarinoff and van den Driessche 1979; von Schulthess and Mazer 1982). A schematic representation of pluripotential stem cell population regulation is shown in Figure 8.9.

Interestingly in (Abkowitz et al. 1988) it was also found that the fraction of these progenitor cells in the DNA-synthesis phase is similar between the grey collie and normal collies, and the fraction does not show any sign of cyclic fluctuation.

A crucial clue to the potential origin of the defect in periodic hematopoiesis is the observation of the effect of continuous cyclophosphamide and busulfan administration in normal dogs (Morley and Stohlman 1970; Morley et al. 1970). Though in most animals these drugs led to a pancytopenia (generalized depression of the numbers of all circulating blood cells) whose severity was proportional to the drug dose, in some dogs low doses led to a mild pancytopenia, intermediate doses gave a periodic hematopoiesis-like behavior with a period between 11 and 17 days, and high drug levels led to either death or gross pancytopenia. When the periodic hematopoiesis-like behavior occurred, it was at circulating white blood cell levels of one-half to one-third normal. To this we must add the observation that patients undergoing hydroxurea therapy sometimes develop periodic hematopoiesis-like symptoms (Kennedy 1970), as do patients receiving cyclophosphamide (Dale et al. 1973).

Both cyclophosphamide and busulfan selectively kill cells within the DNA-synthetic phase of the cell cycle, and the fact that both drugs are capable of inducing periodic hematopoiesis-like behavior strongly suggests that the origin of periodic hematopoiesis as a disease is due to an abnormally large death rate (apoptosis) in the proliferative phase of the cell cycle of a population of pluripotential stem cells, which is at a level more primitive than the granulocyte-macrophage colony-forming cells and the marrow erythroid burst-forming units.

Here we interpret the effects of an increase in the rate of irreversible apoptotic loss from the proliferating phase of the pluripotential stem-cell population (γ in Figure 8.9) on blood-cell production (Mackey 1978).

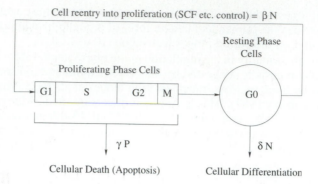

Cell reentry into proliferation (SCF etc. control) = $\beta\,N$

Resting Phase Cells

Proliferating Phase Cells

| G1 | S | G2 | M |

G0

$\gamma\,P$

$\delta\,N$

Cellular Death (Apoptosis)   Cellular Differentiation

Figure 8.9. A schematic representation of the control of pluripotential stem-cell regeneration. Proliferating phase cells (P) include those cells in G_1, S (DNA synthesis), G_2, and M (mitosis) while, the resting-phase (N) cells are in the G_0 phase. Local regulatory influences are exerted via a cell-number dependent variation in the fraction of circulating cells. δ is the **normal** rate of differentiation into all of the committed stem-cell populations, while γ represents a loss of proliferating-phase cells due to apoptosis. See Mackey (1978; 1979a) for further details.

The dynamics of this pluripotential stem-cell population is governed (Mackey 1978; Mackey 1979a) by the pair of coupled differential delay equations (which can be derived using the techniques of (8.4)):

$$\frac{dP}{dt} = -\gamma P + \beta(N)N - e^{-\gamma\tau}\beta(N_\tau)N_\tau, \qquad (8.62)$$

$$\frac{dN}{dt} = -[\beta(N)+\delta]N + 2e^{-\gamma\tau}\beta(N_\tau)N_\tau, \qquad (8.63)$$

where τ is the time required for a cell to traverse the proliferative phase, and the resting- to proliferative- phase feedback rate β is taken to be

$$\beta(N) = \frac{\beta_0\theta^n}{\theta^n + N^n}. \qquad (8.64)$$

An examination of (8.63) shows that this equation could be interpreted as describing the control of a population with a delayed mixed-feedback-type production term $[2e^{-\gamma\tau}\beta(N_\tau)N_\tau]$ and a destruction rate $[\beta(N)+\delta]$ that is a decreasing function of N.

For δ small enough relative to β_0, this model has two possible steady states. There is a steady state corresponding to no cells, $(P_1^*, N_1^*) = (0,0)$, which is stable if it is the only steady state, and which becomes unstable whenever the second positive steady state (P_2^*, N_2^*) exists.

The stability of the nonzero steady state depends on the value of γ, and this is illustrated schematically in Figure 8.10. When $\gamma = 0$, this steady state cannot be destabilized to produce dynamics characteristic of periodic hematopoiesis. On the other hand, for $\gamma > 0$, increases in γ lead to a decrease in the pluripotential stem-cell numbers and a consequent decrease in the cellular efflux (given by δN) into the differentiated cell lines. This diminished efflux becomes unstable when a critical value of γ is reached, $\gamma = \gamma_1$, at which a supercritical Hopf bifurcation occurs (Appendix B). For all values of γ satisfying $\gamma_1 < \gamma < \gamma_2$, there is a periodic solution of (8.63) whose period is in good agreement with that seen in periodic hematopoiesis. At $\gamma = \gamma_2$, a reverse bifurcation occurs and the greatly diminished pluripotential stem-cell numbers as well as cellular efflux again become stable.

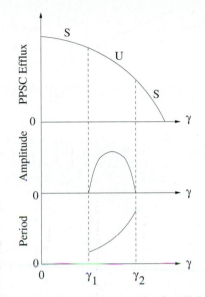

Figure 8.10. Schematic representation of the combined analytic and numerically determined stability properties of the pluripotential stem-cell model. See the text for details.

Separate estimations of the parameter sets for human and grey collie pluripotential stem cell populations give predictions of the period of the oscillation at the Hopf bifurcation that are consistent with those observed clinically and in the laboratory. These results are illustrated in Figures 8.11 and 8.12 for humans and grey collies, respectively.

Numerical simulations of (8.62) and (8.63) confirm the results of the local stability analyses displayed in Figures 8.11 and 8.12. As expected, an increase in γ is accompanied by a decrease in the average number of circulating cells. For certain values of γ an oscillation appears. Over the range of γ in which an oscillation occurs, the period increases as γ increases. However, the amplitude of the oscillation first increases and then decreases. (Similar observations hold for the model of autoimmune hemolytic anemia as the control parameter γ is increased.) When all the parameters in the model are set to the values estimated from laboratory and clinical data, no other types of bifurcations are found. Although these simulations also indicate the existence of multiple bifurcations and chaotic behaviors, these more complex dynamics are observed only for nonphysiological choices of the parameters. Thus, the observed irregularities in the fluctuations in blood-cell numbers in periodic hematopoiesis cannot be related to chaotic solutions of (8.63). These results suggest that periodic hematopoiesis is likely related to defects, possibly genetic, within the pluripotential stem-cell population that lead to an abnormal ($\gamma > 0$) apoptotic loss of cells from the proliferative phase of the cell cycle.

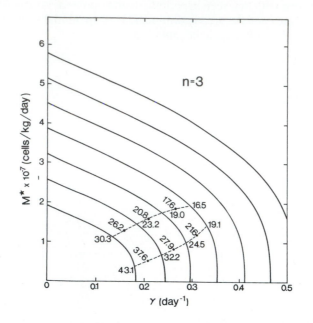

Figure 8.11. Variation of the total steady-state cellular-differentiation efflux ($M^* = \delta N$) as a function of the apoptotic death rate γ from the proliferating cell population in humans ($n = 3$). Parameters in the model were estimated assuming a proliferating fraction of 0.1 and an amplification of 16 in the recognizable erythroid, myeloid, and megakaryocytic precursors populations. See Mackey (1978; 1979a) for details. The pluripotential stem-cell parameters corresponding to each curve from the top down are: $(\delta, \beta_0, \tau, \theta \times 10^{-8}) = (0.09, 1.58, 1.23, 2.52), (0.08, 1.62, 1.39, 2.40), (0.07, 1.66, 1.59, 2.27), (0.06, 1.71, 1.85, 2.13), (0.05, 1.77, 2.22, 1.98), (0.04, 1.84, 2.78, 1.81),$ and $(0.03, 1.91, 3.70, 1.62)$ in units (days^{-1}, days^{-1}, days, cells/kg). The dashed solid lines indicate the boundaries along which stability is lost in the linearized analysis, and the numbers indicate the predicted (Hopf) period (in days) of the oscillation at the Hopf bifurcation.

8.7   UNDERSTANDING LABORATORY-INDUCED CYCLICAL ERYTHROPOIESIS

To illustrate the usefulness of the model presented in Section 8.6 with a second example, we will compare these predictions to experimental observations obtained for ^{89}Sr-induced cyclic erythropoiesis in two congenitally anemic strains of mice, W/W^v and S1/S1^d (Gibson et al. 1985; Gurney et al. 1981). W/W^v mice suffer from a defect in the pluripotential stem-cells, and in S1/S1^d mice the hematopoietic microenvironment is defective. Let us assume that the difference between W/W^v and S1/S1^d mice is related solely to differences in γ. The observation that S1/S1^d mice are more refractory to erythropoietin than W/W^v suggests that γ is higher in S1/S1^d. The results of Section 8.6 predict that a higher γ would increase the likelihood that an oscillation in erythrocyte number occurs. Indeed, in contrast to W/W^v, approximately 40% of S1/S1^d mice have "spontaneous" oscillations in their hematocrit (Gibson et al. 1985). In both strains of mice, a single dose of ^{89}Sr is sufficient to increase γ into a

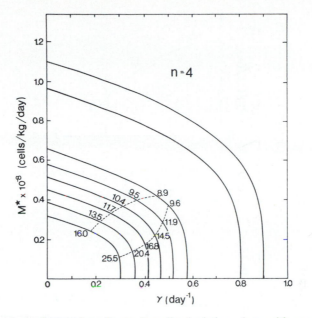

Figure 8.12. As in the previous figure, but all calculations done with parameters appropriate for dogs.

range associated with oscillations in erythrocyte number. Since the value of γ for the S1/S1^d mice is greater than that for W/W^v prior to ^{89}Sr, it is reasonable to expect that it will also be higher following administration of equal doses of ^{89}Sr to both strains of mice, as shown in Figure 8.13. As predicted, experimentally the period of the oscillation is longer, the amplitude larger, and the mean hematocrit lower for S1/S1^d mice than for W/W^v mice.

8.8   UNDERSTANDING COLONY-STIMULATING-FACTOR EFFECTS IN PERIODIC HEMATOPOIESIS

Recent clinical and experimental work has focused on the modification of the symptoms of hematological disorders, including periodic hematopoiesis, by the use of various synthetically produced cytokines (Cebon and Layton 1994; Sachs 1993; Sachs and Lotem 1994), e.g., the recombinant colony stimulating factors rG-colony-stimulating factor and rGM-colony-stimulating factor, whose receptor biology is reviewed in Rapoport et al. (1992), and interleukin-3 (IL-3). These cytokines are now known to interfere with the process of apoptosis, or to lead to a decrease in γ within the context of the pluripotential stem-cell model of Section 8.6.

Human colony-stimulating factors increase both the numbers and proliferation rate of white blood cell precursors in a variety of situations (Bronchud et al. 1987; Lord et al. 1989; Lord et al. 1991). Furthermore, colony-stimulating factor in mice is able to stimulate replication in both stem cells and early erythroid cells (Metcalf et al. 1980).

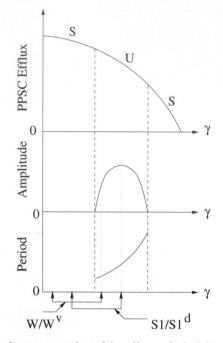

Figure 8.13. Schematic representation of the effects of administering the same dose of ^{51}Sr to W/W^v and S1/S1^d mice. See the text for details.

It is known that in aplastic anemia and periodic hematopoiesis there is an inverse relationship between plasma levels of G-colony stimulating factor and white blood cell numbers (Watari et al. 1989). Further it has been shown (Layton et al. 1989) that the $t_{1/2}$ of G-colony-stimulating factor in the circulation is short — of the order of 1.3 to 4.2 hours — so the dynamics of the destruction of G-colony-stimulating factor are unlikely to have a major role in the genesis of the dynamics of periodic hematopoiesis.

In the grey collie it has been shown that at relatively low doses of G-colony-stimulating factor the mean white blood-cell count is elevated (by 10 to 20 times), as is the amplitude of the oscillations (Hammond et al. 1990), while higher dosages (Hammond et al. 1990; Lothrop et al. 1988) lead to even higher mean white blood cell numbers but eliminate the cycling. Another interesting observation is that in the collie, G-colony-stimulating factor administration results in a decrease in the period of the peripheral oscillation. The elevation of the mean white blood cell levels and the amplitude of the oscillations, as well as an enhancement of the oscillations of platelets and reticulocytes, at low levels of G-colony-stimulating factor has also been reported in humans (Hammond et al. 1990; Migliaccio et al. 1990; Wright et al. 1994), and it has been also noted that the fall in period observed in the collie after G-colony stimulating factor administration occurs in humans with a fall in period from 21 to about 14 days. Finally it should be mentioned that treatment with G-colony stimulating factor in patients with agranulocytosis has also lead to a significant increase in the mean white blood cell counts and, in some patients, to the induction of white blood cell oscillations with periods ranging from 7 to 16 days.

Our major clue to the nature of the effects of G-colony stimulating factor comes from its prevention of apoptosis and from the work of Avalos et al. (1994) who have shown in dogs that there is no demonstrable alteration in the number, binding affinity, or size of the G-colony-stimulating-factor receptor on periodic-hematopoiesis dogs as compared to normal dogs. They thus conclude that periodic hematopoiesis "is caused by a defect in the G-colony-stimulating-factor signal-transduction pathway at a point distal to G-colony-stimulating-factor binding . . . ." The data of Avalos et al. (1994) can be used to estimate that

$$\gamma_{max}^{PH} \simeq 7 \times \gamma_{max}^{norm} \tag{8.65}$$

The results of Hammond et al. (1992) in humans are consistent with these results in dogs.

Less is known about the effect of GM-colony-stimulating factor, but it is known that administration of GM-colony-stimulating factor in humans gives an elevation of the mean white blood cell level but only by relatively modest amounts–1.5 to 3.9 times (Wright et al. 1994), but either dampens the oscillations of periodic hematopoiesis or eliminates them entirely. The same effect has been shown (Hammond et al. 1990) in the grey collie. It is unclear if the period of the peripheral cell oscillations has a concomitant decrease, as is found with G-colony-stimulating factor. The abnormal responsiveness of precursors to G-colony-stimulating factor in grey collies and humans with periodic hematopoiesis (Hammond et al. 1992; Avalos et al. 1994) is mirrored in the human response to GM-colony-stimulating factor (Hammond et al. 1992).

Thus, the available laboratory and clinical data on the effects of colony-stimulating factors in periodic hematopoiesis indicate that: (1) there is extensive intercommunication between all levels of stem cells; and (2) within the language of nonlinear dynamics, colony-stimulating factors may be used to titrate the dynamics of periodic hematopoiesis to the point of inducing a reverse Hopf bifurcation (disappearance of the oscillations). In the course of this "titration" there may also be a shift in the period.

The behavior in periodic hematopoiesis when colony-stimulating factor is administered is qualitatively consistent with the pluripotential stem-cell model discussed in Section 8.6, since it is known that colony-stimulating factor interferes with apoptosis and, thus, administration of colony-stimulating factor is equivalent to a decrease in the apoptotic death rate γ. This is a current active area of research in conjunction with Profs. David Dale and William Hammond of the University of Washington (Seattle). We hope this combined modeling and data-analysis project may give greater insight into the fundamental nature of the regulation of the mammalian cell cycle and, in the future, suggest more rational therapies for patients with periodic hematopoiesis.

8.9 CONCLUDING REMARKS

Delayed feedback mechanisms are important for regulating blood-cell numbers. Under certain conditions, delayed-feedback mechanisms can produce oscillations whose period typically ranges from 2 to 4 times the delay but may be even longer. Thus it is not necessary to search for illusive and mystical entities (Beresford 1988), such as ultradian rhythms, to explain the periodicity of these disorders.

The observations in this chapter emphasize that an intact control mechanism for the regulation of blood-cell numbers is capable of producing behaviors ranging from no oscillation to periodic oscillations to more complex irregular fluctuations, i.e., chaos. The type of behavior produced depends on the nature of the feedback, i.e. negative or mixed, and on the value of certain underlying control parameters, e.g., peripheral destruction rates or maturation times. Pathological alterations in these parameters can lead to periodic hematological disorders.

As an extension to the concept of periodic diseases introduced by Reimann (1963) in 1963, the term "dynamical disease" has been introduced (Glass and Mackey 1979; Glass and Mackey 1988; Mackey and Glass 1977; Mackey and an der Heiden 1982; Mackey and Milton 1987). A dynamical disease is defined as a disease that occurs in an intact physiological control system operating in a range of control parameters that leads to abnormal dynamics. Clearly the hope is that it may eventually be possible to identify these altered parameters and then readjust them to values associated with healthy behaviors. Developments in biotechnology and the analysis of physiological control mechanisms are occurring at so rapid a pace that the feasibility of such an approach may be just around the corner, as illustrated by the material described concerning the effects of the colony-stimulating factors in periodic hematopoiesis.

8.10   ACKNOWLEDGMENTS

These notes closely follow two papers (Mackey and Milton 1990; Milton and Mackey 1989) that I wrote with my friend and colleague John G. Milton of the University of Chicago. My research on the periodic hematological diseases has been supported for over 20 years by the Natural Sciences and Engineering Research Council of Canada (NSERC), and for shorter periods by the FCAR (Quebec), NATO, the Alexander von Humboldt Stiftung, and the Royal Society of London.

REFERENCES

Abkowitz, Janis L., Holly, Richard D., and Hammond, William P. 1988. Cyclic hematopoiesis in dogs: Studies of erythroid burst-forming cells confirm an early stem cell defect. *Experimental hematology*, **16**, 941–945. {*167, 174*}

an der Heiden, Uwe and Mackey, Michael C. 1982. The dynamics of production and destruction: Analytic insight into complex behaviour. *Journal of Mathematical Biology*, **16**, 75–101. {*157, 165, 174*}

an der Heiden, Uwe and Mackey, Michael C. 1987. Mixed feedback: A paradigm for regular and irregular oscillations. Springer series in synergetics, vol. 36. Berlin, Germany / Heidelberg, Germany / London, UK / etc.: Springer Verlag. {*157, 174*}

an der Heiden, Uwe, Mackey, Michael C., and Walther, Hans O. 1981. Complex oscillations in a simple deterministic neuronal network. *Lectures in applied mathematics*, **19**, 355–360. {*157, 174*}

an der Heiden, Uwe, Longtin, Andre, Mackey, Michael C., Milton, John G., and Scholl, R. 1990. Oscillatory modes in a nonlinear second order differential equation with delay. *Journal of dynamics and differential equations*, **2**, 423–449. {*165, 174*}

Avalos, Belinda R., Broudy, Virginia C., Ceselski, Sarah K., Druker, Brian J., Griffen, James D., and Hammond, William P. 1994. Abnormal response to granulocyte colony-stimulating factor (G-CSF) in canine cyclic hematopoiesis is not caused by altered G-CSF receptor expression. *Blood*, **84**, 789–794. {*173, 174*}

Bai-Lin, Hao. 1984. *Chaos*. Singapore; Philadelphia, PA, USA; River Edge, NJ, USA: World Scientific Publishing Co. ISBN 9971-966-50-6, 9971-966-51-4 (paperback). Page 576. {*149, 175*}

Bélair, Jacques and Mackey, Michael C. 1989. Consumer memory and price fluctuations in commodity markets: An integrodifferential model. *Journal of dynamics and differential equations*, **1**, 299–325. {*165, 175*}

Bélair, Jacques, Mackey, Michael C., and Mahaffy, Joseph M. 1995. Age-structured and two delay models for erythropoiesis. *Mathematical Biosciences*, **128**, 317–346. {*165, 175*}

Beresford, C. H. 1988. Time: A biological dimension. *Journal of the Royal College of Physicians of London*, **22**, 94–96. {*173, 175*}

Bronchud, M. H., Scarffe, J. H., Thatcher, N., Crowther, D., Souza, L. M., Alton, N. K., Testa, N. G., and Dexter, T. M. 1987. Phase I/II study of recombinant human granulocyte colony-stimulating factor in patients receiving intensive chemotherapy for small cell lung cancer. *British Journal of Cancer*, **56**, 809–813. {*171, 175*}

Campbell, S. A., Belair, Jacques, Ohira, T., and Milton, J. G. 1995. Complex dynamics and multistability in a damped harmonic oscillator with delayed negative feedback. *Chaos (Woodbury, NY)*, **5**, 640–645. {*157, 165, 175*}

Cebon, Jonathan and Layton, Judith. 1994. Measurement and clinical significance of circulating hematopoietic growth factor levels. *Current opinion in hematology*, **1**, 228–234. {*150, 171, 175*}

Dale, D. C. and Hammond, W. P. 1988. Cyclic neutropenia: A clinical review. *Blood Reviews*, **2**, 178–185. {*166, 175*}

Dale, D. C., Ward, S. B., Kimball, H. R., and Wolff, S. M. 1972a. Studies of neutrophil production and turnover in grey collie dogs with cyclic neutropenia. *Journal of Clinical Investigation*, **51**, 2190–2196. {*166, 175*}

Dale, David C., Alling, David W., and Wolff, Sheldon M. 1972b. Cyclic hematopoiesis: the mechanism of cyclic neutropenia in grey collie dogs. *Journal of Clinical Investigation*, **51**, 2197–2204. {*166, 175*}

Dale, David C., Alling, David W., and Wolff, Sheldon M. 1973. Application of time series analysis to serial blood neutrophil counts in normal individuals and patients receiving cyclophosphamide. *British journal of haematology*, **24**, 57–64. {*167, 175*}

Degn, H., Holden, A. V., and Olsen, L. F. (eds). 1987. *Chaos in Biological Systems*. NATO ASI series. Series A, Life sciences, vol. 138. New York, NY, USA; London, UK: Plenum Press. ISBN 0-306-42685-4. Pages xi + 323. {*149, 175*}

Dunn, C. D. R. 1983. Cyclic hematopoiesis: The biomathematics. *Experimental hematology*, **11**, 779–791. {*149, 150, 167, 175*}

Foss, Jennifer, Longtin, André, Mensour, Boualem, and Milton, John. 1996. Multistability and delayed recurrent loops. *Physical Review Letters*, **76**(4), 708–711. {*157, 175*}

Gibson, C. M., Gurney, C. W., Simmons, E. L., and Gaston, E. O. 1985. Further studies on cyclic erythropoiesis in mice. *Experimental hematology*, **13**, 855–860. {*170, 175*}

Glass, Leon and Mackey, Michael C. 1979. Pathological conditions resulting from instabilities in physiological control systems. *Annals of the New York Academy of Sciences*, **316**, 214–235. {*149, 150, 165, 174, 175*}

Glass, Leon and Mackey, Michael C. 1988. *From Clocks to Chaos: The Rhythms of Life*. Princeton, NJ, USA: Princeton University Press. ISBN 0-691-08495-5 (hardcover), 0-691-08496-3 (paperback). Pages xvii + 248. {*149, 174, 175*}

Glass, Leon, Beuter, Anne, and Larocque, David. 1988. Time delays, oscillations, and chaos in physiological control systems. *Mathematical Biosciences*, **90**, 111–125. {*165, 175*}

Gordon, R. R. and Varadi, S. 1962. Congenital hypoplastic anemia (pure red cell anemia) with periodic erythroblastopenia. *The Lancet (London, England)*, **i**, 296–299. {*155, 176*}

Gurney, Clifford W., Simmons, Eric L., and Gaston, Evelyn O. 1981. Cyclic erythropoiesis in W/W^v mice following a single small dose of ^{89}Sr. *Experimental hematology*, **9**, 118–122. {*170, 176*}

Hammond, William P., Boone, Thomas C., Donahue, Robert E., Souza, Lawrence M., and Dale, David C. 1990. Comparison of treatment of canine cyclic hematopoiesis with recombinant human granulocyte-macrophage colony-stimulating factor (GM-CSF), G-CSF, Interleukin-3, and Canine G-CSF. *Blood*, **76**, 523–532. {*172, 173, 176*}

Hammond, William P., Chatta, Gurkamal S., Andrews, Robert G., and Dale, David C. 1992. Abnormal responsiveness of granulocyte committed progenitor cells in cyclic neutropenia. *Blood*, **79**, 2536–2539. {*173, 176*}

Hayes, N. D. 1950. Roots of the transcendental equation associated with a certain difference-differential equation. *Journal of the London Mathematical Society*, **25**, 226–232. {*163, 164, 176*}

Hodgson, G. and Eskuche, I. 1966. Aplicacion de la teoria de control al estudio de la eritropoyesis. *Archives of Biological and Medical Experiments*, **3**, 85–92. {*157, 176*}

Hoffman, H. J., Guerry, D., and Dale, D. C. 1974. Analysis of cyclic neutropenia using digital band-pass filtering techniques. *Journal of interdisciplinary cycle research*, **5**, 1–18. {*166, 176*}

Jones, J. B. and Lange, R. D. 1983. Cyclic hematopoiesis: Animal models. *Immunology and Hematology Research Monographs*, **1**, 33–42. {*166, 176*}

Jones, J. B., Yang, T. J., Dale, J. B., and Lange, R. D. 1975. Canine cyclic haematopoiesis: Marrow transplantation between littermates. *British journal of haematology*, **30**, 215–223. {*166, 167, 176*}

Kazarinoff, Nicholas D. and van den Driessche, Pauline. 1979. Control of oscillations in hematopoiesis. *Science*, **203**, 1348–1349. {*150, 167, 176*}

Kennedy, B. J. 1970. Cyclic leukocyte oscillations in chronic myelogenous leukemia. *Blood*, **35**, 751–760. {*167, 176*}

Kirk, J., Orr, J. S., and Hope, C. S. 1968. A mathematical analysis of red blood cell and bone marrow stem cell control mechanisms. *British journal of haematology*, **15**, 35–46. {*149, 150, 156, 176*}

Krance, R. A., Spruce, W. E., Forman, S. J., Rosen, R. B., Hecht, T., Hammond, W. P., and Blume, G. 1982. Human cyclic neutropenia transferred by allogeneic bone marrow grafting. *Blood*, **60**, 1263–1266. {*167, 176*}

Lange, Robert D. 1983. Cyclic hematopoiesis: Human cyclic neutropenia. *Experimental hematology*, **11**, 435–451. {*166, 176*}

Layton, Judith E., Hockman, Helen, Sheridan, William P., and Morstyn, George. 1989. Evidence for a novel in vivo control mechanism of granulopoiesis: Mature cell-related control of a regulatory growth factor. *Blood*, **74**, 1303–1307. {*172, 176*}

Longtin, Andre and Milton, John G. 1988. Complex oscillations in the human pupil light reflex with "mixed" and delayed feedback. *Mathematical Biosciences*, **90**, 183–199. {*165, 176*}

Lord, B. I., Bronchud, M. H., Owens, S., Chang, J., Howell, A., Souza, L., and Dexter, T. M. 1989. The kinetics of human granulopoiesis following treatment with granulocyte colony stimulating factor in vivo. *Proceedings of the National Academy of Sciences of the United States of America*, **86**, 9499–9503. {*171, 176*}

Lord, B. I., Molineux, G., Pojda, Z., Souza, L. M., Mermod, J.-J., and Dexter, T. M. 1991. Myeloid cell kinetics in mice treated with recombinant Interleukin-3, granulocyte colony-stimulating factor (CSF), or granulocyte-macrophage CSF in vivo. *Blood*, **77**, 2154–2159. {*171, 176*}

Losson, Jerome, Mackey, Michael C., and Longtin, Andre. 1993. Solution multistability in first order nonlinear differential delay equations. *Chaos (Woodbury, NY)*, **3**, 167–176. {*157, 176*}

Lothrop, Clinton D., Warren, David J., Souza, Lawrence M., Jones, J. B., and Moore, Malcolm A. S. 1988. Correction of canine cyclic hematopoiesis with recombinant human granulocyte colony-stimulating factor. *Blood*, **72**, 1324–1328. {*172, 177*}

MacDonald, N. 1978. Cyclical neutropenia: Models with two cell types and two time lags. *Pages 287–295 of:* Valleron, A.-J. and Macdonald, P. D. M. (eds), *Biomathematics and Cell Kinetics: based on a workshop held at Université Paris 7, Paris, 27–28 February, 1978*. Amsterdam, The Netherlands: Elsevier. {*150, 167, 177*}

Mackey, Michael C. 1978. A unified hypothesis for the origin of aplastic anemia and periodic haematopoiesis. *Blood*, **51**, 941–956. {*149, 150, 164, 167, 168, 170, 177*}

Mackey, Michael C. 1979a. Dynamic haematological disorders of stem cell origin. *Pages 373–409 of:* Vassileva-Popova, J. G. and Jensen, E. V. (eds), *Biophysical and biochemical information transfer in recognition*. New York, NY, USA; London, UK: Plenum Press. {*149, 150, 167, 168, 170, 177*}

Mackey, Michael C. 1979b. Periodic auto-immune hemolytic anemia: An induced dynamical disease. *Bulletin of Mathematical Biology*, **41**, 829–834. {*149, 150, 155, 164, 177*}

Mackey, Michael C. and an der Heiden, Uwe. 1982. Dynamic diseases and bifurcations in physiological control systems. *Funktionelle Biologie und Medizin*, **1**, 156–164. {*149, 174, 177*}

Mackey, Michael C. and an der Heiden, Uwe. 1984. The dynamics of recurrent inhibition. *Journal of Mathematical Biology*, **19**, 211–225. {*157, 177*}

Mackey, Michael C. and Glass, Leon. 1977. Oscillation and chaos in physiological control systems. *Science*, **197**, 287–289. {*149, 150, 165, 174, 177*}

Mackey, Michael C. and Milton, John G. 1987. Dynamical diseases. *Annals of the New York Academy of Sciences*, **504**, 16–32. {*149, 174, 177*}

Mackey, Michael C. and Milton, John G. 1990. Feedback, delays, and the origins of blood cell dynamics. *Comments on modern biology. Part C, Comments on theoretical biology*, **1**, 299–327. {*174, 177*}

Metcalf, D., Johnson, G. R., and Burgess, A. W. 1980. Direct stimulation by purified GM-CSF of the proliferation of multipotential and erythroid precursor cells. *Blood*, **55**, 138–147. {*171, 177*}

Migliaccio, Anna Rita, Migliaccio, Giovanni, Dale, David C., and Hammond, William P. 1990. Hematopoietic progenitors in cyclic neutropenia: Effect of granulocyte colony stimulating factor in vivo. *Blood*, **75**, 1951–1959. {*172, 177*}

Milton, John G. and Mackey, Michael C. 1989. Periodic haematological diseases: Mystical entities or dynamical disorders? *Journal of the Royal College of Physicians of London*, **23**, 236–241. {*174, 177*}

Morley, A. 1979. Cyclic hemopoiesis and feedback control. *Blood Cells*, **5**, 283–296. {*149, 150, 167, 177*}

Morley, A. and Stohlman, F. 1970. Cyclophosphamide induced cyclical neutropenia. *The New England Journal of Medicine*, **282**, 643–646. {*167, 177*}

Morley, A. A., King-Smith, E. A., and Stohlman, F. 1970. The oscillatory nature of hemopoiesis. *Pages 3–14 of:* Stohlman, F. (ed), *Symposium on Hemopoietic Cellular Proliferation; St. Elizabeth's Hospital centennial, 1869–1969, Boston, Massachusetts, November 5–6, 1969*. New York, NY, USA: Grune and Stratton. {*167, 177*}

Nazarenko, V. G. 1976. Influence of delay on auto-oscillations in cell populations. *Biofizika*, **21**, 352–356. {*150, 167, 177*}

Orr, J. S., Kirk, J., Gray, Kathleen G., and Anderson, J. R. 1968. A study of the interdependence of red cell and bone marrow stem cell populations. *British journal of haematology*, **15**, 23–34. {*155, 164, 165, 177*}

Quesenberry, P. J. 1990. Hemopoietic stem cells, progenitor cells, and growth factors.

Pages 129–147 of: Williams, William J., et al.(eds), *Hematology*, fourth edn. New York, NY, USA: McGraw-Hill, Health Professions Division. {*150, 151, 177*}

Quesenberry, Peter J. 1983. Cyclic hematopoiesis: Disorders of primitive hematopoietic stem cells. *Immunology and Hematology Research Monographs*, **1**, 2–15. {*167, 178*}

Ranlov, P. and Videbaek, A. 1963. Cyclic haemolytic anaemia synchronous with Pel-Ebstein fever in a case of Hodgkin's disease. *Acta medica Scandinavica*, **174**, 583. {*155, 178*}

Rapoport, A. P., Abboud, C. N., and DiPersio, J. F. 1992. Granulocyte-macrophage colony stimulating factor (GM-CSF) and granulocyte colony stimulating factor (G-CSF): Receptor biology, signal transduction, and neutrophil activation. *Blood Reviews*, **6**, 43–57. {*171, 178*}

Reimann, Hobart Amsteth. 1963. *Periodic Diseases*. Philadelphia, PA, USA: F. A. Davis Company. Pages vii + 189. {*149, 174, 178*}

Sachs, Leo. 1993. The molecular control of hemopoiesis and leukemia. *Comptes rendus de l'Acad*
'emie des sciences, Paris, **316**, 882–891. {*150, 171, 178*}

Sachs, Leo and Lotem, Joseph. 1994. The network of hematopoietic cytokines. *Proceedings of the Society for Experimental Biology and Medicine*, **206**, 170–175. {*150, 171, 178*}

von Schulthess, N. D. and Mazer, N. A. 1982. Cyclic neutropenia (CN): A clue to the control of granulopoiesis. *Blood*, **59**, 27–37. {*150, 167, 178*}

Watari, Kiyoshi, Asano, Shigetaka, Shirafuji, Naoki, Kodo, Hideki, Ozawa, Keiya, Takaku, Fumimaro, and ichi Kamachi, Shin. 1989. Serum granulocyte colony-stimulating factor levels in healthy volunteers and patients with various disorders as estimated by enzyme immunoassay. *Blood*, **73**, 117–122. {*172, 178*}

Wheldon, T. E. 1975. Mathematical models of oscillatory blood cell production. *Mathematical Biosciences*, **24**, 289–305. {*150, 167, 178*}

Wheldon, T. E., Kirk, J., and Finlay, H. M. 1974. Cyclical granulopoiesis in chronic granulocytic leukemia: A simulation study. *Blood*, **43**, 379–387. {*150, 167, 178*}

Wichmann, H.-Erich and Loeffler, Markus. 1988. *Mathematical Modeling of Cell Proliferation: Stem Cell Regulation in Hemopoiesis*. 2000 Corporate Blvd., Boca Raton, FL 33431, USA: CRC Press. ISBN 0-8493-5503-6 (vol. 1), 0-8493-5504-4 (vol. 2). Page various. Two volumes. {*149, 178*}

Wright, D. G., Dale, D. C., Fauci, A. S., and Wolff, S. M. 1981. Human cyclic neutropenia: clinical review and long term follow up of patients. *Medicine (Baltimore)*, **60**, 1–13. {*166, 178*}

Wright, Daniel G., Kenney, Richard F., Oette, Dagmar H., LaRussa, Vincent F., Boxer, Laurence A., and Malech, Harry L. 1994. Contrasting effects of recombinant human granulocyte-macrophage colony-stimulating factor (CSF) and granulocyte CSF treatment on the cycling of blood elements in childhood-onset cyclic neutropenia. *Blood*, **84**, 1257–1267. {*172, 173, 178*}

9. OSCILLATIONS AND MULTISTABILITY IN DELAYED FEEDBACK CONTROL

John Milton and Jennifer Foss

9.1 INTRODUCTION

Science-fiction writers have long been fascinated with the concept of treating human neurological diseases through the construction of cyborgs, i.e., beings which are part human, part machine. In their stories, lost limbs are replaced by robotic ones (Lucas 1995), and epileptic seizures are prevented by directly interfacing computer circuits with the brain (Crichton 1972). As far fetched as these tales may have once seemed, the use of hybrid computer neural devices to treat disease is surprisingly close to becoming a reality. Already, electrical stimulators are implanted to control seizures (Cooper et al. 1976; Wilder 1990) and pain (Richardson and Goreck 1996), to treat Parkinson's disease (Iacono 1995; Siegfried 1994), and to aid the hearing impaired (Maniglia (ed.) 1995). The recent demonstration that two-way communication can occur between a living neuron and a silicon chip (Fromherz and Stett 1995) suggests that it may eventually be possible to miniaturize hybrid neural computer devices to the cellular level.

Here we discuss the dynamics of closed-loop hybrid neural computer devices. The basic idea is that the variable, x, to be controlled, e.g., membrane potential of a neuron, is monitored by a computer which calculates the required feedback, f, and then outputs an appropriate stimulus, i.e., an electrical pulse, which changes x, thereby closing the loop (Figure 9.1). An intrinsic component of closed-loop hybrid neural computer devices is the presence of a time delay, τ. Time delays arise both within the nervous system (e.g., conduction times along an axon and across the synapse, processing times, rise times of post-synaptic potentials) and within the device (e.g., sensor times, computing time). Consequently, mathematical models take the form of delay differential equations (see Chapter 8); an example is the first-order delay differential equation

$$\dot{x} + \alpha x = f(x_\tau) \tag{9.1}$$

where x, x_τ are, respectively, the values of the variable, x, at times $t, t - \tau$, $x_\tau \equiv x(t - \tau)$, and α is a rate constant.

Possible goals of this neuroelectronic integration range from restoring behaviors lost because of disease (e.g., the ability to move a limb) to the pre-

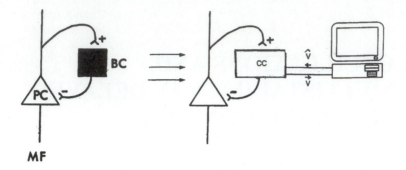

Figure 9.1. On the left a recurrent inhibitory loop in the hippocampus composed of an excitatory neuron, the pyradmidal cell (PC), and an inhibitory interneuron, the basket cell (BC). MF represents the mossy fiber bundle which provides a tonic drive to PC. On the right, a hybrid neural computer device representation in which the role of the inhibitory interneuron is played by a computer.

vention of pathological oscillations (e.g., to stop a seizure) to the controlled delivery of medications (e.g., drug pumps). These goals translate into the mathematical problem of how best to stabilize unstable systems and to stop and start biological oscillations, including chaotic ones. In some cases it may be appropriate to design the feedback to mimic as closely as possible the normal physiological feedback. However, this approach may not always yield the feedback which is most easy to implement electronically and for that matter may not be the feedback most suited to the control problem at hand. For example, piecewise-continuous feedback is optimal for many classes of control problem (Flügge-Lotz 1968) and brief stimuli can be used to stabilize an unstable orbit in a chaotic dynamical system ("OGY" algorithm (Ott et al. 1990)).

Variations on the concept of a closed-loop hybrid neural computer device have been used to construct hybrid neural circuits (Martinez and Segundo 1983; Kohn et al. 1981; Sharp et al. 1992; Sharp et al. 1993b; Sharp et al. 1993a; Vibert et al. 1979; Yarom 1991), to regulate pupil size (Milton et al. 1988; Reulen et al. 1988; Stark 1962), to abort seizures in patients with epilepsy (Forster 1975), to control muscle movement (Andrews et al. 1988; Kostov et al. 1995; Petrofsky and Phillips 1985) and muscle pain (Zhang et al. 1993), to improve the treatment of the obstructive sleep disorder apnea (Behbehani et al. 1995), and to control "chaos" in the brain (Schiff et al. 1994). Other medical applications include the design of automatic drug-delivery systems (Milton et al. 1995) to, for example, deliver insulin to diabetic patients (Mirouze et al. 1977) or control blood pressure during anesthesia (Furutani et al. 1995), and the use of the OGY algorithm to control cardiac chaos (Garfinkel et al. 1992).

In Section 9.2 we discuss the prototype of a hybrid neural computer device, i.e., the clamped pupil-light reflex, and in Section 9.3 we examine a second hybrid neural computer device, the recurrently clamped neuron. Considerable analytical insight into the dynamics of these hybrid neural computer devices can be obtained by replacing continuous-feedback functions by simpler piecewise-constant approximations (compare the feedback functions in

Figures 9.2 and 9.3) (Bayer and an der Heiden 1996; an der Heiden 1985; an der Heiden et al. 1990; an der Heiden and Mackey 1982; an der Heiden and Mackey 1987; an der Heiden and Reichard 1990; Longtin and Milton 1988; Losson et al. 1993; Milton and Longtin 1990). The appearance of multistability, i.e., the co-existence of multiple attractors, is discussed in Sections 9.3 through 9.5. In Section 9.6 we discuss some preliminary observations pointing to multistability in the clamped pupil-light reflex.

9.2 CLAMPED PUPIL-LIGHT REFLEX

The pupil-light reflex regulates the retinal light flux, $\phi = IA$ (lumens), where I (lumens mm^{-2}) is the illuminance and A (mm^2) is the pupil area. It does this by acting like the aperture of a camera: when I increases, the pupil gets smaller (negative feedback) after a time delay of approximately 300 msec (delayed negative feedback).

From a physiological point of view this reflex has only a very modest influence on visual acuity and the total retinal light flux. Its importance is twofold. First, examination of the response of the pupil to light provides the neurologist with an invaluable bedside tool for assessing, for example, the integrity of neural pathways which pass close to regions of the brainstem which regulate consciousness. Second, this reflex can be readily monitored and manipulated using noninvasive techniques. Consequently, biomedical engineers have extensively studied this reflex as an example of a biological servomechanism (Stark 1959). Here we exploit this reflex as THE paradigm of a closed-loop hybrid neural computer device, referred to herein as the clamped pupil light reflex.

It is easy to clamp the pupil-light reflex. The reflex-feedback loop is first "opened" by focusing a small-diameter light beam onto the center of the pupil (Stark and Sherman 1957). This circumvents the shading effects of the iris on the retina; i.e., even though the pupil constricts or dilates, ϕ remains unchanged. The feedback loop is then reclosed with a clamping box (Figure 9.2) which relates measured changes in pupil area to changes in retinal illumination (Milton et al. 1988; Reulen et al. 1988; Stark 1962). By careful design of the clamping box a variety of feedbacks can be inserted into the pupil-light reflex. Since the feedback is known, direct comparisons between theory and observation become possible.

Up to now models of the clamped pupil-light reflex have been of the form (Longtin and Milton 1988; Milton and Longtin 1990; Longtin et al. 1990; Milton et al. 1989; Longtin and Milton 1989b)

$$\dot{A} + \alpha A = F(A_\tau) . \tag{9.2}$$

The lefthand side of (9.2) describes the dynamics of iris musculature, and the right-hand side describes the feedback constructed by the clamping box. In this case we have assumed that, in the presence of constant feedback, the pupil area A will "relax" to a constant value with first-order kinetics given by the rate α (sec^{-1}). Two feedback functions which approximate negative feedback have been studied: piecewise-constant negative feedback (Milton and Longtin

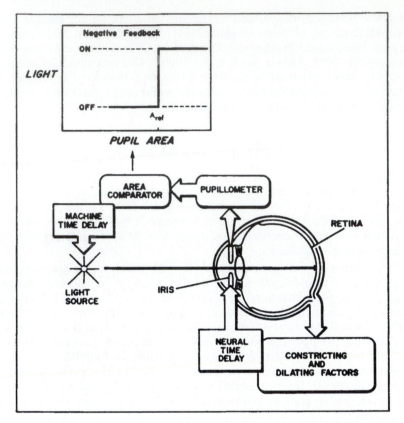

Figure 9.2. Schematic representation of the pupil-light reflex clamped with piecewise constant negative feedback (inset). (Figure from Milton and Longtin (1990).)

1990) (Figure 9.3(a)) of the form

$$F(A_\tau) = \begin{cases} \text{light ON} & \text{if } A_\tau > A_{\text{ref}}, \\ \text{light OFF} & \text{otherwise,} \end{cases} \tag{9.3}$$

where A_{ref} (mm^2) is an adjustable area threshold (cf. equation 8.26), and smooth negative feedback (Longtin et al. 1990) (Figure 9.3(a)):

$$F(A_\tau) = \frac{\theta^n}{A_\tau^n + \theta^n}, \tag{9.4}$$

where n and θ (mm^2) are constants (cf. equation 8.23).

In the case of smooth negative feedback, the gain, G, in the feedback is proportional to n (Longtin et al. 1990).[1] For sufficiently high n, oscillations in pupil area occur (Figure 9.3(b)). The onset of these oscillations corre-

---

[1]G is proportional to dF/dA evaluated at A^*, where A^* is the solution of $\alpha A^* = F(A^*)$. This definition of the gain is inverse to the gain typically considered in engineering applications (Longtin and Milton 1989a).

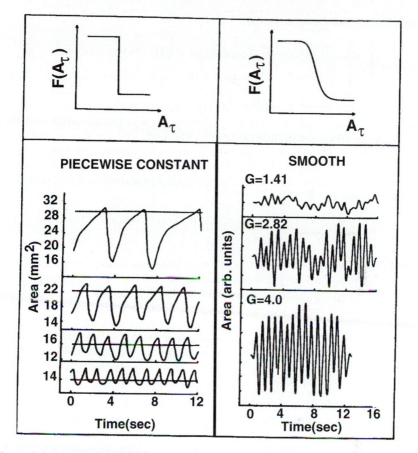

Figure 9.3. Fluctuations in period, T, and amplitude, A, of the pupil oscillations in the pupil-light reflex clamped with piecewise-constant negative feedback (left) and smooth negative feedback (right). (Figure from Longtin et al. (1990).)

sponds to the occurrence of a supercritical Hopf bifurcation (Longtin and Milton 1989b). Linear stability analysis of (9.2) (cf. Section 8.5.1) predicts that the frequency of the oscillation which arises at the point of instability, f, is equal to $\alpha\sqrt{G^2 - 1}$. This implies that the period, T, of the oscillation is bounded by

$$4\tau > T > 2\tau. \qquad (9.5)$$

The observed period of the high-gain oscillations in (Figure 9.3(b)) is approximately 900–1000 msec. Since the latency time (delay τ) is approximately 300 msec, this observed period is in good agreement with that predicted by (9.5), i.e., 600–1200 msec.

Piecewise-constant negative feedback corresponds to the case of infinite gain in the feedback loop (an der Heiden and Mackey 1982; Milton and Longtin 1990). The condition for an oscillation to arise is that the resting pupil area is larger than A_{ref} (otherwise the light never turns ON!). The advantage of studying piecewise-constant negative feedback is that the solution of (9.2) can be constructed by piecing together exponentials (an der Heiden and Mackey

1982; Milton and Longtin 1990),

$$A(t) = \begin{cases} A_{\text{on}} + [A(t_0) - A_{\text{on}}] \exp(-\alpha_c(t - t_0)) & \text{when } A(s - \tau) > A_{\text{ref}}, \\ A_{\text{off}} + [A(t_0) - A_{\text{off}}] \exp(-\alpha_d(t - t_0)) & \text{otherwise}, \end{cases} \quad (9.6)$$

where α_c, α_d are, respectively, the rate constants for pupillary constriction and dilation, and A_{on}, A_{off} (mm^2) are, respectively, the asymptotic values that pupil area tends to when the light is on or off. The form chosen here is equivalent to (8.28) and (8.33), which are derived from the piecewise-constant feedback function (8.26) used to study disorders of the blood in Chapter 8. Employing methods from Section 8.5.1, it is straightforward to show that the period, T, of the oscillations which arise is

$$T = 2\tau + \alpha_c^{-1} \ln \left[\frac{A_{\text{max}} - A_{\text{on}}}{A_{\text{ref}} - A_{\text{on}}} \right] + \alpha_d^{-1} \ln \left[\frac{A_{\text{min}} - A_{\text{off}}}{A_{\text{ref}} - A_{\text{off}}} \right], \quad (9.7)$$

where A_{max}, A_{min} are, respectively, the maximum and minimum amplitudes of the oscillation and we have assumed that the latency for light onset and offset is the same. In contrast to smooth negative feedback, when the pupil is clamped with piecewise-constant negative feedback, T can become much longer than 4τ (Figure 9.4(a)).

The parameters α_c, α_d, A_{on}, A_{off} can be evaluated directly from experimental observations by taking $t_0 = 0$ and $A(t_0) = A_{\text{max}}$. Then from (9.6) we obtain

$$A_{\text{max}} = A_{\text{off}}(1 - \exp(-\alpha_d\tau)) + A_{\text{ref}} \exp(-\alpha_d\tau).$$

Thus we can evaluate α_d from the slope of a plot of A_{max} versus A_{ref} and A_{off} from the intercept. Similarly, α_c and A_{on} can be evaluated from a plot of A_{min} versus A_{ref}.

Figure 9.4 compares pupil cycling observed for different values of A_{ref} to that predicted by (9.6) when the parameters are evaluated using the above procedure. As can be seen, the predicted period and amplitude of the oscillations agree with experiment to within 5% to 10%. Measurements of pupil oscillations when the pupil is clamped with piecewise-constant negative feedback have proven useful for assessing the efferent pathways in the pupil-reflex arc (Milton and Longtin 1990) and for diagnosing pathology within the optic nerve (Milton et al. 1988).

9.3   RECURRENTLY CLAMPED NEURONS

An obvious extension of the ideas in Section 9.2 is to dynamically clamp neurons (Sharp et al. 1993b; Sharp et al. 1993a). Instead of monitoring pupil area, we monitor neuron membrane potential, and the clamping box plays the role of, for example, an interneuron (Figure 9.1). Variations of this technique have been used to study the dynamics of recurrent inhibitory (Kohn et al. 1981; Vibert et al. 1979) and excitatory (Martinez and Segundo 1983) loops involving crayfish slowly-adapting stretch receptors, to form artificial electrical (Sharp et al. 1992) and chemical (Sharp et al. 1993b) synapses, to form simple neural circuits (Sharp et al. 1993b; Sharp et al. 1993a), and to study rhythmogenesis in networks of coupled oscillators (Yarom 1991).

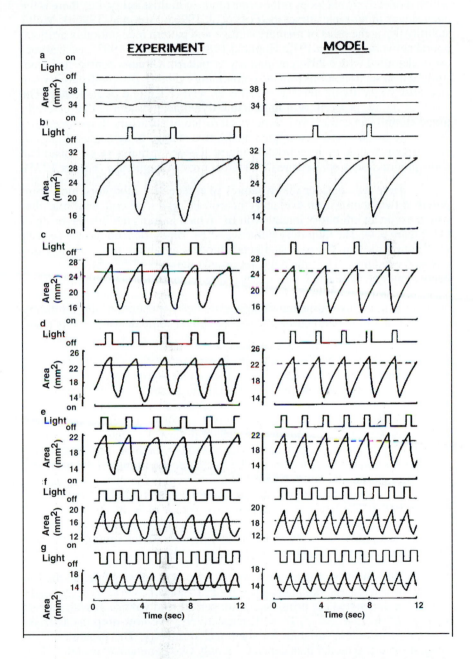

Figure 9.4. Comparison of pupil cycling as a function of A_{ref} to that predicted by (9.6). (Figure from Milton and Longtin (1990).)

A recently emphasized property of the delay differential equations which model recurrent loops is the occurrence of multistability, i.e., there is the coexistence of multiple attractors (Foss et al. 1996a; Foss et al. 1996b). Multistability lies at the basis of memory storage and pattern recognition in artificial neural networks (Cowan 1972; Hopfield 1984; Zipser et al. 1993): each attractor is identified with a different memory or pattern. Circuits constructed from *Aplysia* (sea snail) neurons provide convincing evidence that multistability can also arise in the living nervous systems as well (Kleinfeld et al. 1990). The most intriguing property of a multistable dynamical system is that carefully timed stimuli can cause a switch between different attractors. Experimentally this would be observed as a sudden qualitative change in dynamics. Indeed it is well known that an appropriately timed sensory stimulus (e.g., noise) can abort progression of the seizure in some patients with epilepsy (Forster 1975).

Here we consider the dynamics of a recurrent inhibitory loop composed of two neurons: an excitatory neuron PC gives off collateral branches which excite an inhibitory interneuron BC which in turns inhibits the firing of PC (Figure 9.1). The inhibitory influence of BC on PC depends on the activity of PC at a time τ in the past, and hence models are in the form of delay differential equations. The time delay represents the sum of the conduction-time delay along the axon and dendrites, the time required for neurotransmitter release, processing times in the interneuron, and the rise time of the postsynaptic potentials. In the hybrid-neural-computer-device realization of this inhibitory loop the role of the interneuron BC is played by a computer or possibly a silicon chip.

To illustrate how multistability arises in delayed recurrent loops, consider the simple integrate-and-fire model shown in Figure 9.5. The membrane potential, V, of the neuron increases linearly at a rate, R, until it reaches the firing threshold, θ. When $V = \theta$, the neuron fires and V is reset to its resting membrane potential, V_o. The period is $T = \theta/R$. The firing of the neuron excites the inhibitory interneuron, which in turn at a time τ later delivers an inhibitory postsynaptic potential to the excitatory neuron. The effect of the inhibitory postsynaptic potential will be to change the timing of the next neuronal firing by an amount δ, where δ is a function of the phase at which the inhibitory postsynaptic potential arrives after the neuron has fired (Best 1979; Pinsker 1977). For illustrative purposes we take δ to be independent of the phase. In this simple model, when $R > 0$, changing the period by δ is equivalent to changing V by δ. By convention, when $\delta > 0$, the timing of the next spike is prolonged by δ, and vice versa.

Without loss of generality, we take $V_0 = 0$ volts and define the following dimensionless variables: $\tau^* = \tau/T$, $t^* = t/T$, $v^* = V/\theta$, $\Delta = \delta/\theta$ so that the dimensionless firing threshold, period, and voltage growth rate are $\theta^* = 1$, $T^* = 1$, $R^* = \theta^*/T^* = 1$, respectively. We now drop the asterisks on the dimensionless variables for notational simplicity. The dynamics of the integrate-and-fire model thus depends on only two parameters: τ and Δ. Furthermore we take $0 < \Delta < 1$. The special case $\tau = q$, q a positive integer, has been briefly considered previously (Bressloff and Stark 1990). Here we consider all positive τ. When $\tau < 1$, the recurrent inhibitory model produces only a regular periodic firing pattern. Each excitatory spike produced is followed by an inhibitory pulse at time τ later. The inhibitory pulse decreases v by an

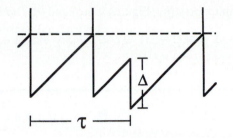

Figure 9.5. Schematic representation of the dynamics of an integrate-and-fire model for recurrent inhibition. The vertical axis is membrane voltage, the horizontal axis is time, and the dotted line is the threshold, θ. See text for discussion.

amount Δ and hence prolongs the period by an amount Δ. Thus all solutions are periodic with period $1 + \Delta$, and we do not consider this case further.

Multistability arises when $\tau > 1$ and Δ is independent of phase. This complex behavior becomes possible, since the inhibitory pulses are not necessarily the result of the immediately preceding excitatory spike (Figure 9.5). It can be shown that the solutions which arise can be constructed from segments of length τ (Foss et al. 1996a): each segment must satisfy an equation of the form

$$\tau = x + m + n\Delta, \tag{9.8}$$

where n, m are positive integers and $0 < x < 1$. For τ, Δ fixed, the total number of (m, n) pairs that satisfy (9.8) is $\lceil \tau/\Delta \rceil$, where the notation $\lceil . \rceil$ denotes the smallest integer greater than τ/Δ. Since the number of (m, n) segments is finite for fixed τ, Δ, it follows that all solutions are periodic. Moreover, since there is a one-to-one relationship between excitatory spikes and inhibitory pulses, and since each inhibitory pulse prolongs the period by Δ, the period of these solutions is $S(1+\Delta)$, where S is a positive integer equal to the number of excitatory spikes per period. The mean interspike interval is therefore $(1 + \Delta)$.

Once the (m, n) pairs have been determined from (9.8), it is possible to construct the nature of the periodic spike trains which arise. The interested reader is referred to Foss *et al.* (1996a) for the method of construction and for a simple-to-construct electronic circuit of a multistable delayed recurrent loop. Here we illustrate the multistability. A periodic neural spike train consisting of j interspike intervals is completely described by the notation $\{n_1, n_2, \ldots, n_j\}$, where n_i is the number of times that the inhibitory neuron fires in the i-th interval. Necessarily $j = \sum_{i=1}^{j} n_i$; i.e., some of the n_i may be zero.

Figure 9.6 illustrates that four qualitatively different periodic spike trains arise when $\tau = 4.1$ and $\Delta = 0.8$: regular spiking ($\{11111\}$), bursting ($\{50000\}$), and two more irregular spiking patterns ($\{13100\}$ and $\{10220\}$). It can be shown that regular spiking (Figure 9.6(a)) occurs when

$$(j - 1)(1 + \Delta) < \tau < (j - 1)(1 + \Delta) + 1, \tag{9.9}$$

bursting patterns (Figure 9.6(b)) occur when

$$j - 1 < \tau < \Delta(j - 1) + 1, \tag{9.10}$$

and the more complex patterns (Figures 9.6(c) and (d)) both occur when

$$3 + \Delta + 5k(1 + \Delta) < \tau < 2 + 3\Delta + 5k(1 + \Delta) \qquad (9.11)$$

(Foss et al. 1996a).

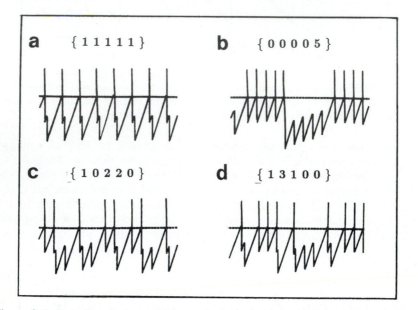

Figure 9.6. Periodic solutions which coexist in the integrate-and-fire model when $\tau = 4.1$ and $\Delta = 0.8$. The vertical axis is membrane voltage, the horizontal axis is time, and the dotted line is the threshold.

The regions in $(\tau - \Delta)$-space for which these four spiking patterns are observed are shown, respectively, in Figures 9.7(a–c). Figure 9.7(d) shows the intersection, $\cap$, of these regions. The fact that $\cap$ has nonzero measure lies at the basis of the multistability of this model. For choices of (τ, Δ) in $\cap$, all four of the above solutions coexist.

9.4   MULTISTABILITY: INITIAL FUNCTIONS

Typically multistability is discussed in the context of dynamical systems described by ordinary differential equations (Winfree 1980). A well-studied example is the bistability associated with the subcritical Hopf bifurcation (Appendix B), which characterizes, for example, the onset of oscillation in excitable systems, such as neurons (Guttman et al. 1980; Hounsgaard et al. 1988). In this case a fixed-point attractor can coexist with a limit, cycle attractor. The dynamics of a multistable system are determined once the initial conditions are chosen: each attractor can be associated with a basin of attraction, i.e., the set of all initial conditions attracted. For an ordinary differential equation it is sufficient to specify the initial conditions at a single instance in time. In contrast, for a delay differential equation it is necessary to specify an initial function, $\Psi(s)$, $s \in [t_0 - \tau, t_0]$ where $t_0 \geq 0$. In other words, multistable

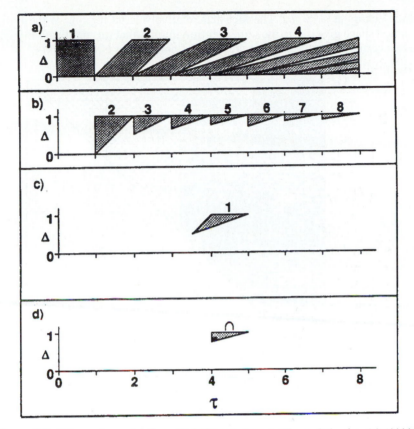

Figure 9.7. Values of τ and Δ for which the solutions in Figure 9.6 exist: (a) {11111}, (b) {50000}, (c) {13100} and {10220}. Subfigure (d) shows the values of (τ, Δ) for which these solutions coexist.

delay differential equations evolve in a functional space in which functions of length τ are mapped into functions of length τ, and the basins of attraction correspond to sets of functions.

The space of initial functions which describe the multiple basins of attraction in a delay differential equation is typically very complex (Losson et al. 1993; Foss et al. 1996b). However, in the case of the integrate-and-fire model it can be determined. Here, the initial function, $\Psi(s)$, consists of a sequence of spike times, $\{t_i\}$, of length τ (Foss et al. 1996b; Foss et al. 1996a). For $t_0 > 0$, Ψ contains two components: one component represents spikes generated intrinsically by the circuit; the other component represents external inputs to the loop introduced by, for example, other neurons or by the experimentalist using a spike generator. The maximum number of t_i which must be specified is that required to define the regular spiking patterns (Figure 9.6(a)), since this pattern has the most evenly distributed firing times, and hence the smallest maximum time between spikes. For the regular spiking pattern when $\tau = 4.1$, $\Delta = 0.8$ we have $j = 3$. Thus $\Psi = \{t_1, t_2, t_3\}$.

Figure 9.8 shows the dependence of the solution of the recurrent inhibitory model for all initial functions composed of three spikes, the last of which occurs at $t_3 = t_0$. The fact that relatively large regions of the initial function space give rise to the same solution implies that this multistability will have a certain robustness in the presence of noise (Foss et al. 1996a).

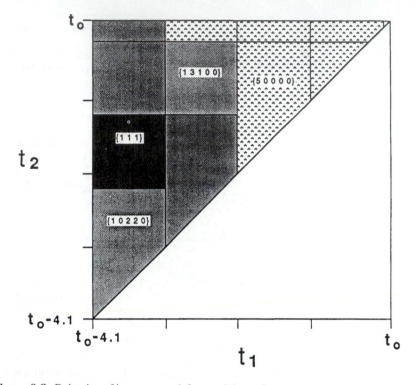

Figure 9.8. Behavior of integrate-and-fire model as a function of initial functions composed of three spikes. t_1 and t_2 are the times of two of the spikes, and the third has been fixed at $t_0 = 0$. (Figure from Foss et al. (1996b).)

9.5   MULTISTABILITY AND MIXED FEEDBACK

More generally, models for recurrent inhibition lead to delay differential equations with nonmonotone feedback (Mackey and an der Heiden 1984; Milton et al. 1990). An der Heiden and Mackey (1982; 1987) were the first to draw attention to the relationship between mixed feedback and the occurrence of multistability in delay differential equations. In particular, they studied delay differential equations in which feedback was piecewise-constant mixed feedback, in particular

$$\dot{x} + ax = \begin{cases} 0 & \text{if } x_\tau < \theta_1, \\ b & \text{if } \theta_1 \le x_\tau < \theta_2, \\ 0 & \text{otherwise.} \end{cases} \tag{9.12}$$

As in the case of piecewise-constant negative feedback (Section 9.2), (9.12) can be solved by piecing together exponentials to obtain

$$
x(t) = \begin{cases} \gamma + (x(t_0) - \gamma)\exp(-a(t - t_0)) & \text{if } \theta_1 \leq x_\tau < \theta_2, \\ x(t_0)\exp(-a(t - t_0)) & \text{otherwise,} \end{cases} \tag{9.13}
$$

where $\gamma = b/a$ is the asymptote of the increasing exponential segment. It is possible to prove analytically the existence of limit cycles (Appendix B), of a homoclinic orbit (Appendix B), and of Li and Yorke-type chaos, and to characterize the global stability of the simple limit cycles (i.e., those with one maximum per period) (an der Heiden and Mackey 1982; an der Heiden and Mackey 1987). Further, multistability can arise. These results have been shown to hold also when the discontinuities in the piecewise-constant mixed feedback are smoothed out (an der Heiden 1985), making the results more relevant to real physical systems. Losson *et al.* (1993) modeled (9.12) using an electronic circuit and were able to demonstrate multistability experimentally.

Equation (9.12) is an idealization of the Mackey-Glass equation

$$
\dot{x} + ax = \frac{\beta x_\tau \theta^n}{x_\tau^n + \theta^n} \tag{9.14}
$$

where a, β, θ, n are constants, which arises in the description of the growth of certain blood-cell populations (Mackey and Glass 1977). Ikeda and Matsumo-tot (1987) were able to demonstrate using numerical simulations that multistability can arise in (9.14). The condition for multistability to occur is that the time delay must be longer than the intrinsic time scale a^{-1}, i.e., $\tau > a^{-1}$. This condition can be identified with the condition $\tau > 1$ for multistability in the simple recurrent inhibitory loop studied in Section 9.3. Multistability in (9.14) has been demonstrated experimentally in experiments involving laser diodes (Aida and Davis 1992). A novel scheme for controlling the multistability in the Mackey-Glass equation in order to store information has been recently proposed (Mensour and Longtin 1995).

9.6 MULTISTABILITY AND NEGATIVE FEEDBACK

Recently it has been realized that multistability can also arise in delay differential equations with negative feedback, i.e., feedback of the type we discussed in Section 9.2 in the context of the pupil light reflex. Multistability cannot arise in a first-order delay differential equation with smooth negative feedback: there is either a stable fixed point or a stable limit cycle. However, multistability can arise in second-order delay differential equations with smooth negative feedback (Campbell et al. 1995), i.e.,

$$
\ddot{x} + b\dot{x} + ax = f(x_\tau)
$$

Models expressed as second-order DDEs arise when Newton's laws are applied to the description of neuromechanical systems, e.g., the control of the movement of a limb. As might be expected, a great deal of insight into the nature of the multistability which arises can be gained by replacing smooth negative feedback by piecewise-constant negative feedback. An der Heiden and his co-workers (Bayer and an der Heiden 1996; an der Heiden et al. 1990; an der Heiden and Reichard 1990) studied the special case of a harmonic oscillator with piecewise-constant negative feedback, i.e.,

$$\ddot{x} + \alpha x = \begin{cases} a & \text{if } x_\tau \leq \theta, \\ b & \text{otherwise.} \end{cases} \tag{9.15}$$

In this case the solution is composed of arcs of circles, and in fact the solutions can be constructed with compass and ruler (an der Heiden et al. 1990)! It is possible to prove that there are six different limit-cycle solutions of (9.15). For certain parameters choices up to three stable solutions coexist (Bayer and an der Heiden 1996; an der Heiden and Reichard 1990). Numerical simulations indicate that multistability also occurs for non zero damping (i.e., $b \neq 0$) provided that b is not too large (W. Bayer and U. an der Heiden, unpublished observations).

By analogy with (9.12) an obvious experimental paradigm to explore for evidence of multistability is the pupil-light reflex clamped with piecewise-constant negative feedback. However, it should be noted that the frequency response of the pupil-light reflex to low-amplitude sinusoidal light stimulation yields an open-loop transfer function which is third order (Stark 1959) This result likely accounts for the observation that the abrupt changes in direction of pupil movement between constriction and dilation predicted by (9.2) are not observed experimentally (Figure 9.4). Thus the inertia of the iris and its musculature is an important consideration. Although at least a third-order DDE is required to model this reflex, we anticipate that some of the same phenomena observed for second-order DDEs will occur.

Figure 9.9 shows a plot of $\dot{A}$ versus A when $A_{\text{ref}} = 30 \text{ mm}^2$ (Milton et al. 1996). This phase portrait has an uncanny resemblance to one of the solutions of (9.15) (compare with portrait c in Figure 9.10). Spontaneous changes in waveforms occur frequently during pupil cycling (compare *a* and *b* in Figure 9.11). It is tempting to speculate that these changes reflect noise-induced transitions in a dynamical system which is inherently multistable. Verification of this hypothesis will require detailed analysis of (9.15) for nonzero damping as well as the experimental demonstration that changes between limit cycles occur in response to an appropriately timed perturbation (e.g., light pulse).

9.7   DISCUSSION

The nervous system is composed of multiple feedback loops which range in size from the two-neuron recurrent inhibitory loop (Figure 9.11) to the multiple polysynaptic loops involved in, for example, the control of posture. An intrinsic component of these loops is the presence of time delays. Once the time delays in these loops become sufficiently large (which must happen at some level of neural organization), the occurrence of multistability becomes inescapable. Thus we are led to the conclusion that multistability is a direct consequence of the way in which the nervous system is constructed. Moreover, since time delays are typically longer in closed-loop hybrid neural computer devices, multistability may even more readily arise.

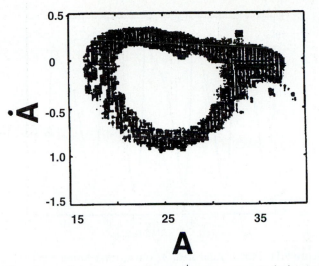

Figure 9.9. Plot of rate of change of pupil area, $\dot{A}$, versus area, A, during pupil cycling.

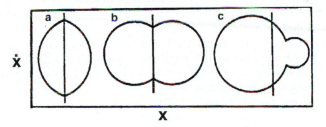

Figure 9.10. Phase-plane portraits, i.e., $\dot{x}$ (vertical axis) versus x (horizontal axis), for the three stable limit cycles that occur for (9.15). The vertical line corresponds to $\dot{x} = 0$. For proof of existence and stability see an der Heiden and Reichard (1990) and Bayer and an der Heiden (1996).

The implications of multistability for delayed-feedback control mechanisms are far-reaching: (1) multistability suggests the possibility of changing or controlling dynamics by using strategies that alter the initial distribution of stimuli Ψ; (2) multistability may underlie some of the complexity observed in neural times series through the effects of noise on Ψ; and (3) multistability may underlie the considerable variability observed between repeated recordings from the same neural population, since, in general, Ψ is not controlled.

Does the multistability which arises in delayed recurrent loops have relevance for the nervous system? Recurrent inhibitory loops involving two or more neurons are ubiquitous in the nervous system and are particularly prevalent in cortical regions important for memory, e.g., the hippocampalmesial temporal lobe complex. Thus multistability may play a role in active short-term memory, i.e., the electrical form of memory that occurs before memories are stored in the form of macromolecules and changes in neural connectivity. It would be anticipated from our model in (9.3) that the code is in the form of temporal patterning of the spike train (Foss et al. 1996b). Indeed, the possibility that the neural code in the nervous system is in part a temporally patterned

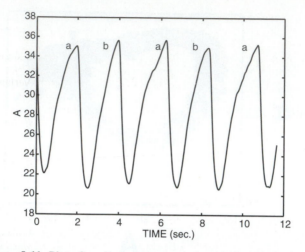

Figure 9.11. Plot of pupil area, A, versus time during pupil cycling.

one has been recently emphasized (Middlebrooks et al. 1994; Mainen and Sejnowski 1995; Ferster and Spruston 1995).

Whether or not multistability in man-made devices, such as hybrid neural computer devices, has practical significance remains to be seen. A fundamental question is whether it is easier to implement a control strategy for a dynamical system that is multistable than for one that is not. Certainly a multistable dynamical system represents a very economical way in which to store multiple dynamical behaviors within the same physical structure: each of these patterns can be brought to the forefront by the application of a single, carefully honed stimulus. As for our science fiction writers ... coexistence of multiple, parallel universes ... multistability ... hmm, we wonder!

9.8   ACKNOWLEDGMENTS

The authors thank Uwe an der Heiden, André Longtin, and Michael Mackey for useful discussions. Work is supported by a grant from the National Institutes of Mental Health and the Brain Research Foundation. J. F. was supported by the Lucille P. Markey Charitable Trust and the Women's Council of the Brain Research Foundation.

REFERENCES

Aida, T. and Davis, P. 1992. Oscillation modes of laser diode pumped hybrid bistable system with large delay and application to dynamic memory. *IEEE Journal of Quantum Electronics*, **28**(3), 686–699. {*191, 194*}

an der Heiden, U. 1985. Stochastic properties of simple differential-delay equations. *Page 351 of:* Meinardus, G. and Nurnberger, G. (eds), *Delay Equations, Approximation and Application: international symposium at the University of Mannheim, October 8–11, 1984*. International series of numerical mathematics, vol. 74. Basel, Switzerland: Birkhäuser Verlag. {*181, 191, 194*} an der Heiden, U. and Mackey, M. C. 1982. The dynamics of production and destruction: Analytic insight into complex behavior. *Journal of Mathematical Biology*, **16**, 75–101. {*181, 183, 190, 191, 194*}

an der Heiden, U. and Mackey, M. C. 1987. Mixed feedback: A paradigm for regular and irregular oscillations. Springer series in synergetics, vol. 36. Berlin, Germany / Heidelberg, Germany / London, UK / etc.: Springer Verlag. {*181, 190, 191, 195*}

an der Heiden, U. and Reichard, K. 1990. Multitude of oscillatory behavior in a nonlinear second order differential-difference equation. *Zeitschrift für Angewandte Mathematik und Mechanik*, **70**, 621–624. {*181, 192, 193, 195*}

an der Heiden, U., Longtin, A., Mackey, M. C., Milton, J. G., and Scholl, R. 1990. Oscillatory modes in a nonlinear second-order differential equation with delay. *Journal of Dynamics and Differential Equations*, **2**(4), 423–449. {*181, 192, 195*}

Andrews, B. J., Baxendale, R. H., Barnett, R., Phillips, G. F., Yamazaki, T., Paul, J. P., and Freeman, P. 1988. Hybrid FES orthosis incorporating closed loop control and sensory feedback. *Journal of biomedical engineering*, **10**, 189–195. {*180, 195*}

Bayer, W. and an der Heiden, U. 1996. Oscillation types and bifurcations of a nonlinear second order differential-difference equation. *Journal of Dynamics and Differential Equations*. In press. {*181, 192, 193, 195*}

Behbehani, K., Yen, F-C., Burk, J. R., Lucas, E. A., and Axe, J. R. 1995. Automatic control of airway pressure for treatment of obstructive sleep apnea. *IEEE Transactions on Biomedical Engineering*, **42**(10), 1007–1016. {*180, 195*}

Best, E. N. 1979. Null space in the Hodgkin-Huxley equations: A critical test. *Biophysical Journal*, **27**, 87–104. {*186, 195*}

Bressloff, P. C. and Stark, J. 1990. Neuronal dynamics based on discontinuous circle maps. *Physics Letters A*, **150**, 187–195. {*186, 195*}

Campbell, S. A., Bélair, J., Ohira, T., and Milton, J. 1995. Complex dynamics and multistability in a damped harmonic oscillator with delayed negative feedback. *Chaos (Woodbury, NY)*, **5**, 640–645. {*191, 195*}

Cooper, I. S., Amin, I., Riklan, M., Waltz, J. M., and Poon, T. P. 1976. Chronic cerebellar stimulation in epilepsy. *Archives of Neurology*, **33**, 559–570. {*179, 195*}

Cowan, J. D. 1972. Stochastic models of neuroelectric activity. *Pages 181–182 of:* Rice, Stuart A., Freed, Karl F., and Light, John C. (eds), *Statistical mechanics; new concepts, new problems, new applications*. Chicago, IL, USA: University of Chicago Press. IUPAP Conference on Statistical Mechanics (6th: 1971: Chicago, IL). {*186, 195*}

Crichton, Michael. 1972. *The Terminal Man*. New York, NY, USA: Ballantine Books. ISBN 0-394-44768-9. Pages xi + 247. {*179, 195*}

Ferster, D. and Spruston, N. 1995. Cracking the neuronal code. *Science*, **270**(5237), 756–757. {*194, 195*}

Flügge-Lotz, Irmgard. 1968. *Discontinuous and Optimal Control*. New York, NY, USA: McGraw-Hill. Pages x + 286. {*180, 195*}

Forster, Francis M. 1975. *Reflex epilepsy, behavioral therapy, and conditional reflexes*. Springfield, IL, USA: Thomas. ISBN 0-398-03614-4. Pages ix + 318. {*180, 186, 195*}

Foss, J., Moss, F., and Milton, J. 1996a. Noise, multistability and delayed recurrent loops. *Physical Review E (Statistical physics, plasmas, fluids, and related interdisciplinary topics)*. Submitted. {*186–190, 195*}

Foss, Jennifer, Longtin, Andre, Mensour, Boualem, and Milton, John. 1996b. Multistability and delayed recurrent loops. *Physical Review Letters*, **76**(4), 708–711. {*186, 189, 193, 195*}

Fromherz, Peter and Stett, Alfred. 1995. Silicon-neuron junction: Capacitive stimulation of an individual neuron on a silicon chip. *Physical Review Letters*, **75**(8), 1670–1673. {*179, 195*}

Furutani, E., Araki, M., Sakamoto, T., and Maetani, S. 1995. Blood pressure control during surgical operations. *IEEE Transactions on Biomedical Engineering*, **42** (10), 999–1006. {*180, 195*}

Garfinkel, A., Spano, M., Ditto, W. L., and Weiss, J. 1992. Controlling cardiac chaos. *Science*, **257**(5074), 1230–1235. {*180, 196*}

Guttman, R., Lewis, S., and Rinzel, J. 1980. Control of repetitive firing in squid axon membrane as a model for a neurooscillator. *Journal of Physiology*, **305**, 377–395. {*188, 196*}

Hopfield, J. J. 1984. Neural networks with graded responses have collective computational properties like those of two-state neurons. *Proceedings of the National Academy of Sciences of the United States of America*, **81**, 3088–3092. {*186, 196*}

Hounsgaard, J., Hultborn, H., Jesperson, B., and Kiehn, O. 1988. Bistability of α-motoneurones in the decerebrate cat and in the acute spinal cat after intravenous 5-hydroxytryptophan. *Journal of Physiology*, **405**, 345–367. {*188, 196*}

Iacono, R. P. 1995. Stimulation of the globus pallidus in Parkinson's disease. *British Journal of Neurosurgery*, **9**, 505–510. {*179, 196*}

Ikeda, I. and Matsumoto, K. 1987. High dimensional chaotic behavior in systems with time-delayed feedback. *Physica D*, **29**, 223–235. {*191, 196*}

Kleinfeld, D., Raccuia-Behling, F., and Chiel, H. J. 1990. Circuits constructed from identified *Aplysia* neurons exhibit multiple patterns of persistent activity. *Biophysical Journal*, **57**(4), 697–715. {*186, 196*}

Kohn, A. F., da Rocha, A. Freitas, and Segundo, J. P. 1981. Presynaptic irregularity and pacemaker inhibition. *Biological Cybernetics*, **41**, 5–18. {*180, 184, 196*}

Kostov, A., Andrews, B. J., Popovic, D. B., Stein, R. B., and Armstrong, W. W. 1995. Machine learning in control of functional electrical stimulation systems for locomotion. *IEEE Transactions on Biomedical Engineering*, **42**(6), 541–551. {*180, 196*}

Longtin, A. and Milton, J. G. 1988. Complex oscillations in the human light reflex with 'mixed' and delayed feedback. *Mathematical Biosciences*, **90**, 183–199. {*181, 196*}

Longtin, A. and Milton, J. G. 1989a. Insight into the transfer function, gain and oscillation onset for the pupil light reflex using delay-differential equations. *Biological Cybernetics*, **61**, 51–58. {*182, 196*}

Longtin, A. and Milton, J. G. 1989b. Modelling autonomous oscillations in the human pupil light reflex using nonlinear delay-differential equations. *Bulletin of Mathematical Biology*, **51**, 605–624. {*181, 183, 196*}

Longtin, A., Milton, J. G., Bos, J. E., and Mackey, M. C. 1990. Noise and critical behavior of the pupil light reflex at oscillation onset. *Physical Review A*, **41**, 6992–7005. {*181–183, 196*}

Losson, J., Mackey, M. C., and Longtin, A. 1993. Solution multistability in first-order nonlinear differential delay equations. *Chaos (Woodbury, NY)*, **3**, 167–176. {*181, 189, 191, 196*}

Lucas, G. 1995. *Return of the Jedi*. New York, NY, USA: Warner Books. {*179, 196*}

Mackey, M. C. and an der Heiden, U. 1984. The dynamics of recurrent inhibition. *Journal of Mathematical Biology*, **19**, 211–225. {*190, 196*}

Mackey, M. C. and Glass, L. 1977. Oscillation and chaos in physiological control systems. *Science*, **197**, 287–289. {*191, 196, 269, 270, 275*}

Mainen, Z. F. and Sejnowski, T. J. 1995. Reliablity of spike timing in neocortical neurons. *Science*, **268**(5216), 1503–1506. {*194, 196*}

Maniglia (ed.), A. J. 1995. Middle and inner ear electronic implantable devices for partial hearing loss. *The Otolaryngologic clinics of North America*, **28**, 1–223. {*179, 196*}

Martinez, O. Diez and Segundo, J. P. 1983. Behavior of a single neuron in a recurrent excitatory loop. *Biological Cybernetics*, **47**, 33–41. {*180, 184, 196*}

Mensour, B. and Longtin, A. 1995. Controlling chaos to store information in delay-differential equations. *Physics Letters A*, **205**(1), 18–24. {*191, 196*}

Middlebrooks, A. E., Xu, L., and Green, D. M. 1994. A panoramic code for sound location by cortical neurons. *Science*, **264**(5160), 842–844. {*194, 197*}

Milton, J., Campbell, S. A., and Bélair, J. 1995. Dynamic feedback and the design of closed-loop drug delivery systems. *Journal of biological systems*, **3**, 711–718. {*180, 197*}

Milton, J., Bayer, W., and an der Heiden, U. 1996. Modeling the pupil light reflex with differential delay equations. *Zeitschrift für Angewandte Mathematik und Mechanik*. In press. {*192, 197*}

Milton, J. G., Longtin, A., Kirkham, T. H., and Francis, G. S. 1988. Irregular pupil cycling as a characteristic abnormality in patients with demyelinative optic neuropathy. *American Journal of Ophthalmology*, **105**, 402–407. {*180, 181, 184, 197*}

Milton, J. G., Longtin, A., Beuter, A., Mackey, M. C., and Glass, L. 1989. Complex dynamics and bifurcations in neurology. *Journal of Theoretical Biology*, **138**, 129–147. {*181, 197*}

Milton, J. G., an der Heiden, U., Longtin, A., and Mackey, M. C. 1990. Complex dynamics and noise in simple neural networks with delayed mixed feedback. *Biomedica Biochimica Acta*, **49**, 697–707. {*190, 197*}

Milton, John G. and Longtin, Andre. 1990. Evaluation of pupil constriction and dilation from cycling measurements. *Vision Research (Oxford)*, **30**(4), 515–526. {*181–185, 197*}

Mirouze, J., Selam, J. L., Pham, T. C., and Cavadore, D. 1977. Evaluation of exogenous insulin homeostasis by the artificial pancreas in insulin-dependent diabetes. *Diabetologia*, **13**, 273–278. {*180, 197*}

Ott, E., Grebogi, C., and Yorke, J. A. 1990. Controlling chaos. *Physical Review Letters*, **64**, 1196–1199. {*180, 197*}

Petrofsky, J. S. and Phillips, C. A. 1985. Closed-loop control of movement of skeletal muscle. *CRC critical reviews in biomedical engineering*, **13**, 35–96. {*180, 197*}

Pinsker, H. M. 1977. Aplysia bursting neurons as endogenous oscillators: I: Phase-response curves for pulsed inhibitory synaptic input. *Journal of Neurophysiology*, **40**, 527–543. {*186, 197*}

Reulen, J. P. H., Marcus, J. T., van Gilst, M. J., Koops, D., Bos, J. E., Tiesinga, G., de Vries, F. R., and Boshuizen, K. 1988. Stimulation and recording of dynamic pupillary reflex: the IRIS technique. *Medical & biological engineering & computing*, **26**, 27–32. {*180, 181, 197*}

Richardson, D. E. and Goreck, J. R. 1996. Deep brain stimulation for pain relief. *Pages 4021–4027 of:* Wilkins, Robert H. and Rengachary, Setti S. (eds), *Neurosurgery*, vol. III. New York, NY, USA: McGraw-Hill. {*179, 197*}

Schiff, S. J., Jerger, K., Duong, D. H., Chang, T., Spano, M. L., and Ditto, W. L. 1994. Controlling chaos in the brain. *Nature*, **370**(6491), 615–620. {*180, 197*}

Sharp, A. A., Abbott, L. F., and Marder, E. 1992. Artificial electrical synapses in oscillatory networks. *Journal of Neurophysiology*, **67**(6), 1691–1694. {*180, 184, 197*}

Sharp, A. A., O'Neil, M. B., Abbott, L. F., and Marder, E. 1993a. Dynamic clamp: Computer-generated conductances in real neurons. *Journal of Neurophysiology*, **69**(3), 992–995. {*180, 184, 197*}

Sharp, Andrew A., O'Neil, Michael B., Abbott, L. F., and Marder, Eve. 1993b. The dynamic clamp: artificial conductances in biological neurons. *Trends in Neurosciences (Regular ed.)*, **16**(10), 389–394. {*180, 184, 197*}

Siegfried, J. 1994. Bilateral chronic electrostimulation of ventroposteriolateral pallidum: a new therapeutic approach for alleviating all parkinsonian symptoms. *Neurosurgery*, **35**(6), 1126–1129. {*179, 197*}

Stark, L. 1959. Stability, oscillations, and noise in the human pupil servomechanism. *Proceedings of the Institute of Radio Engineers*, **47**, 1925–1939. {*181, 192, 197*}

Stark, L. 1962. Environmental clamping of biological systems: Pupil servomchanism.

Journal of the Optical Society of America. A, Optics, image science, and vision, **52**, 925–930. {*180, 181, 197*}

Stark, L. and Sherman, P. M. 1957. A servoanalytical study of the consensual pupil reflex to light. *Journal of Neurophysiology*, **20**, 17–26. {*181, 198*}

Vibert, J. F., Davis, M., and Segundo, J. P. 1979. Recurrent inhibition: Its influence upon transduction and afferent discharges in slowly-adapting stretch receptor organs. *Biological Cybernetics*, **33**, 167–178. {*180, 184, 198*}

Wilder, B. J. 1990. Vagus nerve stimulation for the control of epilepsy. *Epilepsia*. Supplement 2. {*179, 198*}

Winfree, Arthur T. 1980. *The Geometry of Biological Time*. Biomathematics, vol. 8. Berlin, Germany / Heidelberg, Germany / London, UK / etc.: Springer Verlag. ISBN 0-387-09373-7 (New York). 3-540-09373-7 (Berlin).   Pages xiii + 530. {*188, 198, 255–259, 270, 275*}

Yarom, Y. 1991. Rhythmogenesis in a hybrid system interconnecting an olivary neuron to an analog network of coupled oscillators. *Neuroscience*, **44**(2), 263–275. {*180, 184, 198*}

Zhang, X., Ashton-Miller, J. A., and Stohler, C. S. 1993. A closed-loop system for maintaining constant experimental muscle pain in man. *IEEE Transactions on Biomedical Engineering*, **40**(4), 344–352. {*180, 198*}

Zipser, D., Kehoe, B., Littlewort, G., and Fuster, J. 1993. A spiking network model of short-term memory. *Journal of Neuroscience*, **13**(8), 3406–3420. {*186, 198*}

10. CALCIUM AND MEMBRANE POTENTIAL OSCILLATIONS IN PANCREATIC β-CELLS

Arthur Sherman

10.1 INTRODUCTION

We study pancreatic β-cells because of their role in diabetes, a deadly derangement of carbohydrate and lipid metabolism. β-cells are found in micro-organs in the pancreas called islets of Langerhans. Each islet contains of the order of 10^3 β-cells and is up to 0.5 mm in diameter, and there are of the order of 10^6 islets in the pancreas. The β-cell's function is to secrete insulin in response to elevations of glucose in the blood plasma, such as those occurring after a meal. Insulin signals target tissues (muscle, liver, and adipose cells) indicating that glucose is available for use as a fuel or for storage as glycogen or fat. Later, as plasma glucose declines, insulin secretion returns to basal levels, and cells switch back to using stored carbohydrate, fat, or protein as a fuel. Thus, insulin and glucose form a classical negative-feedback loop like a thermostat.

If no insulin is produced, glucose rises to very high levels, and the unrelieved reliance on fat and protein for fuel leads to acidification of the blood and eventual death. This is a result of juvenile (Type I) diabetes, in which the β-cells are destroyed by an autoimmune response. In the more common maturity-onset (Type II) diabetes, chronic low levels of insulin result in elevated glucose levels that cause kidney failure, heart disease, blindness, and premature death. Type II diabetics also exhibit insulin resistance in the target tissues: these tissues do not respond effectively to the signal that the insulin gives. The relative importance of the defects in insulin secretion versus the insulin action in the etiology of Type II diabetes is debated, but most authorities agree that both factors play a role (Kahn and Porte, Jr. 1990).

Type I diabetes is treated by insulin injection, which prevents immediate demise but is a poor substitute for the fine minute-to-minute regulation of a normal pancreas. If Type II diabetes cannot be reversed by diet and exercise, it is sometimes treated by sulfonylurea drugs that enhance insulin secretion and may also ameliorate insulin resistance, and sometimes by insulin injection. The results for both forms of the disease still leave much to be desired. It is hoped that better understanding of the basic mechanisms of insulin secretion and its regulation will lead to better treatment, earlier diagnosis, and prevention of diabetes.

Models for β-cells, like numerous models for cardiac smooth muscle and other excitable cells, are based on the Hodgkin-Huxley equations for neuronal electrical activity (Hodgkin and Huxley 1952). (See Rinzel (1990) and Rinzel and Ermentrout (1989) for a modern perspective.) We will employ a simplified version in the spirit of Morris and Lecar (Rinzel and Ermentrout 1989) with elaborations to account for other aspects of cell biology, including intracellular Ca^{2+} handling, glucose metabolism, and hormonal signaling. These effects will be treated minimally, as parameters that modify ionic fluxes. We will also incorporate gap-junctional coupling of the β-cells. Coupling not only coordinates electrical activity, as in myocardium and other smooth muscle, but modifies the electrical properties.

The feature of β-cell electrical activity on which we will focus most of our attention is bursting, a slow alternation between active (spiking) and silent states (Figure 10.1). Changes in plasma glucose modulate the burst pattern, notably the relative duration of the active and silent phases. Bursting is also ubiquitous in neurons (Wang and Rinzel 1995) and can be thought of as a way to build slow-time-scale (seconds) integrative behavior out of the fast time scales (milliseconds) of channel kinetics and action potentials. We will discuss mechanisms for the generation of bursting and its modulation by glucose metabolism and other biochemical processes. The β-cell models will be seen as exemplars of one form of bursting in a family of patterns that encompasses many neuronal models and, abstractly, a wealth of low-dimensional, singularly perturbed systems.

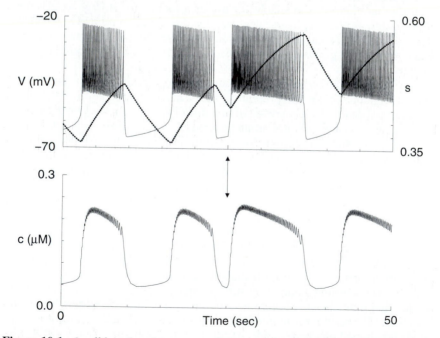

Figure 10.1. β-cell bursting time-course. Top: v solid; s dotted. Glucose increased at arrow by reducing $g_{K(ATP)}$ from 120 to 100 pS. Bottom: Cytosolic calcium concentration.

The following model, which is representative of the great variety of β-cell models (Bertram et al. 1995a; Chay and Keizer 1983; Chay et al. 1995; Keizer and Smolen 1991; Smolen and Keizer 1992), will elucidate β-cell issues and illustrate general mechanisms of cellular electrical activity that are relevant for a wide variety of systems.

$$c_m \frac{dv}{dt} = -I_{\text{ion}}(v, n, s) \tag{10.1}$$

$$= -I_{\text{Ca}}(v) - I_{\text{K(V)}}(v, n) - I_{\text{slow}}(v, s),$$

$$\frac{dn}{dt} = \lambda \left[\frac{n_\infty(v) - n}{\tau_n} \right], \tag{10.2}$$

$$\frac{ds}{dt} = \frac{s_\infty(v) - s}{\tau_s}. \tag{10.3}$$

Here v is the voltage across the cell membrane and n and s are the fraction of the associated ion channel type that is open. The current I_{Ca} (L-type) arises from the inward movement of Ca^{2+}, the current $I_{\text{K(V)}}$ (v-dependent delayed-rectifier) arises from the outward movement of K^+ gated by n, and the current I_{slow} is slowly modulated by its gating variable, s. For v fixed, the gating variables n and s exponentially relax to v-dependent steady-state values, $n_\infty(v)$ and $s_\infty(v)$, with time constants τ_n and τ_s. (In many models the time constants are also v-dependent, but they need not be for our purposes.) The calcium current I_{Ca} is fast relative to $I_{\text{K(V)}}$ and is modeled as instantaneously dependent on v. The time scale of v is determined by the ratio of the typical total conductance and the capacitance, c_m. Here, it is in the range of tens of milliseconds, comparable to τ_n and much faster than τ_s. Thus, the system can be decomposed into a fast v-n subsystem and the slow s equation.

This system contains the minimal features needed to generate bursting oscillations. The fast variables, v and n, generate the spikes during the active phase of a burst, while s is responsible for switching between active and silent phases. Glucose metabolism and other features will be added as we go along. Here, I_{slow} is an inhibitory, K^+ current, but it could just as well conduct Ca^{2+} or a mixture of ions. See the Appendix for details of formulas and parameters. When the voltage is increased, we say that the system is "depolarized," and when it is decreased, we say that the system is "hyperpolarized."

10.2 PHASE PLANES: FAST OSCILLATIONS

We begin with a phase-plane analysis (Appendix B) of (10.1) and (10.2) for v and n with s fixed and used as a bifurcation parameter. The first step is to find conditions that generate instability and the oscillations that will provide the spikes of the active phase of a burst.

The steady states of the system are determined by the zero-crossings of

$$I_{\text{SS}}(v; s) = I_{\text{ion}}(v, n_\infty(v); s) = 0. \tag{10.4}$$

Figure 10.2 shows I_{SS} vs. v for several values of s. The characteristic N-shape of I_{SS} comes from the interaction of I_{K}, which is positive, and I_{Ca}, which is negative, in the interval $v \in [v_{\text{K}}, v_{\text{Ca}}]$ (Appendix). When s is not too small, raising v from v_K first makes I_K and I_{SS} grow. Then, $m_\infty(v)$ turns on, $I_{Ca} = g_{Ca}m_\infty(v)(v - v_{Ca})$ becomes more negative, and the total current decreases. Eventually, however, I_K must dominate because $v - v_{Ca} \to 0$.

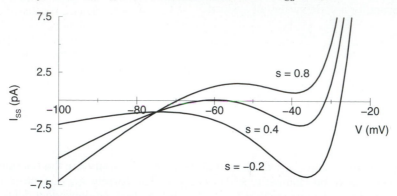

Figure 10.2. I-V curves: $I_{SS}(v; s)$ vs. v for the indicated values of s (10.4).

Figure 10.3 shows the nullclines corresponding to Figure 10.2. Note that changes in s translate the v nullcline up and down, because I_{slow} and $I_{K(V)}$ have the same reversal potential. Planar systems can go unstable, when either the determinant of the Jacobian changes sign (saddle-node bifurcation — SN) or the trace changes sign (Hopf bifurcation — HB) (Appendix B). Both happen here. Figures 10.2 and 10.3 imply that there is an SN for $s \in [-0.2, 0.4]$ and one for $s \in [0.4, 0.8]$, when the v and n nullclines become tangent.

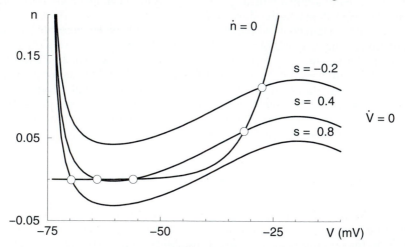

Figure 10.3. Nullclines for fast subsystem (10.1, 10.2) with s fixed. Opaque dots indicate steady states.

For an HB, the n nullcline must intersect the middle branch of the v nullcline with greater slope (Appendix B). This is not sufficient, however. When $s = -0.2$, the (unique) steady state is on the middle branch (Figure 10.3), but it is a stable spiral. By $s = 0.4$, an HB has occurred, and this steady state (now the uppermost of three) has become an unstable spiral. The lowest steady state is a stable node, and the middle is a saddle (Appendix B).

By $s = 0.8$, only the low-voltage steady state remains. A closer look at the phase portrait for $s = 0.4$ (Figure 10.4), including the trajectory and the invariant sets of the saddle, shows how the oscillations die. One unstable manifold (u_1) of the saddle wraps around the limit cycle (LC). Outside the LC is stable manifold s_1. As s is increased, the LC shifts down and to the left and simultaneously merges with u_1, s_1, and the saddle for a unique value of s, $s_{SL} \approx 0.47$, creating an infinite period *homoclinic* orbit or saddle loop (SL). As s increases past s_{SL}, the stable and unstable manifolds cross, with u_1 now lying outside s_1. The system remains excitable but is no longer oscillatory. A topological sketch of these transitions is shown in Figure 10.5.

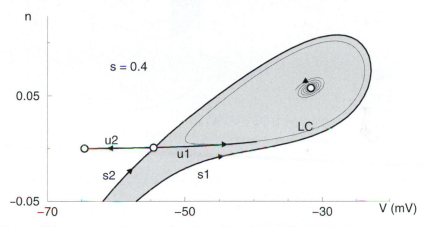

Figure 10.4. Phase plane showing bi-stability between limit cycle (LC) and steady state. All initial conditions in shaded region go to LC; the rest go to the lower steady state.

Figure 10.5. Passage through homoclinic bifurcation.

Figure 10.6 is a bifurcation diagram summarizing the range of behaviors obtained above by varying s. The limit-cycle branch (LC) is born at an HB and dies at an SL (or vice versa). For $s \in [s_{SN}, s_{SL}]$, the system is bistable, with phase planes like that of Figure 10.4.

10.3 SINGULAR PERTURBATION: BURSTING

To realize bursting we can exploit the bistability between spiking and steady-state behavior and add slow s dynamics (10.3). Figure 10.6 shows an

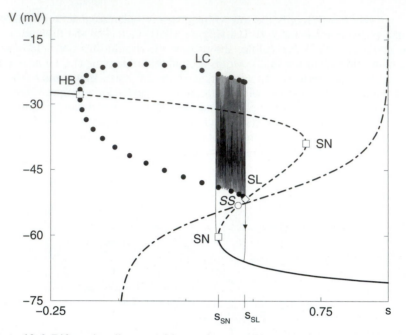

Figure 10.6. Bifurcation diagram with superimposed burst trajectory. On Z-curve, $\dot{v} = \dot{n} = 0$; on dot-dashed sigmoid curve, $\dot{s} = 0$.

overlay of the s nullcline and the burst trajectory on the bifurcation diagram. The Z-shaped curve is now viewed as the slow manifold of the combined, singularly perturbed system, or as the v nullcline in the v-s pseudophase plane. Since s is inhibitory here, all we need do is have s increase for large v and decrease for small v. The active phase ends when s goes through SL; the silent phase ends when s goes through SN. Biophysically, the active phase ends essentially when the spike minimum goes below the rising threshold (the middle, saddle branch of the Z-curve). The bursting, 3-variable system, like the v-n system, oscillates by negative feedback. The v-n subsystem is very fast relative to s, however, so that it is always in quasi-steady state (spiking or silent), except during the brief transitions.

10.3.1   *β*-Cell

We are now ready to incorporate the effects of glucose. At low glucose (up to 7 mM) $β$-cells are quiescent. At high glucose (above 20 mM) they spike continuously. In between those extremes, they burst, and the fraction of time spent in the active phase (*plateau fraction*) increases. Moreover, within the burst regime the spike amplitude and silent-phase potential do not change significantly. The Z-curve picture suggests that all these features can be reproduced by shifting the full-system steady state (the intersection of the s nullcline and the Z-curve (SS in Figure 10.6)). When SS is on the lower branch of the Z-curve, the system is silent; when SS is on the middle branch, the system bursts; and when SS is on the LC branch, the system spikes continuously. For SS on the middle branch, the plateau fraction is low when SS is near SN, because the trajectory slows down leaving the silent-phase and increases monotonically as SS approaches SL. Since these changes do not affect the fast dynamics, the spike amplitude and silent phase potential are invariant throughout the bursting regime.

A biophysical mechanism that moves SS in the desired way is the K-ATP channel, which transduces the glucose-dependent metabolic rate into K^+ conductance. This channel is blocked when it binds adenosine triphosphate (ATP), while adenosine triphosphate (ADP) binding prevents the block. Therefore, its conductance decreases when glucose increases, which increases the ratio of ATP to ADP. The K-ATP channel is also important as the site of action of the sulfonylurea drugs, which block the channel independent of glucose. Therefore, we add a further current term, $-I_{K(ATP)} = -g_{K(ATP)}(v - v_K)$, to (10.1). Geometrically, increasing $g_{K(ATP)}$ shifts the Z-curve in Figure 10.6 to the right. The right portion of Figure 10.1 shows the increase in plateau fraction with no change in spike amplitude when $g_{K(ATP)}$ decreases.

We have made splendid progress analyzing the qualitative behavior of the dynamical system without committing ourselves to a biophysical meaning for s. Numerous $β$-cell models have been published (Chay and Keizer 1983; Keizer and Smolen 1991; Smolen and Keizer 1992) exploring various possibilities for this and also for the glucose-sensing mechanism, but these issues remain unresolved. In the first $β$-cell model (Chay and Keizer 1983) the role of s was played by $c = [Ca^{2+}]_i$, the concentration of free cytoplasmic Ca^{2+}, and the glucose sensor was k_c, the rate of Ca^{2+} removal by pumps and exchangers. The Ca^{2+} balance equation had the form

$$\frac{dc}{dt} = f \left[\alpha I_{Ca}(v) - k_c c \right], \tag{10.5}$$

where α is a factor to convert current to concentration changes, and f is the fraction of free cytoplasmic Ca^{2+}. Since most of the Ca^{2+} that enters the cells is rapidly bound to proteins, f is small and c was a plausible candidate slow variable.

The hypothesis of negative feedback through $[Ca^{2+}]_i$, first proposed by Atwater, Rojas, and colleagues (1980), was appealing because of the known existence of a Ca^{2+}-activated K^+ channel in $β$-cells. The model led to a testable (falsifiable) prediction, that Ca^{2+} would show a sawtooth oscillation, like s in Figure 10.1 (top). This prediction was indeed falsified when the $[Ca^{2+}]_i$ was found by fluorescence measurements to have a time course that was closer to a square wave (Valdeolmillos et al. 1989). It turns out that $[Ca^{2+}]_i$ is not quite slow enough to pace bursts with a period of tens of seconds because f is closer to 0.01 than the 0.001 value needed. Appending (10.5) to our generic model and using the larger value for f, we obtain a roughly square-wave c time course (Figure 10.1, bottom).

Although $[Ca^{2+}]_i$ is ruled out as the slow variable for $β$-cell bursting, it still provides a plausible burst mechanism for neurons that can have burst frequencies of 10 Hz or more, faster than $β$-cell *spike* frequency. The difficulty in finding negative-feedback mechanisms that operate on the long time scale of $β$-cell bursts has been a major barrier to resolving the mechanism. The least problematic candidate currently is the slow, voltage-dependent inactivation of the Ca^{2+} channels observed in patch-clamp experiments on isolated $β$-cells by

Satin and Cook (1989). For recent reviews, see Chay et al. (1995), Satin and Smolen (1994), and Sherman (1995).

The importance of Ca^{2+} is the link it provides between electrical activity and insulin secretion. Glucose concentration is transduced through metabolism and the relative levels of ATP and ADP into $g_{K(ATP)}$ conductance (Appendix), which determines the plateau fraction. Since the silent and active phase levels of c, like those of v, are nearly invariant within the bursting regime, $g_{K(ATP)}$ also determines the mean Ca^{2+} concentration, averaged over many bursts. This suggests that the secretory machinery is slow and responds to average, rather than instantaneous, c. This story may be incomplete, as there is evidence that both electrical activity and secretion are regulated by factors other than $g_{K(ATP)}$.

In addition to its biophysical successes, this family of models has generated a great deal of mathematical activity that we can only point to. Miura and colleagues have used Melnikov's method to calculate semianalytically the location of homoclinic orbits (Pernarowski et al. 1992). Combined with the method of averaging, this gives an efficient way to calculate the plateau fraction and its dependence on putative glucose-sensing parameters. Terman (1991) has proved that bursting solutions exist using geometric singular perturbation methods borrowed from nonlinear wave propagation theory. The proof also confirmed a conjecture of Rinzel that in the limiting case $\tau_s \to \infty$, continuous spiking ensues precisely when the full system steady state ($\mathcal{SS}$) coincides with SL. When τ_s is finite, chaotic bursting and spiking occur (Chay et al. 1995; Terman 1992). Chaos can also occur during the transition from N to $N + 1$ spikes per burst (Terman 1991).

10.3.2   Other types of bursters

So far, we have varied only one parameter of the fast subsystem, s. By varying a second parameter we unveil a large family of topologically distinct bursters. Some of these have large-amplitude spikes and look more like neuronal bursters. A convenient choice for the second parameter is λ. Because λ modifies only the time constant of n, it does not change the shape of the Z-curve, only the stability of the steady states and the characteristics of the periodic branch(es). Increasing λ causes the LC amplitude to decrease until no oscillations exist for the fast subsystem; the full system is then reduced to a relaxation oscillator (Appendix). Biophysically, the oscillations result from the slow response of I_K, and if I_K is activated as rapidly as I_{Ca}, the opposite ion fluxes cancel. Decreasing λ, on the other hand, facilitates the emergence of Hopf bifurcation and increases the amplitude of the oscillations. The bifurcation diagram smoothly changes, with both HB and SL moving to the left (Figure 10.7). The range of s values traversed during a burst, and hence the burst period, decreases (see horizontal cut labeled Type Ia in the two-parameter bifurcation diagram, Figure 10.8). The range of $g_{K(ATP)}$ values that supports bursting also shrinks.

The SL eventually merges with the SN at a saddle-node loop (SNL) (Figure 10.8), and approximately curve C, Figure 10.7. This is a codimension-two bifurcation because two constraints must be satisfied (Appendix B), but the homoclinic orbit will generally persist at the knee for a finite interval of

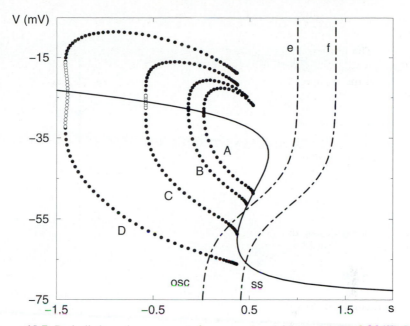

Figure 10.7. Periodic branches grow as λ decreases. λ = 1.05 (A), 1.0 (B), 0.86 (C), 0.6 (D). Open circles indicate an unstable portion of the branch and closed circles indicate a stable portion of the branch. Curves *e* and *f* are projected *s* nullclines. See text in Section 10.5.

λ values as the minimum *v* continues to drop (curve *D*, Figure 10.7). This is called a *saddle node on an invariant circle* or SNIC. The fast subsystem is no longer bistable and bursting can no longer occur by the mechanism of Section 10.3.1. If the slow nullcline is like curve *f*, Figure 10.7, the full system has a stable steady state on the lower branch of the Z-curve, while with nullcline *e*, a continuous spiking solution is the global attractor. One could get bursting in this case by adding a second slow variable that would have the effect of sweeping the *s* nullcline back and forth across the SNIC (cut labeled Type II in Figure 10.8) or, equivalently, shifting the Z-curve. A necessary condition to avoid getting stuck in either the silent or oscillatory region is that the second variable be excitatory, rather than inhibitory. One can predict that firing frequency will be low at both the beginning and the end of the active phase as the trajectory passes near the SNIC. This is called parabolic bursting because of the shape of the spike frequency profile, first realized in a model of Plant (1978) for the R-15 neuron. There the inhibitory slow variable was Ca^{2+} acting on a K^+ channel and the excitatory slow variable was a slowly activated I_{Ca}. Rinzel and colleagues (Baer et al. 1995; Rinzel and Lee 1987) have analyzed this situation by constructing the averaged phase-plane dynamics for the slow variables.

Also, note that as λ decreases, the HB in Figure 10.7 becomes subcritical (Appendix B). This is another form of bistability in which the limit cycle surrounds a stable steady state, with an unstable limit cycle as separatrix (Appendix B). Bursting (Type III) can also happen in this situation, with a single slow variable (Wang and Rinzel 1995). Type III, unlike I and II, does not require N-shaped I_{SS}. An alternative way to get Type III bursting is to desta-

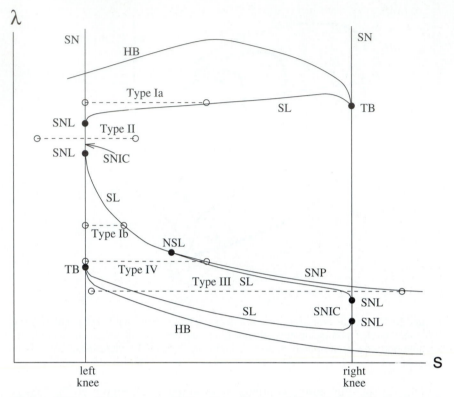

Figure 10.8. Two-parameter bifurcation diagram (adapted from Bertram *et al.* (1995b)). SN = Saddle-Node; HB = Hopf Bifurcation; SL = Saddle-Loop; SNL = Saddle-Node-Loop; SNIC = Saddle-Node-on-Invariant Circle; NSL = Neutral Saddle-Loop; TB = Takens-Bogdanov bifurcation;SNP = Saddle-Node of Periodics.

bilize the bottom branch of the Z-curve by further reducing λ. See the bottom portion of Figure 10.8. This scenario is found in a lobster cardiac ganglion model (Av-Ron et al. 1993).

Our minimal model has gone far indeed, and, although Figure 10.8 may not be quite the end of the line (G. deVries, manuscript in preparation), future developments will probably come from more complex models. The first generation models are all essentially modifications of the primeval cubic, the van der Pol equation, to include an oscillatory excited state. However, as modelers try to keep up with the electrophysiologists, models with more currents and variables (10 or more) are emerging. In some cases, cells seem to redundantly parameterize simple behaviors, perhaps due to genetic constraints, but new behaviors and coexistence of different old behaviors also result. Here are some examples worth investigating. The thalamic neuron model of Rush and Rinzel (1994) can exhibit both subthreshold oscillations and bursting with spikes, owing to a quintic slow manifold. A recent R-15 neuron model (Canavier et al. 1993) exhibits multistability of burst and spike patterns. The analysis by Smolen et al. (1993) of a β-cell-derived model with two *inhibitory* slow variables is interesting for both the results and the techniques used. Finally, some models of bursting arise from interaction of multiple neurons in a network (Rinzel and Frankel 1992) or spatially segregated compartments in a single neuron (Pinsky and Rinzel 1994).

10.4   ROLE OF INTERNAL Ca²⁺ STORES

Although glucose is the primary stimulus for β-cells, secretion can be potentiated by acetylcholine (ACh), provided adequate glucose is present. The *in vivo* signal originates in cognitive stimulation via the hypothalamus (e.g., seeing food or knowing that it is lunchtime) and is transmitted to the islet by the vagus nerve. This preempts excessive rises in plasma glucose by increasing insulin in advance. *In vitro* application of ACh leads to depolarization (Santos and Rojas 1989) and increased $[Ca^{2+}]_i$ (Bertram et al. 1995a). Unlike increased glucose, ACh increases the absolute levels of both v and $[Ca^{2+}]_i$, rather than the plateau fraction (Figure 10.1). In the absence of glucose, ACh has little effect on v and only transiently increases $[Ca^{2+}]_i$.

ACh works by binding to muscarinic receptors on the β-cell plasma membrane, leading to the production of inositol 1,4,5-trisphosphate (IP_3) and diacylglycerol (DAG). The latter activates protein kinase C (PKC), which appears to sensitize the secretory machinery to Ca^{2+} but does not affect electrical activity and will not be considered here. IP_3 diffuses to the endoplasmic reticulum (ER) where it activates ligand-gated Ca^{2+} channels, releasing Ca^{2+} into the cytosol.

We augment our simple model with ER equations from a model for pituitary gonadotrophs (Li and Rinzel 1994) where IP_3 mediates $[Ca^{2+}]_i$ oscillations. We add fluxes into (J_{in}) and out of (J_{out}) the ER to (10.5) for $[Ca^{2+}]_i$ and add an equation for ER calcium concentration, c_{ER}:

$$\frac{dc}{dt} = f\left[\alpha I_{Ca}(v) - k_c c\right] + \frac{1}{\mu}\left(J_{out} - J_{in}\right) \tag{10.6}$$

$$\frac{dc_{ER}}{dt} = \frac{1}{\mu\sigma}\left(-J_{out} + J_{in}\right) \tag{10.7}$$

The factor σ accounts for the difference in ER and cytosolic volumes and Ca^{2+} buffering capacity, and μ sets the ER time scale. See the Appendix for details.

With the gonadotroph parameters, oscillations are generated by the combination of rapid activation and slow inhibition of the IP_3 receptors by Ca^{2+}. See also Chapter 6, this volume. Here, ER-driven oscillations are unimportant, so we choose parameters where they do not occur.

Two new ionic currents are needed to link the events at the ER to the membrane potential: Ca^{2+}-activated K^+ channel, I_{K-Ca}, which is known to exist in β-cells, and a calcium-release-activated-current, I_{CRAC}, whose conductance increases as the ER empties and whose existence is less clearly established by direct measurement. The power of I_{CRAC} to explain a variety of phenomena, however, provides circumstantial evidence for its existence.

In Figure 10.9, ACh is added during normal glucose-induced bursting. The complex biochemistry above is reduced to an increase in IP_3. First the bursting is interrupted by release of Ca^{2+}, which activates I_{K-Ca}. This is followed by gradual depolarization, ending in high-frequency, "muscarinic" bursting with depolarized silent phases. The depolarization results from both the removal of Ca^{2+} from the cell, reducing $I_{K\text{-}Ca}$, and the depletion of Ca^{2+} in the ER, increasing the magnitude of I_{CRAC}. The agreement between the simulations and experiment (Bertram et al. 1995a; Santos and Rojas 1989), including the complicated transients, supports the hypothesis that ACh works through I_{CRAC}.

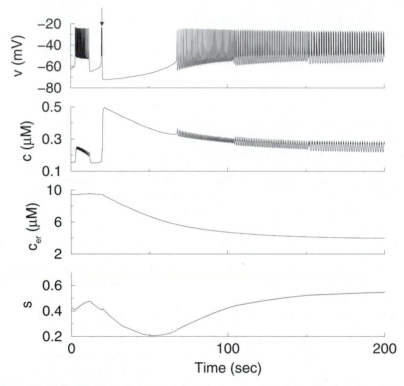

Figure 10.9. Simulation of addition of ACh at arrow by raising IP$_3$ from 0 to 0.6μM.

The model also makes strong predictions about so-called "biphasic" transients when glucose is raised above the threshold for bursting, in which a minute or more of intense spiking is seen before steady-state bursting begins. The model (Figure 10.10) suggests that the ER is depleted and I_{CRAC} is turned on in low glucose. Activity is blocked, however, because the larger $I_{K(ATP)}$ is also on. This also explains why ACh has little effect on electrical activity in the absence of glucose. Glucose relieves this block by turning off $I_{K(ATP)}$, but bursting cannot begin until the ER fills sufficiently to suppress I_{CRAC}. The model is consistent with experiments showing that when Ca^{2+}-uptake into the ER is blocked with thapsigargin, preventing ER refilling, the initial transient does not progress to bursting (Worley, III et al. 1994).

Now, even in the absence of an ER, abrupt changes in $g_{K(ATP)}$ can jerk the slow manifold in the v-s pseudo phase plane, in a manner similar to post-inhibitory rebound (Rinzel and Ermentrout 1989), and result in a longer first burst. This is seen when raising glucose in Figure 10.1. However, to get a transient that is significantly longer than the burst period, one needs a process that is significantly slower than the slow variable. Any inhibitory process that is

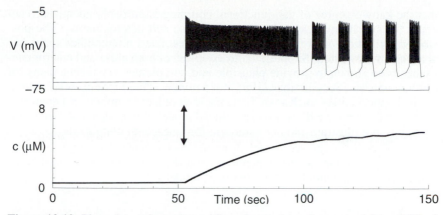

Figure 10.10. Phase 1 transient when adding glucose (stepping $g_{K(ATP)}$) from 5000 pS to 120 pS at arrow. (From Bertram et al. (1995a).)

activated by high glucose or electrical activity will do; the model shows that I_{CRAC} can do the job. In contrast, ACh needs an excitatory process that is correlated with ER dumping (e.g., a channel gated directly by ACh, a G-protein, the rise in $[Ca^{2+}]_i$, or, as here, the ER Ca^{2+} concentration). Elegantly, I_{CRAC} can play both roles in different circumstances.

Of mathematical interest, the muscarinic bursts in the right portion of Figure 10.9 are no longer driven by s, which is nearly constant, but rather by oscillations of c acting through g_{K-Ca}. Although c is too fast to drive glucose-induced bursting, it is just slow enough to drive the more rapid muscarinic bursting. The poor separation of time scales between c and the v-n subsystem actually helps the performance of the model. The silent-phase depolarization is enhanced because the trajectory does not go all the way down to the bottom branch of the Z-curve. From Terman's analysis of chaotic bursting one would also expect to find chaos more readily when the slow variable is not very slow. This is indeed the case in the model and perhaps also in experiments with ACh (Santos and Rojas 1989).

10.5   ELECTRICAL COUPLING

Electrical coupling by gap-junctions within the islets of Langerhans is important for two reasons. First, it synchronizes the activity of the β-cells. Thus, our simple model really describes the behavior of one representative cell in a synchronized population. Second, in isolated cells one rarely observes bursting, but more commonly erratic spiking. This has led to several hypotheses for how bursting might be an emergent phenomenon dependent on coupling. The noise hypothesis (Atwater et al. 1983) proposes that single-channel fluctuations are significant because of low channel densities and disrupt bursting in single cells, but are damped in islets by the conductance load of the network. The heterogeneity hypothesis holds that individual cells are unlikely to have parameters that fall in the narrow bursting regime, whereas islets effectively average the parameters.

Computational studies involving hundreds of cells have established that the known degree of gap-junctional coupling is probably adequate to account for synchrony, taking into account noise and heterogeneity. The possible contribution of diffusion of K$^+$ in the restricted intercellular space has also been investigated. Suppression by coupling of both noise and heterogeneity has been shown to provide plausible and complementary mechanisms for rhythmogenesis. See Sherman (1995) for a review. These models offer mathematical opportunities, such as studying the laws of large numbers in a nonlinear context, but here we will examine a reduced model of two identical, deterministic cells and illustrate another paradigm for emergent bursting.

Gap junctions are modeled, as in other electrically coupled tissues such as myocardium, as v-independent conductances between pairs of cells. The membrane potentials v_i for a two-cell model satisfy

$$c_m \frac{dv_i}{dt} = -I_{\text{ion}}(v_i, n_i, s_i) - g_c(v_i - v_j) \tag{10.8}$$

where I_{ion} includes all the currents in (10.1).

Weakly coupled oscillators generically have the possibility to oscillate out of phase. When the coupling is diffusive, as here, the out-of-phase solutions tend to have smaller amplitude, because each cell is pulled away from its extreme values by the other. This simple effect has two consequences for Type I bursters. The top panel of Figure 10.11 shows that coupling can substantially increase burst period. The bottom panel shows that coupling can convert continuously spiking (beating) cells into bursters.

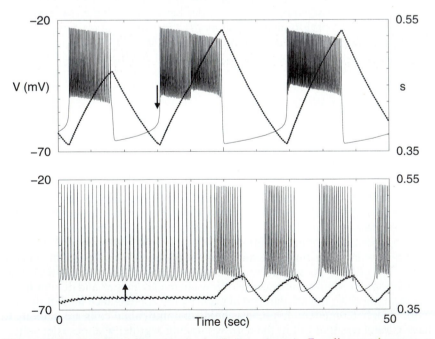

Figure 10.11. Coupling of two identical cells (one shown). Coupling conductance g_c increased from 0 to 10 pS at arrows. Top: $\lambda = 1.0$; bottom: $\lambda = 0.86$.

To explain this, we make Figure 10.7 do double duty. For the period increase, suppose that the LC branch labeled B applies to an isolated cell and that curve e is its s nullcline. This cell can, of course, burst. The same diagram can be used for the coupled pair, because the cells are identical and their s values are approximately the same. Branch B now represents the solution where the two cells are in phase (IP) and have the same time course as when uncoupled; the IP solution is always a solution of the coupled equations, but it may not be stable. For weak coupling the bifurcation diagram of our model includes a branch like A of antiphase (AP) oscillations (180° out of phase). This results from a second HB on the upper branch. That is, a second pair of eigenvalues will cross the imaginary axis, retarded by an amount that increases with g_c. Note that its amplitude is smaller, and hence, that its SL lies further to the right than that of B. Moreover, in the region near the left knee, A is usually stable while B is unstable. Therefore, the burst trajectory will follow the antiphase branch and will tend to have a larger period because the s amplitude is larger. As g_c increases, one expects the cells to synchronize in phase, and in fact the AP branch shrinks and loses stability while the IP branch regains stability. For details of this complex transition, which goes through quasi-periodic and asymmetric periodic solutions, see Sherman (1994).

For the beat-to-burst case, suppose that branch C in Figure 10.7 corresponds to the uncoupled and IP solutions, B to the AP, and e again to the s nullcline. Without coupling, the trajectory is trapped on branch C, and the cell spikes continuously. With weak coupling, it can happen that branch C is unstable near the intersection with nullcline e, while branch B is stable. Then the two cells will burst with antiphase spikes. This arrangement is delicate and breaks down as g_c is increased even modestly, because the IP branch becomes stable and trajectories coming off the bottom branch of the Z-curve are trapped. However, if there is still a stable out-of-phase branch, adding noise paradoxically restores bursting at the high coupling strength, because the trajectory is perturbed onto the out-of-phase branch. Simulations (unpublished) suggest that this can be a robust mechanism for emergent bursting in large networks with channel noise where none of the individual cells can burst.

10.6 FINAL REMARKS

The β-cell models have achieved some notable successes, for example in explaining the regulation of bursting and Ca^{2+} levels by glucose and ACh. A general mathematical framework has been developed in which to understand these phenomena. Ironically, the very universality of the mechanisms, which makes them applicable to other cell types as well, has left us with a superfluity of models that match at least some aspects of the experimental data. How do we know which if any are correct? Ultimately, the models have the same status as all biological hypotheses: their predictions must be tested by experiment and the models revised over and over. In the meantime, they provide a more quantitative and disciplined way to conceptualize the phenomena and find the right questions to ask.

There remain many areas that have barely been touched by modeling. Another receptor-mediated potentiator of insulin secretion, currently of clinical interest, is glucagon-like peptide, which activates cyclic adenosine monophosphate (cAMP). Oscillations much slower than bursts (with periods of several minutes) are seen in both isolated cells and islets (Sherman 1995). The relation between these phenomena and islet bursting is unclear. We have discussed stimulus-secretion coupling only up to the point of Ca^{2+} entry. There exist more detailed models of glucose metabolism (Sweet and Matschinsky 1995) and also phenomenological models of secretion (Grodsky 1972) that need to be tied mechanistically to the channel-based models.

For pedagogical reasons, we have concentrated on the simplest models that illuminate given phenomena. For example, we have stressed reduced, two-variable models or subsystems isolated by range of activation or time scale. In addition to giving insight, this often reduces dynamic problems to algebraic ones of determining shape and position of nullclines. Thus, although nullclines are sometimes deprecated mathematically because, unlike bifurcations, they are not invariant features under coordinate transformations, they are very useful for modeling purposes.

Many of the biophysical models, however, are considerably more complicated, because they have followed the data rather than a preconceived template of how the dynamics should work. They have thus led to previously overlooked possibilities. We have also worked here mostly with limiting cases, such as well-separated time scales, to facilitate analysis. What is a virtue to the theorist, however, may not be a virtue to the cell. Indeed, in the I_{CRAC} model, imperfect separation of time scales helped realize the modeling goal of raising silent-phase potential in the presence of ACh. This was discovered serendipitously by adding to the model an I_{K-Ca} current that was thought to be irrelevant. It was originally excluded for "simplicity" but turned out to be the key. Thus, a combination of detailed and simplified models can be more effective than either alone.

10.7   APPENDIX: MODEL PARAMETERS

Exercises and input files for the models suitable for use with Bard Ermentrout's xpp program can be found at http://mrb.niddk.nih.gov/sherman; follow links for this chapter.

10.7.1   Basic model (equations (10.1), (10.2), (10.3), (10.5))

- Ionic currents:

$$
\begin{aligned}
I_{Ca} &= g_{Ca} m_\infty(v)(v - v_{Ca}), \\
I_K &= g_K n(v - v_K), \\
I_{slow} &= g_s s(v - v_K), \\
I_{K(ATP)} &= g_{K(ATP)}(v - v_K).
\end{aligned}
$$

Parameters: $g_{Ca} = 1000$ pS, $g_K = 2700$ pS, $g_s = 200$ pS, $v_{Ca} = 25$ mV, $v_K = -75$ mV, $g_{K(ATP)} = 120$ pS except where indicated.

- Gating variables:

$$x_\infty(v) = 1/(1 + \exp((v_x - v)/s_x)).$$

Parameters: $x = m, n, s$. $v_m = -20$ mV, $s_m = 12$ mV, $v_n = -16$ mV, $s_n = 5.6$ mV, $v_s = -52$ mV, $s_s = 5$ mV. $\lambda = 1.0$ except where indicated otherwise.

- Other parameters: $\tau_n = 20$ ms, $\tau_s = 20,000$ ms, $f = 0.01$, $k_c = 0.2$ ms^{-1}, and $c_m = 5300$ fF (femtoFarads). In order to convert Ca^{2+} current in fA (femtoamps = pS $\cdot$ mV) to Ca^{2+} concentration in μM, we multiply by $\alpha = 1/(2FV)$, where F is the Faraday constant and V is the cell volume. For a cell of radius 6.5 μm, $\alpha \approx -4.5 \times 10^{-6} \mu$M fA^{-1} ms^{-1}.

10.7.2 Extended model with ER (Equations (10.1), (10.2), (10.3), (10.6), (10.7))

- Modified parameters: $k_c = 0.12$ ms^{-1}, $s_s = 10$ mV.
- Additional currents:

$$I_{\text{K-Ca}} = g_{\text{K-Ca}} \left[c^5/(k_d^5 + c^5) \right] (v - v_K),$$

with $g_{\text{K-Ca}} = 1000$ pS and $k_d = 0.6 \, \mu M$;

$$I_{\text{CRAC}} = g_{\text{CRAC}} z_\infty(c_{\text{ER}})(v - v_{\text{CRAC}})$$

with $g_{\text{CRAC}} = 40$ pS, $v_{\text{CRAC}} = -30$ mV;

$$z_\infty(c_{\text{ER}}) = 1/(1 + \exp(c_{\text{ER}} - \bar{c}_{\text{ER}}))$$

with $\bar{c}_{\text{ER}} = 4 \, \mu$M.

- ER fluxes (see also Bertram et al., (1995a) and Li and Rinzel, (1994)):
  - $\mu = 250$ ms and $\sigma = 5$.
  - ER influx (pump:)

$$J_{\text{in}} = v_p c^2/(k_p^2 + c^2),$$

where $v_p = 0.24 \, \mu$M ms^{-1}, $k_p = 0.1 \, \mu$M.

  - ER efflux (J_{out}) is the sum of leak, $p_l(c_{\text{ER}} - c)$, and flux through the IP$_3$ receptor channel, $p_{\text{ip3}}(c_{\text{ER}} - c)$. $p_l = 0.02$ (non-dimensional), and

$$p_{\text{ip3}} = a(c)^3 b(\text{IP}_3)^3 h_\infty(c)^3,$$

where

$$a(c) = c/(c + d_{\text{act}})$$

represents activation of the receptor by cytosolic Ca^{2+};

$$b(\text{IP}_3) = \text{IP}_3/(\text{IP}_3 + d_{\text{ip3}})$$

represents activation by IP$_3$; and

$$h_\infty(c) = d_{\text{inh}}/(c + d_{\text{inh}})$$

represents inhibition by [Ca^{2+}]$_i$. $d_{\text{act}} = 0.1 \, \mu$M, $d_{\text{ip3}} = 0.2 \, \mu$M, $d_{\text{inh}} = 0.4 \, \mu$M.

REFERENCES

Atwater, I., Dawson, C. M., Scott, A., Eddlestone, G., and Rojas, E. 1980. The nature of the oscillatory behavior in electrical activity for pancreatic β-cell. *Hormone and metabolic research. Supplement series*, **Suppl. 10**, 100–107. {*205, 216*}

Atwater, I., Rosario, L., and Rojas, E. 1983. Properties of calcium-activated potassium channels in the pancreatic β-cell. *Cell Calcium*, **4**, 451–461. {*211, 216*}

Av-Ron, E., Parnas, H., and Segel, L. 1993. A basic biophysical model for bursting neurons. *Biological Cybernetics*, **69**, 97–95. {*208, 216*}

Baer, S. M., Rinzel, J., and Carrillo, H. 1995. Analysis of an autonomous phase model for neuronal parabolic bursting. *Journal of Mathematical Biology*, **33**, 309–333. {*207, 216*}

Bertram, R., Smolen, P., Sherman, A., Mears, D., Atwater, I., Martin, F., and Soria, B. 1995a. A role for calcium release-activated current (CRAC) in cholinergic modulation of electrical activity in pancreatic β-cells. *Biophysical Journal*, **68**, 2323–2332. {*201, 209–211, 215, 216*}

Bertram, R., Butte, M., Kiemel, T., and Sherman, A. 1995b. Topological and phenomenological classification of bursting oscillations. *Bulletin of Mathematical Biology*, **57**, 413–439. {*208, 216*}

Canavier, C. C., Baxter, D. A., Clark, J. W., and Byrne, J. H. 1993. Nonlinear dynamics in a model neuron provide a novel mechanism for transient synaptic inputs to produce long-term alterations of postsynaptic activity. *Journal of Neurophysiology*, **69**, 2252–2257. {*208, 216*}

Chay, T. R. and Keizer, J. 1983. Minimal model for membrane oscillations in the pancreatic β-cell. *Biophysical Journal*, **42**, 181–190. {*201, 205, 216*}

Chay, T. R., Fan, Y. S., and Lee, Y. S. 1995. Bursting, spiking, chaos, fractals, and universality in biological rhythms. *International journal of bifurcation and chaos in applied sciences and engineering*, **3**, 595–635. {*201, 206, 216*}

Grodsky, G. 1972. A threshold distribution hypothesis for packet storage of insulin and its mathematical modeling. *Journal of Clinical Investigation*, **51**, 2047–2059. {*214, 216*}

Hodgkin, A. L. and Huxley, A. F. 1952. A quantitative description of membrane current and its application to conduction and excitation in nerve. *Journal of Physiology*, **117**, 205–249. {*200, 216*}

Kahn, S. E. and Porte, Jr., D. 1990. The Pathophysiology of Type II (Noninsulin-Dependent) Diabetes Mellitus: Implications for Treatment. *Pages 436–456 of:* Rifkin, Harold and Porte, Jr., Daniel (eds), *Ellenberg and Rifkin's diabetes mellitus: theory and practice*, fourth edn. Amsterdam, The Netherlands: Elsevier. {*199, 216*}

Keizer, J. and Smolen, P. 1991. Bursting electrical activity in pancreatic β-cells caused by Ca^{2+}- and voltage-inactivated Ca^{2+} channels. *Proceedings of the National Academy of Sciences of the United States of America*, **88**, 3897–3901. {*201, 205, 216*}

Li, Y.-X. and Rinzel, J. 1994. Equations for $InsP_3$ receptor-mediated Ca^{2+} oscillations derived from a detailed kinetic model: A Hodgkin-Huxley like formalism. *Journal of Theoretical Biology*, **166**, 461–473. {*209, 215, 216*}

Pernarowski, M., Miura, R. M., and Kevorkian, J. 1992. Perturbation techniques for models of bursting electrical activity in the pancreatic β-cells. *SIAM Journal on Applied Mathematics*, **52**, 1627–1650. {*206, 216*}

Pinsky, P. F. and Rinzel, J. 1994. Intrinsic and network rhythmogenesis in a reduced Traub model. *Journal of Computational Neuroscience*, **1**, 39–60. {*209, 216*}

Plant, R. E. 1978. The effects of calcium++ on bursting neurons. A modeling study. *Biophysical Journal*, **21**, 217–237. {*207, 216*}

Rinzel, J. 1990. Electrical Excitability of Cells, Theory and Experiment: Review of the

Hodgkin-Huxley Foundation and an Update. *Bulletin of Mathematical Biology*, **52**, 5–23. {*200, 216*}

Rinzel, J. and Ermentrout, G. B. 1989. Analysis of Neural Excitability and Oscillations. *Pages 135–169 of:* Koch, Christof and Segev, Idan (eds), *Methods in Neuronal Modeling*. Cambridge, MA, USA: MIT Press. {*200, 210, 217*}

Rinzel, J. and Frankel, P. 1992. Activity patterns of a slow synapse network predicted by explicitly averaging spike dynamics. *Neural Computation*, **4**, 535–545. {*209, 217*}

Rinzel, J. and Lee, Y. S. 1987. Dissection of a model for neuronal parabolic bursting. *Journal of Mathematical Biology*, **25**, 653–675. {*207, 217*}

Rush, M. and Rinzel, J. 1994. Analysis of bursting in a thalamic neuron model. *Biological Cybernetics*, **71**, 281–291. {*208, 217*}

Santos, R. M. and Rojas, E. 1989. Muscarinic receptor modulation of glucose-induced electrical activity in mouse pancreatic B-cells. *FEBS Letters*, **249**, 411–417. {*209–211, 217*}

Satin, L. and Cook, D. 1989. Calcium current inactivation in insulin-secreting cells is mediated by calcium influx and membrane depolarization. *Pflügers Archiv: European Journal of Physiology*, **414**, 1–10. {*206, 217*}

Satin, L. and Smolen, P. 1994. Electrical bursting in β-cells of the pancreatic islets of Langerhans. *Endocrine*, **2**, 677–687. {*206, 217*}

Sherman, A. 1994. Anti-phase, asymmetric, and aperiodic oscillations in excitable cells — I. Coupled bursters. *Bulletin of Mathematical Biology*, **56**, 811–835. {*213, 217*}

Sherman, A. 1995. Theoretical aspects of synchronized bursting in β-cells. *Pages 323–337 of:* Huizinga, Jan D. (ed), *Pacemaker Activity and Intercellular Communication*. 2000 Corporate Blvd., Boca Raton, FL 33431, USA: CRC Press. {*206, 212, 214, 217*}

Smolen, P. and Keizer, J. 1992. Slow voltage inactivation of Ca^{2+} currents and bursting mechanisms for the mouse pancreatic β-cell. *Journal of Membrane Biology*, **127**, 9–19. {*201, 205, 217*}

Smolen, P., Terman, D., and Rinzel, J. 1993. Properties of a bursting model with two slow inhibitory variables. *SIAM Journal on Applied Mathematics*, **53**, 861–892. {*208, 217*}

Sweet, I. R. and Matschinsky, F. M. 1995. Mathematical model of beta-cell glucose metabolism and insulin release. I. Glucokinase as glucosensor hypothesis. *American Journal of Physiology*, **268**, E775–E788. {*214, 217*}

Terman, D. 1991. Chaotic spikes arising from a model of bursting in excitable membranes. *SIAM Journal on Applied Mathematics*, **51**, 1418–1450. {*206, 217*}

Terman, D. 1992. The transition from bursting to continuous spiking in excitable membrane models. *Journal of Nonlinear Science*, **2**, 135–182. {*206, 217*}

Valdeolmillos, M., Santos, R., Contreras, D., Soria, B., and Rosario, L. 1989. Glucose-induced oscillations of intracellular Ca^{2+} concentration resembling bursting electrical activity in single mouse islets of Langerhans. *FEBS Letters*, **259**, 19–23. {*205, 217*}

Wang, X.-J. and Rinzel, J. 1995. Oscillatory and Bursting Properties of Neurons. *Pages 686–691 of:* Arbib, Michael A. (ed), *The Handbook of Brain Theory and Neural Networks*. Cambridge, MA, USA: MIT Press. {*200, 207, 217*}

Worley, III, Jennings F., McIntyre, Margaret S., Spencer, Benjamin, Mertz, Robert J., Roe, Michael W., and Dukes, Iain D. 1994. Endoplasmic reticulum calcium store regulates membrane potential in mouse islet β-cells. *The Journal of Biological Chemistry*, **269**(May), 14359–14362. {*210, 217*}

PART III: PHYSIOLOGY

Hans G. Othmer

The chapters in Part III cover physiological processes from the cellular to the organismic level. E. Pate begins with a detailed treatment of how to formulate a model for muscle dynamics. Force in contracting muscle is generated by the interaction of the contractile proteins actin and myosin, and Pate shows how to model the chemomechanical interaction of these proteins. Muscle modeling is a very classical problem in biophysics and the model Pate develops is an extension of an earlier model due to A. F. Huxley. The mathematical model gives insight into a number of experimentally observed phenomena, and Pate shows how experimental results can be used to successively refine a model.

The second chapter, by L. Glass, deals with biological rhythms, a topic that is at least superficially familiar to all. The periods of naturally-occurring rhythms range from fractions of a second to a year or more, and the question of interest is how these intrinsic oscillations interact with external forcing. Stimuli delivered to the oscillators usually induce a resetting of the oscillation, so that the timing of the oscillation after stimulation is different than it would have been if the stimulus was not applied. Determining the response of oscillators to perturbations administered at different phases of the cycle can give important information about the oscillator and also may be useful in determining the behavior of the oscillator in a fluctuating environment. Two major points emerge in this chapter: (i) mathematical modeling and analysis are essential for understanding how an intrinsic oscillator interacts with external stimuli, and (ii) some relatively simple mathematical models of forced oscillators produce a great deal of insight into observed phenomena.

Cardiac muscle, like other excitable tissues that consist of electrically connected cells, exhibits phenomena both at the scale of a single cell and at the multicellular or tissue level. Of course, the latter behavior stems from that of individual cells, but for some phenomena this level of detail is not required. However, it is essential that processes that occur at the cellular level be correctly included in a multicellular description, and in the third chapter W. Krassowska shows how this can be done for cardiac tissue. Using a mathematical technique called averaging or homogenization, she is able to show how the behavior of individual cells is reflected in the partial differential equations that govern phenomena at the tissue level. Equally important, this analysis shows when the simplest tissue-level continuum description is not appropriate and suggests how it can be generalized.

In the fourth chapter C. Peskin treats a complementary cardiac problem, that of modeling and analyzing blood flow during a cycle of contraction and relaxation. Peskin treats the ventricular muscle wall as a fluid with actively contracting fibers immersed in it. This description simplifies the mathematical description of the system and leads to numerical and computational techniques for handling such problems. The outcome of this is a detailed simulation of blood flow in a chamber of the heart and a better understanding of how fluid flow interacts with the heart valves that control it. This chapter provides a good example of how mathematical modeling, analysis, and computation can be integrated to provide new insight into a significant physiological problem.

Bioconvection is the term used to describe the phenomenon of spontaneous pattern formation in suspensions of swimming microorganisms. In the last chapter N. Hill shows how to develop and analyze a mathematical model of this phenomenon, which is commonly seen in the laboratory and produced by protozoa, bacteria, and single-celled algae. Typically the microorganisms that generate such patterns are more dense than the fluid in which they swim (essentially water), and a variety of different directional responses to their environment cause them to aggregate in specific regions. The resulting spatial variations in the density of the suspension drive a bulk flow with downwelling where the concentration of cells is high and upwelling where the concentration is low, and the result is a spatially nonuniform distribution of the cell density.

In these chapters, as in the earlier chapters, there is a strong interplay between experimental observations and theoretical models developed to explain them. The models developed here give rise to ordinary or partial differential equations, and in several instances numerical computations play an essential part in understanding the behavior of the models. This is especially true in Peskin's chapter, where little progress can be made toward understanding blood flow in the heart without large-scale numerical simulations.

11. MATHEMATICAL MODELING OF MUSCLE CROSSBRIDGE MECHANICS

Edward Pate

11.1 INTRODUCTION

Force and motion in actively contracting muscle results from the interaction of the contractile proteins, actin and myosin. In this chapter we will discuss one of the more successfully employed techniques for modeling the chemomechanical interaction of these proteins (Huxley 1957). This will require developing mathematical models describing the force-producing biochemical interactions in contracting muscle at the fundamental level of the individual contractile molecules. In order to motivate this effort, we will first discuss the relevant structural, biochemical, and biophysical properties of vertebrate skeletal muscle. Building upon these fundamental, experimentally-observed constraints, we will subsequently develop mathematical models of the interactions which produce biologically useful force and motion. As always, a fundamental challenge for mathematical modelers is to then use models to augment our understanding of experimental results. In this regard, the second part of the chapter will be devoted to combining our theoretical framework with recent biochemical and mechanical experimental studies. We will suggest that this interdisciplinary approach can indeed provide unexpected insights into muscle function, along with potential resolutions of currently controversial questions in muscle chemomechanics.

11.2 MUSCLE STRUCTURE AND FUNCTION

Figure 11.1 shows a schematic representation of a sarcomere, the fundamental contractile unit in vertebrate skeletal muscle. The sarcomere is composed of two interdigitating, parallel arrays of filaments. The thin filaments project from structures termed the Z-lines and are composed primarily of actin. The bipolar, thick filaments are composed primarily of myosin. The sarcomeres are connected in series along the length of the muscle via the Z-lines. Thus in Figure 11.1, actin filaments for the next sarcomere project to the right (left) from the rightmost (leftmost) Z-line, interdigitating with another set of thick filaments. The thick filaments also interdigitate with mirror-image sets of actin

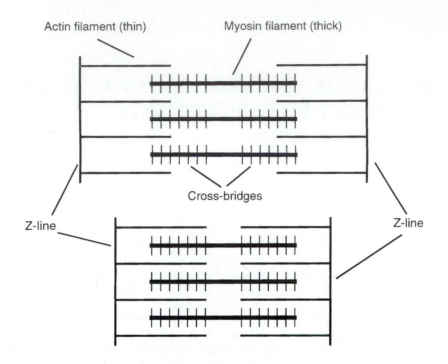

Figure 11.1. Schematic representation of the relationship between actin and myosin filaments in a sarcomere.

filaments connected to yet another Z-line. The sarcomeres are also connected in parallel across the width of the muscle. The actin filaments are arranged in a close-packed hexagonal array, with a myosin thick filament at the center of each hexagon in the filament-overlap zone. Thus, each actin filament has three nearest-neighbor myosin filaments with which it interacts. Muscle contraction results from the cyclic interaction of crossbridges extending from the thick, myosin filaments with binding sites on the adjacent six actin filaments. The crossbridges are staggered along the thick filament so that an equal number interact with each of the six adjacent actin filaments. A crossbridge attaches to a binding site and executes a powerstroke, causing a relative shear displacement between the parallel filaments. When continued attachment would start to resist active sliding, the crossbridge detaches and can begin a new cycle. This process has frequently been termed the "rowing-oar model." Individual actin and myosin filaments are thought to be relatively inextensible. Thus, muscle shortening results not from major changes in lengths of the filaments themselves, but instead from changes in the thick-filament–thin-filament overlap resulting from relative sliding generated by the polar nature of each half-sarcomere. This is demonstrated in Figure 11.1, where the shorter sarcomere has filaments which are identical in length to the longer sarcomere, but the former are more highly overlapped than the latter.

In order to provide some feeling for the length scales involved, we will take typical values from vertebrate skeletal muscle. The myosin thick filaments are approximately 1.6μm in length (1μm $= 1 \times 10^{-6}$m), the actin fil-

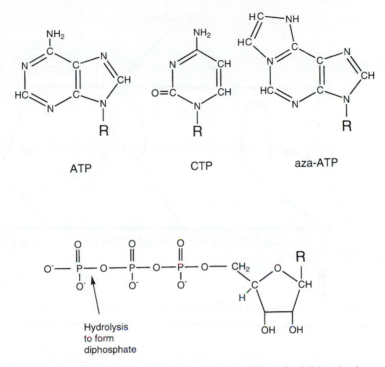

ATP   CTP   aza-ATP

Hydrolysis
to form
diphosphate

Figure 11.2. Structures of the physiological substrate, ATP, and additional substrates used in the subsequent experimental and modeling studies. The bottom half shows ribose triphosphate, a component of all substrates. The top line gives the differing "ring" moieties for each substrate. The ring structures and ribose triphosphate are connected at the locations marked "R" (a ribose-to-ring nitrogen bond in each case). The arrow indicates the bond which is hydrolyzed for the conversion of the triphosphate form to the diphosphate form and free phosphate during the contractile cycle.

aments (one side of the Z-line only) are approximately $1\mu m$ long. At rest a typical sarcomere is $2.5\mu m$ in length (Z-line to Z-line). The maximum sarcomere length at which force can be developed is $3.6\mu m$, since for lengths greater than this there will be no overlap of the thick and thin filaments. The distance in cross section from a myosin filament to an adjacent actin filament spanned by an attached crossbridge is approximately 25nm ($1nm = 1 \times 10^{-9}m$.) Thus a "representative" cylindrical muscle 4cm in length and 1cm in diameter would be made up of approximately 16,000 sarcomeres connected in series along its length and 4.8×10^{10} hexagonal arrays of filaments (i.e., myosin thick filaments) across its width. Ancillary structures reduce these numbers in practice (see Woledge et al. (1985); Bagshaw (1993)). Nonetheless, the numbers remain useful order-of-magnitude estimates.

To produce work, thermodynamics requires that there be a free energy source. Under physiological conditions, chemical free energy is supplied by the hydrolysis of the chemical species adenosine 5'-triphosphate (ATP) to the diphosphate form (ADP) and a free phosphate (P). The structure of ATP is given in Figure 11.2. It will be considered in greater detail as subsequently required.

One of the first things one faces in study of muscle is the wealth of

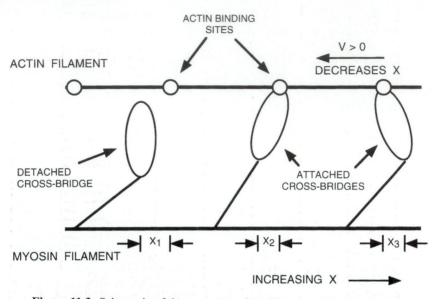

Figure 11.3. Schematic of the geometry of the Huxley crossbridge model.

data on structure, biochemistry, and mechanics which has been collected over the past 50 years. A complete survey is beyond the scope of the present work, but a large number of surveys of the field are available. For example, the book by Bagshaw (1993) provides an excellent introduction for the non-specialist to muscle structure, mechanics and biochemistry. Much more complete reviews of the field are contained in Cooke (1986) and Sellers and Goodson (1995). Muscle mechanics and mathematical models are discussed in considerable detail in the book of McMahon (1984). Woledge et al. (1985) provides an excellent discussion of muscle mechanics as related to energetics. In addition, all of the above citations contain extensive bibliographies to other relevant works.

11.3   THE CROSSBRIDGE MODEL OF A. F. HUXLEY

Having described the basic structure of skeletal muscle, we will begin the process of constructing mathematical models for crossbridge function. In this regard, the question becomes: how can one develop mathematical relationships describing the chemical binding of myosin crossbridges to actin filaments, and the subsequent generation of mechanical force and motion? By far, the most influential solution to this problem has been the crossbridge model of Huxley (1957). We will consider it in some detail, since its analytical tractability facilitates quantitative understanding of so many of the ideas we will subsequently employ.

Following Huxley, we consider infinitely long, parallel, inextensible actin and myosin filaments as shown schematically in Figure 11.3. Crossbridges project from the thick filament. They can exist in either (1) a state detached from actin or (2) a state attached to actin. We assume that a crossbridge can interact with only the nearest actin binding site (termed the single-site assumption). Let the variable, x, termed the distortion, measure the distance along the thick filament between some reference location on a myosin crossbridge and an actin binding site. Precise identification of such a reference location on a myosin crossbridge remains a subject of intense experimental study. The choice in Figure 11.3 should be viewed as serving pedagogical purposes only. We assume that x increases to the right. Thus in Figure 11.3, x_1 and $x_2 > 0$, while $x_3 < 0$. Let $V > 0$ represent the sliding velocity of the actin binding sites to the left relative to the fixed myosin filament. Denote by $n(x, t)$ the proportion of myosin which see the nearest actin binding site at spatial location x at time t *and* are bound to actin. Thus, by our sign conventions, $V > 0$ decreases x for an attached crossbridge. It is an experimental observation that the periodicity of the actin filament and the myosin crossbridges are not constant multiples of each other. Thus, to a first approximation, we can consider $n(x, t)$ to be continuously distributed in x for a large ensemble of crossbridges, as would exist in a functioning muscle.

We let ρ_m be the number of myosin crossbridges per unit length (density). As noted, crossbridges are assumed to bind to and detach from actin sites. Let $f(x)$ be the spatially-dependent, first-order rate constant (units, s^{-1}) for the binding of a detached crossbridge to actin at distortion x. In other words, we assume a Poisson process with $f(x) \Delta t$ approximating the probability of attachment over a short time Δt. Let $g(x)$ be the corresponding detachment-rate function. Since bound crossbridges produce force and motion, our goal is to develop an expression for the time evolution of the attached fraction, $n(x, t)$. This will allow us to describe the chemomechanical properties of contracting muscle in terms of the underlying crossbridge dynamics. Consider those crossbridges which at time t are attached to actin with distortions in the spatial region $[x_0, x_0 + \Delta x]$. By the assumptions above, there are only three ways this fraction can change with time: (1) attached crossbridges can detach; (2) detached crossbridges which see an actin binding site in the region $[x_0, x_0 + \Delta x]$ can attach; and (3) with $V > 0$, sliding of the actin filament relative to the myosin filament will cause a reduction of distortion for attached crossbridges. Thus attached crossbridges with distortions slightly greater than $x_0 + \Delta x$ will be carried into the region $[x_0, x_0 + \Delta x]$ at a rate proportional to V. Conversely, attached crossbridges with distortions slightly greater than x_0 may be carried out of the region $[x_0, x_0 + \Delta x]$. The flux density into the region will be $J(x_0 + \Delta x, t) = \rho_m V n(x_0 + \Delta x, t)$; the density out will be $J(x_0, t) = -\rho_m V n(x_0, t)$. If $n(x, t)$ is the attached crossbridge distribution, then $1 - n(x, t)$ is the detached distribution. At a given time, the crossbridge population in the region $[x_0, x_0 + \Delta x]$ is proportional to the integral of $n(x, t)$ over the spatial domain. Thus considering the above three factors, we have the following balance law expressing the time rate of change of crossbridges in $[x_0, x_0 + \Delta x]$

$$\frac{\partial}{\partial t} \int_{x_0}^{x_0 + \Delta x} \rho_m n(x, t) \, dx = \int_{x_0}^{x_0 + \Delta x} -g(x) \rho_m n(x, t) \, dx$$

$$+ \int_{x_0}^{x_0 + \Delta x} f(x) \rho_m [1 - n(x, t)] \, dx + J(x_0 + \Delta x, t) - J(x_0, t).$$

(11.1)

Using the definition of J, dividing out the common factor, ρ_m, and applying the mean value theorem for integrals, we obtain

$$\frac{\partial n(\beta_1, t)}{\partial t} \Delta x = -g(\beta_2)n(\beta_2, t)\Delta x + f(\beta_3)[1 - n(\beta_3, t)]\Delta x$$
$$+ Vn(x_0 + \Delta x, t) - Vn(x_0, t), \tag{11.2}$$

where $x_0 < \beta_1, \beta_2, \beta_3 < x_0 + \Delta x$. Divide by Δx, let $\Delta x \to 0$, and obtain

$$\frac{\partial n(x_0, t)}{\partial t} = -g(x_0)n(x_0, t) + f(x_0)[1 - n(x_0, t)] + V\frac{\partial n(x_0, t)}{\partial x}. \tag{11.3}$$

Noting that there is nothing special about the point x_0, we may replace it with x. Rearranging terms gives our final evolution equation,

$$\frac{\partial n(x, t)}{\partial t} - V\frac{\partial n(x, t)}{\partial x} = f(x)[1 - n(x, t)] - g(x)n(x, t). \tag{11.4}$$

This, along with initial conditions and boundary conditions, describes the evolution of the attached fraction. It should be noted that the lefthand side is the standard material derivative, modulo our somewhat nonstandard sign convention for V.

It is useful to choose the attachment and detachment rate functions, $f(x)$ and $g(x)$ prior, to further analysis. After considerable trial and error, Huxley chose the following:

$$f(x) = \begin{cases} 0, & x \le 0, \\ f_1 x/h, & 0 \le x \le h, \\ 0, & x > h, \end{cases} \qquad g(x) = \begin{cases} g_2, & x \le 0, \\ g_1 x/h, & x > 0, \end{cases} \tag{11.5}$$

where f_1, g_1, g_2, and $h > 0$. These are shown in Figure 11.4. In terms of active shortening ($V > 0$), actin binding sites enter from the right. The rate functions imply that detached crossbridges begin to attach to actin binding sites only after the binding sites come to within a distance $x = h$. We shall call h the "powerstroke" length. Attachment is biased toward the beginning of the powerstroke. Detachment is possible during the powerstroke, but at a rate much lower than attachment. Continued sliding (without detachment) results in a further reduction in crossbridge distortion. After sufficient sliding while attached, the distortion for a crossbridge becomes negative. Here attachment is not allowed, and the detachment rate, g_2, increases dramatically. Subsequent discussion will provide additional justification for these particular choices.

11.4   ANALYSIS OF ISOMETRIC CONTRACTION

The derived balance law (11.4) is a hyperbolic partial differential equation. Two frequently employed experimental protocols allow simplification to ordinary differential equations. When muscle actively contracts at a fixed length, no net shortening occurs, and hence $V = 0$ (termed isometric contraction). This is the situation, for example, when one tries to pick up a weight

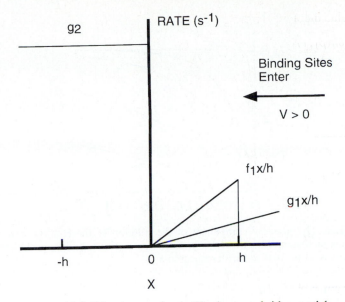

Figure 11.4. Kinetic rates for the Huxley crossbridge model.

from a surface and the weight is too heavy to lift. Equation (11.4) becomes an ordinary differential equation in time at each spatial location

$$\frac{dn(x,t)}{dt} = f(x)[1 - n(x,t)] - g(x)n(x,t). \tag{11.6}$$

To gain insight into the implications of the model, we will solve (11.6) for the case where all crossbridges are detached initially. The rate functions require considering three separate regions in order to solve for $n(x,t)$.

Region 1: $x > h$, $g(x) = g_1 x/h$, $f(x) = 0$, with equation and initial condition

$$\frac{dn(x,t)}{dt} = \frac{f_1 x}{h}[1 - n(x,t)] - \frac{g_1 x}{h}n(x,t), \qquad n(x,0) = 0, \tag{11.7}$$

and solution $n(x,t) = 0$.

Region 2: $0 \le x \le h$, $g(x) = g_1 x/h$, $f(x) = f_1 x/h$, with equation and initial condition

$$\frac{dn(x,t)}{dt} = \frac{f_1 x}{h}[1 - n(x,t)] - \frac{g_1 x}{h}n(x,t), \qquad n(x,0) = 0, \tag{11.8}$$

and solution

$$n(x,t) = \frac{f_1}{f_1 + g_1}[1 - e^{-(f_1+g_1)xt/h}]. \tag{11.9}$$

Region 3: $x < 0$, $g(x) = -g_2$, $f(x) = 0$, with equation and initial condition

$$\frac{dn(x,t)}{dt} = g_2 n(x,t), \qquad n(x,0) = 0, \tag{11.10}$$

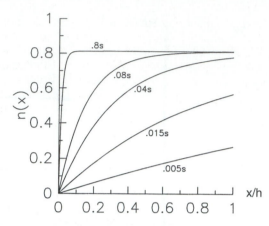

Figure 11.5. The nonzero component of the solution of the Huxley model (Equation 11.9) at increasing times with $f_1 = 65\text{s}^{-1}$, $g_1 = 15\text{s}^{-1}$, and $g_2 = 313.5\text{s}^{-1}$. The spatial variable is normalized with respect to h.

and solution $n(x, t) = 0$.

Figure 11.5 shows the solution as a function of time assuming $f_1 = 65\text{s}^{-1}$, $g_1 = 15\text{s}^{-1}$ and $g_2 = 313.5\text{s}^{-1}$. These values, along with $h = 10\text{nm}$, were suggested by Brokaw (1976) to be appropriate for frog muscle contracting at $0°C$. The only nonzero component of the solution is in the spatial Region 2, $0 \leq x \leq h$. The spatial coordinate in the figure is normalized with respect to h. As $t \to \infty$, the Region 2 solution tends to the uniform solution $n(x, t) = f_1/(f_1 + g_1) = 0.81$. In addition, our previous suggestion that crossbridges at the beginning of the powerstroke should attach more rapidly is confirmed. As $x \to 0^+$, $f(x) \to 0$, and thus even at $t = 0.8s$, the steady-state solution has not been reached for sufficiently small x.

The physiological function of contracting muscle is to produce force and motion. We have solved for the attached fraction and we now extend the analysis to relate crossbridge attachment to force. The experimentally observed linear decrease in force with linearly decreasing filament overlap has been taken as evidence that crossbridges within a sarcomere produce force by summing in parallel (Gordon et al. 1966). The change in muscle tension following very rapid lengthening or shortening of muscle is linear in strain (Huxley and Simmons 1971), which suggests that the individual crossbridges produce force as linearly elastic elements (Hookean springs). Thus we assume that for an attached crossbridge with distortion x, the force produced will be equal to κx, where κ is the elastic force constant. To express the tension $T(t)$ produced by a given muscle, let A be the cross-sectional area, m be the number of crossbridges per unit volume which are capable of acting with actin, s be the sarcomere length, and ℓ be the distance between myosin sites. Then $T(t) = A(s/2) \times$ (average force per crossbridge), or

$$T(t) = \frac{Asm}{2\ell} \int_{-\ell/2}^{\ell/2} n(x, t)\kappa x \, dx. \tag{11.11}$$

The myosin thick filaments are bipolar, with the crossbridges on each end producing forces which act in series. Thus, under isometric conditions, each half of each myosin thick filament must produce a force of equal magnitude. Otherwise there would be filament movement due to the force imbalance. This is the reason for the factor $s/2$ (note that we are ignoring a small zone at the center of the thick filament which is devoid of myosin crossbridges). The single-site assumption implies that $n \approx 0$ outside of $[-\ell/2, \ell/2]$. Thus we shall here, and elsewhere, take advantage of the simplification of replacing the above integral by one over $(-\infty, \infty)$. Let $P(t) = T(t)/A$ be the tension per unit cross-sectional area. Then

$$P(t) = \frac{m\kappa s}{2\ell} \int_{-\infty}^{\infty} n(x,t) x \, dx. \tag{11.12}$$

For the solution to the Huxley model with $n(x,t)$ nonzero only for $0 \le x \le h$, this becomes

$$P(t) = \frac{m\kappa s}{2\ell} \int_{0}^{h} \left(\frac{f_1}{f_1 + g_1} \right) [1 - e^{-(f_1+g_1)xt/h}] x \, dx, \tag{11.13}$$

which is equivalent to

$$\begin{aligned} P(t) = \ & \frac{m\kappa s h^2}{4\ell} \left(\frac{f_1}{f_1 + g_1} \right) \left\{ 1 + \frac{2}{(f_1 + g_1)t} e^{-(f_1+g_1)t} \right. \\ & - \left. \frac{2}{[(f_1 + g_1)t]^2} (1 - e^{-(f_1+g_1)t}) \right\}. \end{aligned} \tag{11.14}$$

As $t \to \infty$,

$$P(t) \to \frac{m\kappa s h^2}{4\ell} \left(\frac{f_1}{f_1 + g_1} \right). \tag{11.15}$$

We shall term this P_0, the isometric tension. Tension normalized with respect to the isometric value is given by the term in braces in (11.14).

We now consider the idealized situation in which an active muscle is very rapidly stretched or released by a small amount. Here, "rapidly" will mean a small stretch or release over a small enough time to allow us to assume that no attachment or detachment of crossbridges occurs during the length change. Then we define mechanical stiffness, S, to be the change in force divided by the change in length, or more precisely

$$S(t) = \lim_{\Delta x \to 0} \left\{ \frac{m\kappa s}{2\ell} \left[\int_{0}^{h} n(x,t)(x + \Delta x) \, dx - \int_{0}^{h} n(x,t) x \, dx \right] \right\} / \Delta x. \tag{11.16}$$

Simplifying, canceling the remaining terms in Δx, and trivially taking the limit,

$$S(t) = \frac{m\kappa s}{2\ell} \int_{0}^{h} n(x,t) \, dx. \tag{11.17}$$

Because $n(x,t)$ is integrated over space, S is a measure of the fraction of attached crossbridges. Using the isometric crossbridge distribution, isometric stiffness can easily be shown to be $S_0 = ms\kappa h f_1/[2\ell(f_1 + g_1)]$. Note also that infinite limits of integration would yield the same result for S. Using equations (11.12) and (11.14), it is straightforward to show that P/P_0 rises more rapidly than S/S_0 as t increases from 0.

Taken together, Figure 11.5 and the time course of P/P_0 and S/S_0 highlight one of the fundamental new insights into muscle physiology provided by the Huxley model — in active contraction, all crossbridges are not equal. Owing to the continuum of spatial distances between crossbridges and binding sites, coupled with the spatial dependence of the actin-myosin crossbridge kinetics, some crossbridges will be situated to attach more rapidly than others. This is demonstrated in the spatial distribution of the attached fraction as a function of time in Figure 11.5. Furthermore, those that do attach more rapidly (larger x), produce more force. Thus normalized force rises more rapidly than the normalized fraction attached. The first crossbridges to attach produce more than their fair share of tension. Although subsequent models have used different spatial properties of crossbridges, the fundamental concept of distortion-dependent mechanical and kinetic properties of crossbridges has remained a basic property of subsequent crossbridge models.

An additional example emphasizing this interplay is shown in Figure 11.6, which shows tension recovery following a rapid increase in muscle length. We take f_1, g_1, and g_2 as before. Figure 11.6(a) shows the initial, isometric crossbridge distribution. At time $t = 0$ the muscle is assumed to be instantaneously stretched by a length of $0.5h$ per half-sarcomere. All attached crossbridges are thus strained by an equivalent amount, resulting in the nonzero crossbridge distribution in the region $0.5h \leq x \leq 1.5h$ shown in Figure 11.6(b). Following the stretch, some crossbridges are now attached in the region $x > h$. Here $f(x) \equiv 0$, while $g(x) = g_1 x/h$. Thus the attached fraction in the region $x > h$ decreases with time. The solutions for $n(x, t)$ in this region at three times (0.005s, 0.03s, and 0.08s) are given in Figure 11.6(c). This decrease will contribute to a decrease in muscle force. Conversely, immediately following the stretch, the region $0 \leq x \leq 0.5h$ is devoid of crossbridges. Since $f_1, g_1 \neq 0$ in this region, new crossbridges will be recruited into the attached pool, and $n(x, t)$ increases with time, contributing to an increase in force. Solutions are shown at times 0.005s, 0.03s, 0.08s, and 0.8s. In the invariant region, $0.5h \leq x \leq h$, the equilibrium crossbridge distribution remains the same. It is important to remember, however, that the crossbridges which maintain this equilibrium are the crossbridges which maintained an equivalent equilibrium, via different, slower transition rates, in the region $0 \leq x \leq 0.5h$ prior to the stretch. Hence, as regards the tension transient following stretch, there are two competing effects, one decreasing and the other increasing force. It is the detachment of the more highly strained ($x > h$) crossbridges which dominates, however, and force decreases with time as shown in Figure 11.6(d).

The reequilibration of tension is qualitatively similar to that experimentally observed by Huxley and Simmons (1971) [Figure 2]. Several features in the experimental transients are missed by the simple two-state model. This has led to more complex, multistate, elastic crossbridge models. However, what we want to emphasize with Figure 11.6 is the rather complex interplay between spatially dependent crossbridge kinetics, elastic crossbridges, and the actual force production. This interplay remains at the heart of the original Huxley analysis and these subsequent models.

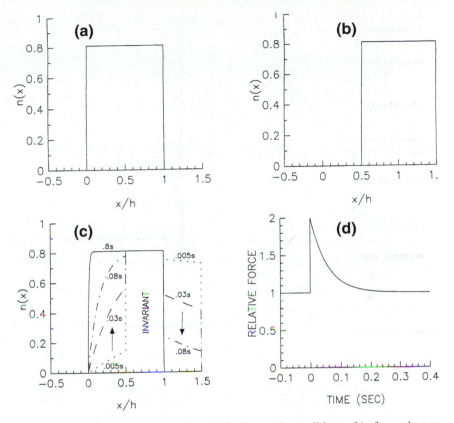

Figure 11.6. Crossbridge distributions (a) for isometric conditions, (b) after an instantaneous strain of $0.5h$ at time $t = 0$, and (c) at various times during recovery. Panel (d) shows the ensuing tension transient following stretch.

11.5   ANALYSIS OF ISOTONIC CONTRACTION

The balance law (11.4) derived for $n(x, t)$ was a partial differential equation. By considering isometric contraction, we could reduce the analysis to that of an ordinary differential equation. An alternative experimental protocol which has frequently been employed is to allow a muscle to shorten at a constant tension. Figure 11.7 shows a representative case. In this type of experiment, a muscle fiber is mounted between a force transducer for determining the generated tension and a rapid motor for manipulating muscle length. Using a computer-controlled feedback loop between the transducer and motor, a muscle is allowed to shorten at a tension equal to a predetermined fraction of P_0. The plot on the right gives the tension against which the muscle shortens as a function of time. Two different isotonic releases are shown at approximately 5% and 20% of P_0. The plot on the left shows the position of a motor arm to which the shortening muscle is attached. The lower release tension on the right corresponds to the upper position trace on the left. After an initial transient period following initiation of the release, the position-vs.-time plot on the left is well approximated as linear in time. The dashed line gives a least-squares

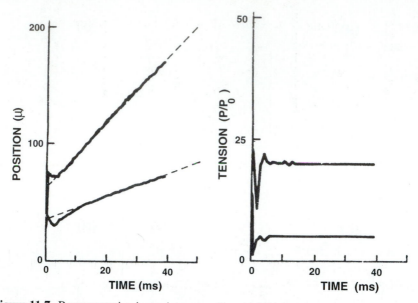

Figure 11.7. Representative isotonic contractions. Additional detail of this experimental protocol is in (Pate et al. 1993a) and references therein.

linear fit to the position plot. Thus the slope of the dashed line is shortening velocity. Figure 11.13(a) (to be discussed subsequently in more detail) shows more complete data sets of normalized shortening velocity (muscle lengths per second) as a function of normalized release tension, P/P_0. Each data point is from a single isotonic release such as in Figure 11.7. Based upon experimental observations, (Hill 1938) originally suggested that the force-velocity relationship for shortening muscle was hyperbolic. The solid lines are least-squares fits to the Hill equation

$$V = V_{max} \frac{(a/P_0)(1 - P/P_0)}{a/P_0 + P/P_0}. \tag{11.18}$$

Here V_{max} is the maximum velocity of shortening which is obtained at zero release tension, and a/P_0 is another parameter for the fit.

From our modeling perspective, it is important to remember that a linear position-vs.-time plot for shortening muscle is not mandated a priori. However, given the experimental observation, we may now take advantage of the substantial analytical simplification resulting from reducing the full partial differential equation to an ordinary differential equation for steady-state, constant-velocity shortening:

$$-V \frac{dn(x)}{dx} = f(x) - [f(x) - g(x)]n(x) \tag{11.19}$$

with appropriate boundary condition. To solve for the case of isotonic shortening, we let $V > 0$ and solve in the direction of positive V in three regions as before. To simplify the notation, we drop the explicit time dependence in n:

Region 1: $x > h$, $g(x) = g_1 x / h$, $f(x) = 0$, boundary condition $n(x) \to 0$ as $x \to \infty$,

$$-V \frac{dn(x)}{dx} = 0 \tag{11.20}$$

and solution $n(x) = 0$.

Region 2: $0 \le x \le h$, $g(x) = g_1 x / h$, $f(x) = f_1 x / h$, and boundary condition $n(h^-) = n(h^+)$, i.e., continuity across the boundary, with equation

$$-V \frac{dn(x)}{dx} = \frac{f_1 x}{h} - \left(\frac{f_1 x}{h} + \frac{g_1 x}{h} \right) n(x) \tag{11.21}$$

and solution

$$n(x) = \frac{f_1}{f_1 + g_1} \{ 1 - e^{(f_1 + g_1)(x^2 - h^2)/(2hV)} \}. \tag{11.22}$$

Region 3: $x < 0$, $g(x) = g_2$, $f(x) = 0$ and boundary condition $n(0^-) = n(0^+)$, with equation

$$-V \frac{dn(x)}{dx} = -g_2 n(x) \tag{11.23}$$

and solution

$$n(x) = \frac{f_1}{f_1 + g_1} \left[1 - e^{-(g_1 + f_1)h/2V} \right] e^{g_2 x / V}. \tag{11.24}$$

Note that in region 3, since $x < 0$, and g_2, $V > 0$, then $n(x) \to 0$ as $x \to \infty$, as expected.

Let v be the shortening velocity normalized as muscle lengths per second (more generally used as the measure of velocity by physiologists). Then $V = (s/2)v$, where s is the sarcomere length. Define $\varphi = (f_1 + g_1)h/s$. The constant velocity solution can be written as

$$n(x) = \begin{cases} 0, & x > h, \\[2mm] \dfrac{f_1}{f_1 + g_1} [1 - e^{(x^2/h^2 - 1)\varphi/v}], & 0 \le x \le h, \\[2mm] \dfrac{f_1}{f_1 + g_1} [1 - e^{-\varphi/v}] e^{(2xg_2/sv)}, & x < 0. \end{cases} \tag{11.25}$$

Figure 11.8 shows the distribution of $n(x)$ for isometric contraction and three increasing values of shortening velocity. For $v = 0$, all attached crossbridges are in the region $0 \le x \le h$. As v increases, the value at $n(0)$ decreases, since the passage time through the region $0 \le x \le h$ decreases. Simultaneously, an increasing fraction of the crossbridges which did attach in the region $x > 0$ are carried by active sliding into the region $x < 0$. Owing to assumption of linearly elastic crossbridges, the crossbridges in region $x < 0$ produce a resistive, drag force on active sliding. As is evident, this dragging

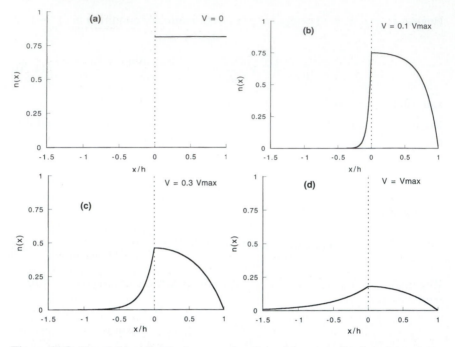

Figure 11.8. Crossbridge distributions as a function of the normalized spatial variable x/h for (a) isometric conditions, (b) $V = 0.1\,V_{\max}$, (c) $V = 0.3\,V_{\max}$, and (d) $V = V_{\max}$.

effect becomes more pronounced as v increases, and $V_{\max}$, the maximum velocity of shortening, is thus determined by that velocity for which the positive force produced by crossbridges in the region $x > 0$ is exactly balanced by the drag force from the region $x < 0$. This is consistent with the experimental observation (Figure 11.13(a)) that at $V_{\max}$, muscle produces no tension. The observation from the mathematical model that $V_{\max}$ was determined by a balance of positive and negative crossbridge elastic forces was another of the really fundamental conceptual insights that Huxley provided. For the first time, there was a mechanical explanation for the existence of a maximum velocity of shortening.

Using the integral relationship for tension (11.12) and the steady-state expression for $n(x)$ from (11.18), the relationship between force and velocity becomes

$$P = \frac{mksh^2}{4\ell}\left(\frac{f_1}{f_1+g_1}\right)\left\{1 - \frac{v}{\varphi}(1 - e^{-\varphi/v})\left[1 + \frac{1}{2}\left(\frac{f_1+g_1}{g_2}\right)^2\frac{v}{\varphi}\right]\right\}$$

(11.26)

where the term outside the braces is the previously derived expression for isometric tension, P_0 (11.15). Hence

$$P/P_0 = 1 - \frac{v}{\varphi}(1 - e^{-\varphi/v})\left[1 + \frac{1}{2}\left(\frac{f_1+g_1}{g_2}\right)^2\frac{v}{\varphi}\right]$$

(11.27)

a transcendental equation relating shortening velocity and tension. Maximum shortening velocity, V_{max}, is determined by setting the righthand side equal to zero and solving for v. We will develop an expression for energy turnover prior to analyzing this transcendental equation.

An alternative way of looking at the expression for P_0 is to consider the maximum work, w (work = force × distance) that a single crossbridge can perform. The maximum occurs when the crossbridge attaches at $x = h$, traverses the powerstroke to $x = 0$, and then detaches. Then

$$w = \int_0^h \kappa x \, dx = \kappa h^2/2, \tag{11.28}$$

and

$$P_0 = \frac{msw}{2\ell}\left(\frac{f_1}{f_1 + g_1}\right). \tag{11.29}$$

As previously noted, the physiological energy source for muscle contraction is the hydrolysis of ATP. Assuming that each attachment-detachment cycle of a functioning crossbridge requires a molecule of ATP, we now will develop an expression for the rate of energy liberation. At steady state, this will be equivalent to the rate of crossbridge attachment (or, equivalently, detachment). Following McMahon (1984), let V equal the filament sliding velocity and ℓ equal the spatial separation of myosin sites. Then the number of times per second that an actin site gets within range of a myosin is V/ℓ. The probability of attachment during some short time dt when a myosin crossbridge is located at spatial location x is $f(x)(1 - n(x, t))dt$. For the case of steady-state, active sliding, the probability of coming within range and attaching is $(V/l)f(x)(1 - n(x))dt$. Let $t = 0$ correspond to the time when an actin site first comes within range for attachment (i.e., $x = h$). The region of attachment, $0 \le x \le h$, will be traversed in time $t = h/V$. Thus the average number of times per second that a crossbridge in the spatial region $0 \le x \le h$ attaches is given by

$$\frac{V}{\ell} \int_{t=0}^{t=h/V} f(x)[1 - n(x)] \, dt. \tag{11.30}$$

Observing that $v = -dx/dt$, and transforming to an integral in space, the energy liberation per unit volume can be given as

$$E = \frac{me}{\ell} \int_0^h f(x)[1 - n(x)] \, dx \tag{11.31}$$

where m is the crossbridge density and e is the amount of energy (units = Joules, ergs, etc.) liberated per site in one cycle. Integrating the expression for $n(x)$ in the region $0 \le x \le h$, (see (11.25)), we obtain

$$E = \frac{meh}{2\ell}\left(\frac{f_1}{f_1 + g_1}\right)\left\{g_1 + \frac{f_1 v}{\varphi}(1 - e^{-\varphi/v})\right\}. \tag{11.32}$$

The first law of thermodynamics requires that the energy, E, be in the form of heat or work. Under isometric conditions ($v = 0$) only heat is produced. Then the "maintenance heat," E_0, is given by

$$E_0 = \frac{meh}{2\ell}\left(\frac{f_1 g_1}{f_1 + g_1}\right). \tag{11.33}$$

The experimental verification of the existence of a maintenance heat is the reason Huxley chose g_1 to have a nonzero value, even though a more efficient isometric tension could be generated with $g_1 = 0$. The latter choice would mean that crossbridges initially attach in the region $0 \leq x \leq h$, but then could not detach via the g_1 pathway with concurrent ATP hydrolysis. Substituting, we have the normalized "excess energy" as a function of velocity, *i. e.* that above the maintenance value

$$\frac{E - E_0}{P_0 v_{max}} = \frac{e}{w} \left(\frac{f_1}{f_1 + g_1} \right) \left(\frac{v}{V_{max}} \right) \left[1 - e^{-(\frac{\varphi}{v_{max}})(\frac{v_{max}}{v})} \right]. \tag{11.34}$$

In developing the Huxley model, we have introduced nine parameters: $f_1, g_1, g_2, m, e, \kappa, h, s,$ and ℓ. Of these, the sarcomere length, s, the myosin crossbridge spacing, ℓ, and the number of crossbridges which can interact with actin, m, can be experimentally observed. We have devoted considerable attention to the analysis of isometric contraction and mechanical transients following rapid fiber-length changes, using the parameters suggested by Brokaw (1976) in order to highlight the mechanical implications of Huxley's fundamental assumptions regarding crossbridge behavior. The original analysis by Huxley, however, concentrated not on transient data but primarily on (1) the steady-state force-velocity relationship and (2) energy liberation [(11.27), (11.33), and (11.34)]. Restricting attention to these quantities, we can show that instead of requiring nine parameters, the model can be specified by considering only three dimensionless groups. By a trial-and-error process, Huxley found excellent agreement with the Hill fit for the experimentally determined force-velocity and the observed energy-liberation relationship for frog muscle ($0°C$) by choosing $g_2/(f_1 + g_1) = 3.919$, $g_1/(f_1 + g_1) = 3/16$, and $w/e = \kappa h^2/(2e) = 3/4$. The previously employed parameters of Brokaw are consistent with these values. Solving the transcendental (11.27), the Huxley analysis gives $V_{max} = 4\varphi$, which may now be used to simplify the exponential in (11.34). For the Brokaw parameters, this gives $v_{max} = 1.28$ muscle lengths/s, or in terms of actual filament sliding velocity, $V_{max} = 1600$ nm/s. These are representative values for frog satorius muscle Hill (1938). Figure 11.9 shows the hyperbolic Hill fit ($a/P_0 = 0.25$ for frog muscle) and the Huxley model prediction for the force-velocity relationship. I reemphasize that the Hill fit (11.4) is in the form of a hyperbola; the Huxley model fit is the solution to a seemingly unrelated transcendental equation. Figure 11.10 gives a fit to the experimental observations of Hill (1938) and the predictions of the Huxley model for normalized excess energy liberation as a function of shortening velocity. As can be seen, the fits to both the experimental force-velocity data and the Hill (1938) energy data are really quite exceptional for such a simple crossbridge model. Indeed, the ability of the model to fit these data was viewed as a fundamental triumph of the Huxley analysis.

Hill (1938) observed that energy liberation increased monotonically with shortening velocity. Aubert (1956), and others subsequently, questioned this observation, suggesting that energy liberation increased only initially with velocity. For sufficiently large velocities ($> 0.5 V_{max}$), they suggested that energy liberation actually began to decrease with increasing velocity. If one thinks in terms of driving an automobile, this alternative observation doesn't really make that much physical sense — we could all increase fuel economy

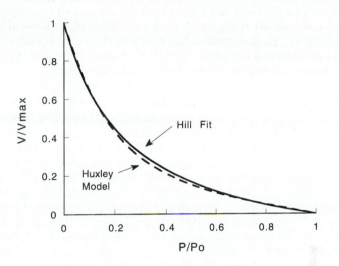

Figure 11.9. The Huxley crossbridge model fit using the parameters as described in the text for the relationship between isotonic release tension, normalized with respect to P_0, and shortening velocity normalized with respect to V_{max} (dashed line) and the Hill fit (Equation (11.4)) with $a/P_0 = 0.25$ (solid line).

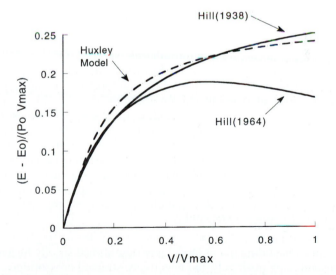

Figure 11.10. The Huxley model prediction for energy liberation as a function of shortening velocity (dashed line) along with the experimentally determined values of Hill (1938) and the revised values of Hill (1964).

by simply driving faster. However, Hill (1964) re-did his original experiments using improved equipment and agreed with the alternative conclusion. His fit to the data is also shown in Figure 11.10. Huxley had fallen victim to the "modeler's nightmare." He had produced an excellent fit to the existing experimental data, only to have the experimentalists then change the data on him! This remains an occupational hazard for mathematical modelers.

Despite the failure of the two-state Huxley model to fit the energy liberation data, the basic properties of his model – (1) spatially dependent rate functions and (2) linearly elastic crossbridges have remained the driving principles behind most crossbridge models for the past 30 years. Indeed, modifications of the original model by Huxley (1973) and others have been able to suggest potential resolutions of the energy liberation data. In the second part of this chapter we will continue to use the fundamental crossbridge ideas originally proposed by Huxley, attempting at the same time to more carefully relate actomyosin biochemistry and muscle mechanics in the context of the Huxley framework. This will allow us to address a fundamental and controversial question in the study of muscle physiology — the magnitude of the power-stroke length, h. Surprisingly, this analysis will ultimately bring us back to the energy-liberation question.

11.6   ADDITIONAL CONSIDERATIONS IN MODELING CROSSBRIDGE MECHANICS

With any model, careful attention must be paid to the underlying assumptions. With this in mind, it is important to consider several details regarding our own basic assumptions in formulating our crossbridge model. One initial assumption in deriving (11.4) was that the actin and myosin filaments are inextensible. This has been a widely held interpretation of the experimental data since the original experimental observations of sliding filaments by Huxley and Niedergerke (1954) and by Huxley and Hanson (1954). Recent interpretations of experimental data have made this assumption more controversial, with suggestions that there is a significant compliance in the actin filaments themselves (Higuchi et al. 1995) and references therein. In our model development we have retained the mathematical simplicity which inextensible filaments provide. Future analyses (especially, for example, the modeling of tension transients following rapid length changes) may require inclusion of the complication of filament extensibility.

Filament sliding in our model was determined strictly by a balance of forces generated by positively and negatively strained crossbridges. We have neglected to include two additional, obvious, potential influences on sliding velocity: (1) the inertial terms resulting from the mass of the filaments themselves, and (2) the viscous forces resulting from the surrounding aqueous environment. Analyses of magnitudes of these effects and justification of their omission are given in Pate and Cooke (1989).

11.7 USING BIOCHEMISTRY, MECHANICS, AND MATHEMATICAL MODELING TO DETERMINE THE POWERSTROKE LENGTH

We have analyzed the actomyosin interaction in terms of attachment-detachment cycles. A working crossbridge traverses the powerstroke $h > x > 0$ producing useful work, and then detaches in the region $x < 0$ when continued attachment produces a negative strain, resisting sliding velocity. The length of the fundamental crossbridge working distance remains controversial, with values ranging from 5–10nm to distances in excess of 100nm (see Table 11.1). The smaller values are compatible with the 20nm length physical size of the myosin crossbridge. The larger values would appear to require a major rethinking of our concept of the working powerstroke. In this section we will extend the fundamental ideas of the Huxley model to include more realistic biochemistry for the actin-myosin-ATP interaction. This will allow us to address the question of the value for h in the Huxley model and to suggest a possible resolution of the short- and long-interaction-length observations.

Table 11.1. Summary of crossbridge distance measurements.

Reference	Value	Units
This work	5–7	nm
Molloy et al. (1995)	4–5	nm
Finer et al. (1994)	11	nm
Huxley and Simmons (1971)	10	nm
Huxley and Kress (1985)	4–12	nm
Uyeda et al. (1990)	5–20	nm/ATP
Higuchi and Goldman (1991)	> 60	nm/ATP
Yanagida et al. (1985)	120	nm/ATP
Harada et al. (1990)	>200	nm/ATP

The primary technique for studying the biochemical pathway in the interaction of actin, myosin, and ATP has been to disrupt the sarcomere structure in order to individually isolate the contractile proteins. The proteins are then resuspended in solution at known concentrations, and standard biochemical techniques are employed to identify and characterize specific states in the actomyosin biochemical cycle. The following basic cycle has been identified, as shown in Figure 11.11. Here we use the abbreviations M = myosin, A = actin, T = ATP, D = ADP, and P = phosphate. The individual steps are reversible. The circle with arrows shows the primary kinetic direction. Starting arbitrarily at the A•M state, ATP binds to the A•M state with second-order rate constant, k_t (units = $M^{-1}s^{-1}$); the first order rate in units of s^{-1} is determined by $k_t[ATP]$[1]. Myosin then dissociates from actin and ATP is hydrolyzed on the detached myosin head. The M•D•P crossbridge reattaches to actin. The hydrolysis product **P** is released first. ADP is subsequently released with first-

---

[1]Here and hereafter [·] denotes a concentration.

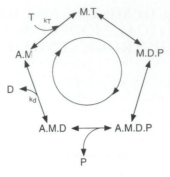

Figure 11.11. Actomyosin biochemical cycle as discussed in text.

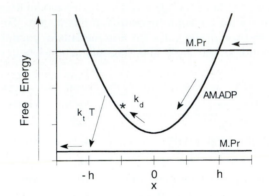

Figure 11.12. Free energy diagram of an expanded crossbridge cycle incorporating individual biochemical states identified in solution studies. Additional details provided in text.

order rate constant k_d. ATP then rebinds, and the cycle repeats. Each cycle requires one molecule of ATP. Note that the hydrolysis step occurs when myosin is detached from actin. Additional kinetic states have been identified (reviewed in Cooke (1986)). However, the above pathway is a fundamental subset of virtually all kinetic analyses and, except as subsequently noted, will be sufficient for our modeling. The above kinetic scheme has been identified from studies on the isolated proteins suspended and free to diffuse in solution. Thus the steric effects found in the intact filament array of functioning muscle are absent. In the Huxley model, these steric effects appear as the spatial (i.e., x) dependence of both protein kinetics and crossbridge forces. In our prior analysis of the Huxley model we did not consider actual biochemistry; indeed, the kinetic cycle was only very poorly understood when Huxley originally proposed his model. To proceed further, it is necessary to correlate the solution biochemistry with the working crossbridge cycle. Figure 11.12 summarizes current ideas. Here we have plotted chemical free energy as a function of the spatial variable, x, in the Huxley model. We are dealing with linearly elastic crossbridges. Chemical free energy can be viewed as somewhat similar to mechanical potential energy. The force produced by a crossbridge with an extension of magnitude x and with elastic force constant κ will be κx; the potential energy is parabolic in strain and equal to $\kappa h^2/2$. Force is the derivative of chemical free energy.

In Figure 11.12, binding sites enter at the right. At spatial location $x = h$, detached M•Pr (Pr = ATP or ADP•P) crossbridges attach to actin. Initial attachment and transition into the powerstroke is associated with release of phosphate from A•M•D•P crossbridges. Continued sliding reduces the value of the spatial coordinate, as the attached A•M•D crossbridges traverse the powerstroke. In this process, work is produced at the expense of chemical free-energy as the elastic spring is relaxed. At $x = 0$, a crossbridge is at its minimum attached free energy and produces no force. For $x < 0$, continued sliding results in a negative force, counterproductive to active sliding. Thus we assume that for $x < 0$, ADP is released with rate constant k_d. ATP can then bind with rate k_t[ATP]. The crossbridge detaches and ATP is hydrolyzed. The detached crossbridge can now participate in a subsequent cycle. In solution, without steric constraints, an A•M complex will be at its free energy minimum. This corresponds to $x = 0$ in Figure 11.12, the value of x for which ADP release and ATP binding begin. Hence, at the start of detachment, $x = 0$, and through the detachment region $x < 0$ we will take k_d and k_t to be the values measured in solution. Additional justification of these assumptions is provided in Cooke (1986), Pate and Cooke (1989) and Pate et al. (1993b).

We expand our original two-state model to a model including three states in order to determine the crossbridge distributions in the above cycle for the case of isotonic shortening. We let $u(x)$ be the unattached fraction (M• Pr = M•T and M•D•P crossbridges are lumped together). Further assume that phosphate release is associated with transition into the powerstroke, letting $r(x)$ be the attached fraction in the A•M•D state. Let $s(x)$ be the fraction in the A•M state. To simplify the mathematics involved, we assume that a fraction of crossbridges, p, attach into the A•M•D state and traverse the powerstroke without detachment. Then

$$u_+(x) = 1 - p \tag{11.35}$$
$$r_+(x) = p \qquad 0 \le x \le h \tag{11.36}$$
$$s_+(x) = 0. \tag{11.37}$$

For $x < 0$, the balance law becomes

$$-V\frac{dr(x)}{dx} = -k_d r(x), \tag{11.38}$$

$$-V\frac{ds(x)}{dx} = -k_t T s(x) + k_d r(x), \tag{11.39}$$

$$u(x) = 1 - r(x) - s(x). \tag{11.40}$$

with boundary conditions $r(0) = p$, $s(0) = 0$, and $u(0) = 1 - r(0) - s(0)$. The solution is

$$r_-(x) = p e^{k_d x/V}, \tag{11.41}$$

$$s_-(x) = p k_d [e^{k_t T x/V} - e^{k_d x/V}]/(k_d - k_t T), \tag{11.42}$$

$$u_-(x) = 1 - r_-(x) - s_-(x). \tag{11.43}$$

Here the subscripts $+$ and $-$ have been added to denote those populations in the positive and negative force region and $T = [ATP]$. Then $V_{\max}$, is determined by the V solving the negative and positive force balance

$$\int_{-\infty}^{0} [r_-(x) + s_-(x)]\kappa x \, dx + \int_{0}^{h} [r_+(x) + s_+(x)]\kappa x \, dx = 0. \tag{11.44}$$

This results in an algebraic equation for velocity. Defining $K_m = k_d/k_t$, and $V_\infty = k_d h/\sqrt{2}$, the solution can be written as

$$V = \frac{V_\infty[ATP]}{K_m + [ATP]} M(K_m, [ATP]) \tag{11.45}$$

where

$$M(K_m, [ATP]) = \left\{ \frac{1 - K_m[ATP]}{(K_m + [ATP])^2} \right\}^{-1/2} \tag{11.46}$$

It is straightforward to show that $M \in (1, 2/\sqrt{3} \approx 1.15)$ for all [ATP]. Then with only a maximum 15% error, we can set $M = 1$ in (11.45). The expression for $V_{\max}$ as a function of [ATP] is now seen to be the well-known expression for Michaelis-Menten saturation behavior. V_∞ is the velocity at infinite [ATP]; K_m is the [ATP] yielding half-maximal velocity. A frequently used form for Michaelis-Menten behavior is obtained by converting the hyperbola to double-reciprocal format

$$1/V_{\max} = 1/V_\infty + (K_m/V_\infty)(1/[ATP]). \tag{11.47}$$

Here the plot of $1/V_{\max}$ as a function of $1/[ATP]$ is linear with vertical intercept $1/V_\infty$ and slope K_m/V_∞.

This is a rather surprising result, since Michaelis-Menten behavior is generally associated with a specific type of enzymatic chemical reaction in solution. Here we have Michaelis-Menten behavior (to within 15%) resulting from spatially dependent crossbridge kinetics and force generation, with our end product being velocity instead of a chemical species. At present, this is of course, only a model prediction, but it constitutes an experimentally testable hypothesis. We now test this hypothesis.

Figure 11.7 showed representative experimental data used to determine shortening velocity as a function of release tension. A fit of such force-velocity data to the hyperbolic Hill equation allows one to determine $V_{\max}$. In the present case, we want to vary the concentration of substrate, and determine $V_{\max}$ as a function of substrate. Then we can ask whether it exhibits Michaelis-Menten behavior. Figure 11.13(a) shows force-velocity data collected at three different substrate concentrations, along with the hyperbolic Hill fits. The intercepts of the Hill fits at $P/P_0 = 0$ give $V_{\max}$. For future reference, an important observation is the scatter in the data which is an unavoidable consequence of dealing with this muscle preparation. Figure 11.13(b) gives a plot of $1/V_{\max}$ vs. $1/[substrate]$, and a least-squares linear fit to the data. The linear plot does indeed confirm our model prediction of Michaelian behavior with $K_m = 1.92$mM and $V_\infty = 1.13$ muscle lengths/sec.

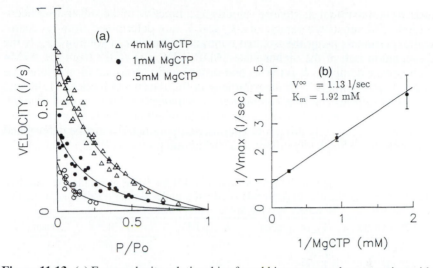

Figure 11.13. (a) Force-velocity relationships for rabbit psoas muscle contracting with three different concentrations of MgCTP as substrate. The solid lines are least-squares hyperbolic Hill fits (11.18) to the data. Velocities are normalized with respect to muscle length, tensions with respect to the isometric value. (b) A least-squares plot of $1/V_{max}$ vs. $1/[MgCTP]$. Linearity confirms our model prediction of Michaelian behavior. Similar linear plots are observed using aza-ATP and ATP as substrates. For additional details see Pate et al. (1993a) and references therein.

We have previously noted that the physiological substrate for the actomyosin interaction is ATP, or to be more precise, MgATP (magnesium complexed with ATP). Note, however, the substrate for the data of Figure 11.13 was actually MgCTP (CTP = cytidine triphosphate). Myosin is a promiscuous enzyme that will hydrolyze a variety of nucleotide triphosphates but with varying chemomechanical efficacy. In addition to CTP, we have also shown that ATP and aza-ATP also yield Michaelian behavior for V_{max} as a function of substrate concentration Pate et al. (1993a). The nucleotide structures are shown in Figure 11.2. All substrates have similar structures, the only difference being the "rings" connected to the ribose and triphosphate moieties. The differing values for K_m and V_∞ in the Michaelis-Menten relationship for V_{max} (11.45) for the above three substrates are given in Table 11.2. We will now exploit the fact that different substrates produce different values in order to estimate the powerstroke length, h.

The parameter ratios used to define K_m and V_{max} allow us to develop two expressions for h. Using our definition $V_\infty = k_d h/\sqrt{2}$, we can solve for h, obtaining

$$h = \sqrt{2}V_\infty/k_d. \tag{11.48}$$

Using the fact that $K_m = k_d/k_t$, we can substitute for k_d in (11.48), obtaining

$$h = (\sqrt{2}V_\infty)/(k_t K_m). \tag{11.49}$$

It is useful to keep in mind where the parameters in (11.48) and (11.49) come from. V_∞ and K_m are derived from experiments with active muscle fibers, looking at maximum shortening velocity as a function of the substrate concentration. The other two parameters, k_d and k_t, are determined from biochemical experiments using the isolated contractile proteins in solution. k_d is the dissociation rate of the diphosphate (ADP, CDP, aza-ADP) from the A•M• diphosphate complex. k_t is the second-order binding constant for the substrate to the $A • M$ complex. With the parameter definitions in our model, (11.48) and (11.49) are equivalent from a *mathematical* perspective. From an *experimental* perspective, however, they are completely different approaches for determining h since completely different experimental parameters must be determined by employing completely different experimental protocols (White et al. 1993).

Table 11.2. Calculated step length, h, from (11.49) for rabbit psoas muscle with biochemical and mechanical experiments conducted under similar conditions, using different substrates with widely varying kinetics and mechanics.

Substrate	V_∞(nm/sec)	$K_m(\mu M)$	$k_t(M^{-1}s^{-1})$	h(nm)
ATP	2040	150	2.7×10^6	7.1
CTP	1360	1900	1.4×10^5	7.1
aza-ATP	470	375	3.4×10^5	5.2

Under physiological conditions, the specific rate-limiting step in the actomyosin cycle remains a subject of controversy. However, for sufficiently low substrate concentration, the rate-limiting step can be unambiguously identified: it becomes the step involved in substrate binding, and the rate is k_t [substrate]. We have two expressions for h. The question now becomes whether they give consistent, plausible values for h. We will consider (11.49) first, since it involves experimental data taken as a function of the substrate concentration, i.e., data for which we know the rate limiting step. Table 11.2 gives the experimentally determined values of V_∞, k_t, and K_m for the three substrates tested, using rabbit psoas muscle. Note that by varying the substrates employed, we have been able to vary the mechanical and kinetic parameters by a factor of up to almost 20. Nonetheless, using (11.49), Table 11.2 shows that the experimental parameters and model analysis give remarkably consistent values of $h = 7.1$nm, 7.1nm, and 5.2nm. These values are identical to within the experimental accuracy with which the physiology (e.g., see Figure 11.13(a)) and the biochemistry can be done. More importantly, the values are "small." They are consistent with the physical size of a myosin crossbridge. Despite the experimental scatter, the model analysis is clearly inconsistent with a fundamental powerstroke length of > 100nm.

It is useful to compare models as they evolve. How does our current analysis compare with our previous analysis of the Huxley model? In the current analysis, we have added a much more detailed consideration of the biochemistry involved. From a mathematical perspective, this resulted in the necessity to consider an additional crossbridge state and thus an additional differential equation describing the population as a function of x. Nonetheless, at the end, we determined V_{max} by balancing linearly elastic, positive and negative crossbridge forces resulting from spatially dependent kinetics, just as in the prior analysis of the Huxley model. Our fundamental explanation for the existence of V_{max} remained unchanged — only more biochemically complicated. Additionally, we assumed a constant attached fraction in the power-stroke. Mathematically, this simplification resulted in an algebraic equation expressing the balance of positive and negative forces which determined V_{max} as a function of [ATP]. In the original Huxley analysis, the equivalent relationship was (11.27), a transcendental equation relating force and velocity. Have we lost anything important in this simplification? One can show that more plausible assumptions, e.g., the Huxley kinetics for $x > 0$, still yield a "short" powerstroke compatible with the size of a crossbridge (Pate and Cooke (1989); Pate et al. (1993b)).

11.8 AN ALTERNATIVE ANALYSIS FOR h SUGGESTS A NEW CROSSBRIDGE CYCLE

Our original motivation for using (11.49) to determine h was that the equation involved data associated with the ATP binding step. For sufficiently large [ATP], the ATP binding step can no longer be rate limiting. Some step preceding ATP binding, e.g., ADP release, must limit crossbridge detachment. The release of ADP from A•M has been shown to be a two-step process involving an isomerization between two conformations of the A•M•ADP complex, followed by dissociation of ADP (Taylor 1991). In fast skeletal muscle, the rate of ADP dissociation is very rapid (too fast to experimentally measure, $> 1000s^{-1}$, $10°C$). The rate of the isomerization step is slow (30 to 90 s^{-1}, $10°C$). An alternative procedure to determine h is to use (11.48). Taking $V_\infty = 2040$nm/sec (Table 11.2), and using the rapid dissociation rate, we obtain $h < 2.8$nm. Using a midrange value of $50s^{-1}$ for the isomerization rate, we obtain $h = 57$nm. Our problem becomes obvious. The latter value is larger than the physical size of a myosin crossbridge. The former value, while small, is now only an upper limit and is becoming too small. Neither kinetic rate yields a value for h consistent with that previously obtained, when we could identify the rate-limiting step.

Accepting $h = 7$nm as the powerstroke length using (11.49), one can ask what value does (11.48) require for k_d? Substituting the velocity above, one obtains a k_d of approximately $400s^{-1}$. Figure 11.14 shows the normalized attached crossbridge distribution at $V_{max} = 2040$nm/s with $k_d = 400s^{-1}$, and $h = 7$nm. For $x > 0$, the model results in a uniform distribution with the powerstroke attached fraction, p, arbitrarily chosen to be 1. For $x < 0$, the attached fraction decays as $n(x) = \exp[k_d x / V_{max}]$. This is the dashed line in Figure 11.14. It is useful to determine the distribution of positive and negative force generated by the crossbridges. Taking $n(x) = 1$, the positive force over the region $0 < x < h$ is, of course, $\kappa h^2 / 2$. (To be precise, it is this factor times $ms/(2\ell)$ as in (11.18).) This proportionality factor appears in all terms and will be neglected to simplify notation. At V_{max}, the total negative force will be of equal magnitude. How much negative force is produced by crossbridges in the region symmetric to the powerstroke, $-h \leq x \leq 0$? Integrating $n(x)kx$ over [-7,0], and using the parameters above, one obtains the result that 40% of the negative force is produced by crossbridges in the region $-h \leq x \leq 0$. This means that 60% of the negative force results from crossbridges which have been strained into the region $x < -h$. Note from Figure 11.14 that the population of crossbridges is small in the region $x < -h$, only about. 25% of the total

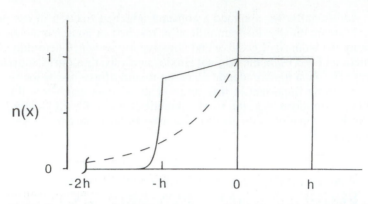

Figure 11.14. Crossbridge distributions in the model used to initially analyze the step length (Figure 11.12) and in the subsequent model (Figure 11.15) developed to analyze the implications of a weakly coupled detachment pathway.

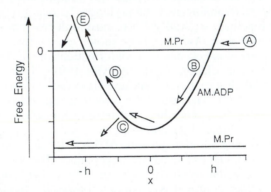

Figure 11.15. A modified crossbridge cycle which allows for crossbridge detachment without substrate hydrolysis for crossbridges bearing large negative strains. Additional details are provided in the text.

population of negatively strained crossbridges. Nonetheless, their large strains compensate, and they produce more aggregate force. This is yet another example of our previous observation that in the Huxley model, "all crossbridges are not created equal."

The preceding analysis implies that more than half of the negative force is provided by crossbridges strained beyond $x = -h$. Equivalently, and with regard to Figure 11.12, these crossbridges are bearing so much strain that their free-energy actually exceeds that of the detached state. This suggests the prospect of an alternative dissociation pathway into the detached state, driven by mass action and free energy differences, as opposed to the chemomechanically less efficient pathway requiring triphosphate hydrolysis. This modified crossbridge cycle is demonstrated in Figure 11.15. The initial portion is identical to that of the Huxley model (open arrows). Binding sites enter from the right (**A**) and attached crossbridges traverse the working powerstroke (**B**). In the dragstroke region, the detachment via ADP dissociation and ATP binding (**C**) may still occur. However, for crossbridges which do not detach via path (**C**), an alternative pathway (**D**, closed arrows) allows crossbridges to be additionally strained by other attached, working, powerstroke crossbridges. Cross-bridges strained into the region $x < -h$ have free energies that rapidly rise above that of the detached state. Energetically, these crossbridges are unstable with respect to the detached state, and it would appear reasonable to believe that such crossbridges would make the transition back to a detached state (**E**), essentially being pulled off the actin ("mechanically detached") as their free energy rises and the attached state becomes less stable. We now consider the implications of this modified crossbridge scheme.

Our analysis to determine the conditions for V_{max} will be as before, except that now the negative force must be handled as two components, one for $-h \leq x \leq 0$ (F_{1-}) and one for $x < -h$ (F_{2-}). For more memorable terminology, we will denote the dissociation pathway requiring ATP hydrolysis as "strongly coupled to ATP hydrolysis" or the "strongly coupled" pathway. The mechanical dissociation pathway followed by highly strained ($x < -h$) crossbridges as they detach without ATP hydrolysis will be termed the "weakly coupled" pathway. To simplify notation, we will also revert to the Huxley notation and let g represent the dissociation rate via the strongly coupled pathway.

The positive force will again be $F_+ = \kappa h^2/2$. In the region $-h \leq x \leq 0$, the attached fraction decays as $n(x) = exp[gx/V]$. Our now-standard integration shows that the negative force contribution is

$$F_{1-} = \kappa \left(\frac{V^2}{g^2} \right) \left\{ 1 - \left[\frac{gh}{V} + 1 \right] e^{-gh/V} \right\}. \qquad (11.50)$$

The calculation of F_{2-}, the negative force generated in the region $x \leq -h$, is more difficult, and it depends upon the nature of the rapid equilibration between attached and detached states. The following calculation provides an estimate of the magnitude of the negative force produced during the mechanical dissociation of a myosin head. We assume that for $x \leq -h$, attached crossbridges can be mechanically detached with a rate, k_{off}, equilibrating rapidly with the detached state. We further assume that the equilibration occurs more quickly than the release of ADP and subsequent binding of ATP. Hence, we have only one process occurring in this region. This greatly simplifies the mathematics without significantly changing the final result. This restriction will be eliminated in the more detailed, multistate model considered subsequently. For $x \leq -h$, again let $n(x)$ be the probability of being in the attached fraction. The probability of being attached as a function of x is now determined by solving the differential equation

$$-V \frac{dn(x)}{dx} = -n(x) \cdot k_{off}$$
$$+ (1 - n(x)) \cdot k_{off} \cdot e^{-\kappa[x^2-h^2]/kT}, \qquad (11.51)$$

where k is the Boltzmann constant and T is the absolute temperature.

The exponential factor in the last term results from the thermodynamic-consistency requirement that the forward (detachment) and reverse (attachment) rates obey a Gibbs relationship, in that they are related by the free-energy difference between the attached and detached states (Hill 1974). In other words, of the forward rate, the reverse rate, and the free-energy difference, only two can be specified independently. Note that we have not "enforced" this requirement in our analyses to this point. However, for plausible free-energy differences, its inclusion in the Huxley model or in the analysis of Section 11.7 does not significantly affect the model results (for discussion, see Brokaw (1976)).

To fix parameters, we take $k_{off} = 10^4 \text{s}^{-1}$, in the range suggested for a rapid equilibration at the beginning of the working powerstroke (Schoenberg 1988). Setting $V = 2000\text{nm/sec}$, one must solve (11.51) using standard numerical methods, and then integrate to determine the negative force, obtaining $F_{2-} \approx -2kT$. Obviously, simplifying assumptions have been used in obtaining this result. However, the major point with this calculation is to show that the additional energetic cost of mechanically dissociating a crossbridge in this fashion is small, with the negative work amounting to only a few kT. This compares with the $23kT$ of free energy available from the hydrolysis of ATP. In the discussion below we take $F_{2-} = -2kT$.

V_{max} is again determined by balancing forces and solving $F_{1-} + F_{2-} + F_+ = 0$. Let $E(> 0)$ be the magnitude of the difference between the free energy of the detached state and the minimum free energy of the attached state in units of kT (i.e., the free energy at the minimum of the parabola in Figure 11.15). Then $E = \kappa h^2/2$. Using the above expressions and rearranging, one obtains a new expression which relates V_{max}, g and h:

$$2E(V_{max}^2/g^2h^2)[1 - exp(-gh/V_{max})(gh/V_{max} + 1)] + 2kT = E. \tag{11.52}$$

This equation has two variables, $F_+ = E = \kappa h^2/2$ and the combination gh/V_{max}. F_+ can be estimated from the efficiency of actively shortening muscle. Actively shortening muscle exhibits high efficiency, 50% to 60%. Taking the free energy available from ATP hydrolysis as $23kT$, $F_+ = 15kT$ defines an efficiency within the above limits. Thus with knowledge of F_+ (or equivalently E), (11.52) becomes transcendental in the single variable, gh/V_{max}. Numerical methods then yield a root for (11.52) of

$$gh/V_{max} = 0.2 \tag{11.53}$$

The crucial observation is that this value is dramatically different from the value we obtained in first-step length analysis where detachment was strongly coupled to ATP hydrolysis. In that case (see (11.48)) we obtained $gh/V_{max} = \sqrt{2}$. The righthand side is more than a factor of 7 greater than the right-hand side of 0.2 above. Thus, for given values of h and V_{max}, the detachment rate, g, can be considerably slower in the region $0 \geq x \geq -h$ in the current model, incorporating an additional weakly coupled pathway for dissociation. Although the two models are applicable at different substrate regimes, we assume that they share the same powerstroke length. Using the values for $V_{max} = 2040\text{nm/s}$ and the previously taken value for h of 7nm from our previous powerstroke analysis, (11.53) gives a value for g of 57s^{-1}. This value is in the range of the rate constant for the slow isomerization which has been experimentally observed to precede the much more rapid ADP release step in rabbit fast muscle. The slower step should be rate limiting at high substrate concentration. Thus, unlike the previous model with only a strongly coupled dissociation pathway, the slow isomerization rate can now be part of a crossbridge cycle including an additional weakly coupled dissociation pathway. In fact, the isomerization rate fits very nicely into the rate predicted by the model.

The weakly coupled model has the property that not every crossbridge which attaches to actin requires an ATP in order to detach. The fraction of crossbridges which do not hydrolyze ATP is given by the population of crossbridges which traverse the dragstroke and remain attached at $x = -h$. This will be equal to

$$e^{-gh/V_{max}}. \tag{11.54}$$

For our root of (11.52), $gh/V_{max} = 0.2$, we find that about 80% of the attached crossbridges do not detach via the ATP hydrolysis step, but instead must be mechanically detached when $x < -h$. Figure 11.14 shows the fraction of attached crossbridges as a function of x for our first model, in which detachment was strongly coupled to ATP hydrolysis (dashed line), and in the new model, allowing for the weakly coupled detachment pathway (solid line). As is evident in the figure, as a crossbridge passes through the dragstroke at V_{max} in the weakly coupled model, very little detachment occurs for $-h < x < 0$. Significant, and very rapid, detachment occurs only after the free energy of the attached state exceeds that of the detached state ($x < -h$).

Alternatively, we may consider the fraction f of crossbridges which detach via ATP hydrolysis pathway (rate g) as a function of normalized shortening velocity, $V_n = V/V_{max}$. From (11.53) and (11.54) $f(V_n) = 1 - \exp[-0.2/V_n]$. For low shortening velocities, the strongly coupled pathway dominates. As V_n increases, an increasing fraction of crossbridges choose the weakly coupled pathway. Finally, at V_{max} i.e., ($V_n = 1$), only one crossbridge in five binds an ATP during each interaction, with four crossbridges in five recycling into another attachment cycle without ATP hydrolysis. This 80% of the total crossbridges have been termed "passenger" crossbridges (Malcolm 1991).

If as the model suggests, a crossbridge cycles approximately five times before hydrolyzing an ATP in a muscle shortening at V_{max}, then the distance traveled per ATP hydrolyzed can be approximated as five times the sum of the powerstroke and dragstroke lengths, or $10h$ ($= 70$nm). The distance per ATP is clearly velocity dependent and will increase roughly linearly with velocity. Higuchi and Goldman (1991) experimentally observed precisely this type of behavior in skinned muscle fibers. The limiting value of approximately 70nm per ATP that our model predicts at V_{max} is furthermore compatible with their estimate of 60nm.

The original problem which has led us to our comparison of weakly and strongly coupled crossbridge models was the question of the step length h. Some investigators have suggested short interaction distances; others have suggested quite long distances. Table 11.1 gives representative values. The crucial observation from Table 11.1 is that the longer interaction distances are in reality distances *per ATP hydrolyzed*. Thus our modeling suggests a thermodynamically plausible resolution of the otherwise apparently contradictory observations of widely disparate interaction distances. Our analysis of the step size via (11.49) (valid for ATP binding as the rate-limiting step, i.e., when the strongly coupled pathway dominates) suggests a short fundamental step length of 5 to 7nm. At more rapid shortening velocities, the weakly coupled pathway dominates. Multiple attachment and detachment cycles per ATP hydrolyzed become thermodynamically plausible, and long interaction distances can be observed as the passenger crossbridges cycle through. This occurs while the fundamental powerstroke length remains small.

11.9   A MODEL APPROPRIATE FOR INTERMEDIATE VELOCITIES AND SUBSTRATE CONCENTRATIONS

As noted, the previous strongly and weakly coupled models have been developed in order to explain the properties of muscle contracting at V_{max} for two distinctly different regimes — those of low and high substrate concentrations, respectively. Our explanation for the presence of both short and long interaction distances requires that there be a model valid for all substrate concentrations as a function of shortening velocity. We briefly show here that such a model does exist, and that with this model we can also potentially shed light on the question of energy liberation as a function of shortening velocity. This was one of the failings of the original Huxley model. A model incorporating more realistic assumptions about the individual states is required. The following additional assumptions are made. In addition to a detached state (State 1), both the powerstroke and dragstroke in the model are assumed to be composed of three attached states: a preisomerization A•M*•D state (State 2), a postisomerization A•M•D state (State 3), and an A•M state (State 4), which is reached following ADP release. The binding of ATP to the A •M state detaches the crossbridge. Precise free energies and transition rates are given in Figure 11.16. Their justification on the basis of solution biochemistry is discussed in Cooke et al. (1994). Two points are important, however. The isomerization rate (State 2 to State 3) is taken to be the previously suggested $60s^{-1}$. We also now include a finite attachment rate at the beginning of the powerstroke.

Figure 11.16 shows results from numerical solution of the model. Figure 11.16(a) shows roughly hyperbolic, isotonic-velocity-vs.-tension plots as are experimentally observed for simulated substrate concentrations of 5 mM and 50 μM. Furthermore, the modeled V_{max} exhibits Michaelian behavior with respect to substrate concentration (simulations not shown). At low velocities, the basic interaction length is short, $\sim$ 7nm. For higher velocities, passenger bridges appear in the model and give an average interaction distance approaching 70nm per ATP hydrolyzed. One additional observation comes from the analysis. Figure 11.16(b) shows relative ATPase and mechanical stiffness as a function of normalized tension during releases, assuming a physiological, 5 mM ATP concentration. Mechanical stiffness decreases in the simulations with increasing velocity. This is consistent with the experimental observations of Julian and Sollins (1975).

In terms of the original Huxley model, decreasing stiffness implies a decreasing fraction of attached crossbridges with increasing velocity. In the current model, this results from our assumption of a finite, spatially dependent attachment rate in the powerstroke. However, the decreasing attached fraction, coupled with the increasing numbers of passenger crossbridges, results in a surprising model prediction. As shortening velocity increases from zero, ATPase initially increases. However, with a continued increase in ve-

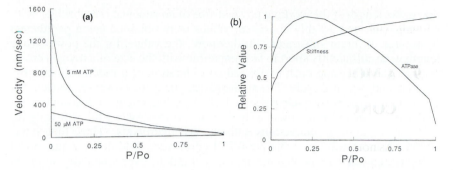

Figure 11.16. Simulation results from a composite model. (a) Velocity (nm/sec) at simulated [ATP] of 5 mM (upper plot) and 50 μM (lower plot) as a function of normalized tension. (b) Simulated relative ATPase and stiffness for 5 mM ATP as a function of relative tension. Parameters for the States 1–4 as defined in text: Let $G_i(x)$ be the relative free energy of State i in units of kT, $\kappa = 0.3\ kT/\text{nm}^2$ be the crossbridge elastic-force constant, and $R_{ij}(x)$ be the transition rate from State i to State j. $G_1(x) = 0$, $G_2(x) = -15 + \kappa x^2$, $G_3(x) = -19 + \kappa x^2$, $G_4(x) = -20 + \kappa x^2$, $\Delta G_{ATP} = -23$. For $0\ \text{nm} < x < 7\ \text{nm}$, $R_{12} = 1.5\text{s}^{-1}$, $R_{23} = 6\text{s}^{-1}$, $R_{34} = 3\text{s}^{-1}$, $R_{41} = 3 \times 10^6\text{M}^{-1}\text{s}^{-1}$; for $x \geq 7\ \text{nm}$, rates are identical to $0 < x < 7\ \text{nm}$, except $R_{12} = 200\text{s}^{-1}$; for $-7.5\ \text{nm} \leq x \leq 0\ \text{nm}$, $R_{12} = 0$, $R_{23} = 60\text{s}^{-1}$, $R_{34} = 2000\text{s}^{-1}$, $R_{41} = 3 \times 10^6\text{M}^{-1}\text{s}^{-1}$; for $x \leq -7.5\ \text{nm}$, rates are identical to $-7.5\ \text{nm} \leq x \leq 0\ \text{nm}$, except $R_{23} = 10^5\text{s}^{-1}$. In all cases the reverse transition rates are determined as $R_{ji}(x) = R_{ij}(x)\exp[G_j(x) - G_i(x)]$. Additional details of these simulations are provided in Cooke et al. (1994).

locity, eventually the decrease in the attached fraction, coupled with the increased fraction of crossbridges which use the weakly coupled pathway for detachment, results in an overall decrease in the ATPase rate! This was the observation by Hill (1964) that the original Huxley model failed to duplicate. A potential explanation has come quite naturally, and unexpectedly, out of our consideration of the relationships between weakly and strongly coupled crossbridge models. Observe that it truly is the combination of both a decreased attached fraction and the presence of passenger crossbridges which allows the model to exhibit a decreased ATPase at high velocities. In the model of Section 11.8 which contained passenger crossbridges but a velocity-independent attached fraction through the powerstroke, ATPase was proportional to $V_n f(V_n) = V_n[1 - \exp(-0.2/V_n)]$. This function is monotonically increasing for $0 < V_n \leq 1$, and there is no decrease in ATPase with increasing velocity.

In summary, working at the level of the individual motor proteins, we have developed a mathematical model for the generation of force and motion by muscle. The analysis has relied heavily upon the original crossbridge model of A. F. Huxley. We have extended his fundamental ideas, however, to incorporate a more realistic representation of the specific biochemical and chemomechanical states which occur in the contractile cycle. Our analysis led us to the model hypothesis that muscle-shortening velocity should follow Michaelis-Menten saturation behavior as a function of substrate concentration. The model prediction was experimentally tested and verified. The parameters in the Michaelis-Menten relationship between the substrate concentration and velocity suggested two approaches by which a combination of biochemistry, muscle mechanics, and mathematical modeling could address the question of the length of the working powerstroke of an attached crossbridge. With one approach, remarkably consistent values were obtained using parameters from three different substrates. Furthermore, the values for the powerstroke length were sufficiently short to be compatible with the size of a myosin crossbridge. The other approach motivated consideration of a modification of the fundamental working cycle. The mathematical analysis, however, showed that the revised crossbridge cycle presented a thermodynamically consistent crossbridge mechanism for retaining a short, basic powerstroke, while allowing for multiple actomyosin interactions without ATP hydrolysis. Thus, long interaction distances were possible per ATP hydrolyzed. Additionally, the model analysis suggested a potential mechanism behind the experimentally observed decrease in ATP hydrolysis at large velocities of shortening.

11.10   ACKNOWLEDGMENTS

This work was supported by USPHS grant AR39643.

REFERENCES

Aubert, Xavier. 1956. *Le couplage énergetique de la contraction musculaire.* Brussels, Belgium: Éditions Arscia. Page 315. Thèse d'agrégation — Université Catholique de Louvain. {*236, 252*}

Bagshaw, C. R. 1993. *Muscle Contraction.* Second edn. London, UK: Chapman and Hall, Ltd. ISBN 0-412-40370-6. Pages x + 155. {*223, 224, 252*}

Brokaw, C. J. 1976. Computer simulation of movement generating cross-bridges. *Biophysical Journal*, **16**, 1013–1027. {*228, 236, 248, 252*}

Cooke, R. 1986. The mechanism of muscle contraction. *CRC Critical Reviews in Biochemistry*, **21**, 53–118. {*224, 240, 241, 252*}

Cooke, R., White, H. D., and Pate, E. 1994. A model of the release of myosin heads from actin in rapidly contracting muscle. *Biophysical Journal*, **66**(3), 778–788. {*250–252*}

Finer, J. T., Simmons, R. M., and Spudich, J. A. 1994. Single myosin molecule mechanics: PicoNewton forces and nanometre steps. *Nature*, **368**(6467), 113–119. {*239, 252*}

Gordon, A. M., Huxley, A. F., and Julian, F. J. 1966. The variation in isometric tension with sarcomere length in vertebrate muscle fibers. *Journal of Physiology*, **184**, 170–192. {*228, 252*}

Harada, Y., Sakurada, T., Aoki, T., Thomas, D. D., and Yanagida, T. 1990. Mechanochemical coupling in actomyosin energy transduction studies in an in vitro movement assay. *Journal of Molecular Biology*, **216**, 49–68. {*239, 252*}

Higuchi, H. and Goldman, Y. E. 1991. Sliding distance between actin and myosin filaments per ATP molecule hydrolysed in skinned muscle fibres. *Nature*, **352**(6333), 352–354. {*239, 249, 252*}

Higuchi, H., Yanagida, T., and Goldman, Y. E. 1995. Compliance of thin filaments in skinned fibers of rabbit skeletal muscle. *Biophysical Journal*, **69**(3), 1000–1010. {*238, 252*}

Hill, A. V. 1938. The heat of shortening and dynamic constants of muscle. *Proceedings of the Royal Society of London. Series B. Biological sciences*, **126**, 136–195. {*232, 236, 237, 252*}

Hill, A. V. 1964. The effect of load on the heat of shortening muscle. *Proceedings of the Royal Society of London. Series B. Biological sciences*, **159**, 297–318. {*237, 238, 251, 252*}

Hill, T. L. 1974. Theoretical formalism for the sliding filament model of contraction of striated muscle, Part I. *Progress in Biophysics and Molecular Biology*, **28**, 267–340. {*247, 253*}

Huxley, A. F. 1957. Muscle structure and theories of contraction. *Progress in Biophysics and Biophysical Chemistry*, **7**, 255–318. {*221, 224, 253*}

Huxley, A. F. 1973. A note suggesting that cross-bridge attachment during muscle contraction may take place in two stages. *Proceedings of the Royal Society of London. Series B. Biological sciences*, **183**, 83–86. {*238, 253*}

Huxley, A. F. and Niedergerke, R. 1954. Interference microscopy of living muscle fibres. *Nature*, **173**, 971–973. {*238, 253*}

Huxley, A. F. and Simmons, R. M. 1971. Proposed mechanism of force generation in striated muscle. *Nature*, **233**, 533–538. {*228, 230, 239, 253*}

Huxley, H. E. and Hanson, J. 1954. Changes in the cross-striations of muscle during contraction and stretch and their structural interpretation. *Nature*, **173**, 973–976. {*238, 253*}

Huxley, H. E. and Kress, M. 1985. Cross-bridge behaviour during muscle contraction. *Journal of Muscle Research Cell Motility*, **6**, 153–161. {*239, 253*}

Julian, F. J. and Sollins, M. R. 1975. Variation of muscle stiffness with force at increasing speeds of shortening. *Journal of General Physiology*, **66**, 287–302. {*250, 253*}

Malcolm, I. 1991. Motor proteins: Biomechanics goes quantum. *Nature*, **352**(6333), 284–286. {*249, 253*}

McMahon, T. A. 1984. *Muscles, Reflexes, and Locomotion*. Princeton, NJ, USA: Princeton University Press. ISBN 0-691-08322-3, 0-691-02376-X (paperback). Pages xvi + 331. {*224, 235, 253*}

Molloy, J. E., Kendrick-Jones, J., Burns, J. E., Tregear, R. T., and White, D. C. S. 1995. Movement and force produced by a single myosin head. *Nature*, **378**(6553), 209–212. {*239, 253*}

Pate, E. and Cooke, R. 1989. A model of crossbridge action: the effects of ATP, ADP, and Pi. *Journal of Muscle Research Cell Motility*, **10**, 181–196. {*238, 241, 245, 253*}

Pate, E., Franks-Skiba, K., White, H. D., and Cooke, R. 1993a. The use of differing nucleotides to investigate cross-bridge kinetics. *The Journal of Biological Chemistry*, **268**(14), 10046–10053. {*232, 243, 253*}

Pate, Edward, White, Howard D., and Cooke, Roger. 1993b. Determination of the myosin step size from mechanical and kinetic data. *Proceedings of the National Academy of Sciences of the United States of America*, **90**(6), 2451–2455. {*241, 245, 253*}

Schoenberg, M. 1988. Characterization of the myosin adenosine triphosphate (M•ATP) crossbridge in rabbit and frog skeletal muscle fibers. *Biophysical Journal*, **54**, 135–148. {*248, 253*}

Sellers, James R. and Goodson, Holly V. 1995. *Protein Profile. Motor Proteins 2: Myosin*. Protein profile, vol. 2(12). New York, USA: Academic Press. Page 1423. {*224, 253*}

Taylor, Edwin W. 1991. Kinetic studies on the association and dissociation of myosin subfragment 1 and actin. *The Journal of Biological Chemistry*, **266**(1), 294–302. {*245, 253*}

Uyeda, T. Q. P., Kron, S. J., McNalley, E. M., Niebling, K. R., Toyoshima, C., and Spudich, J. A. 1990. The myosin step size: estimation from slow sliding movement of actin over low densities of heavy meromyosin. *Journal of Molecular Biology*, **214**, 699–710. {*239, 253*}

White, H. D., Belknap, B., and Jiang, W. 1993. Kinetics of binding and hydrolysis of a series of nucleoside triphosphates by actomyosin-S1: relationship between solution rate constants and properties of muscle fibers. *The Journal of Biological Chemistry*, **268**(14), 10039–10045. {*244, 253*}

Woledge, R., Curtin, N., and Homsher, E. 1985. *Energetic Aspects of Muscle Contraction*. New York, USA: Academic Press. ISBN 0-12-761580-6. Pages xiii + 359. {*223, 224, 254*}

Yanagida, T., Arata, T., and Oosawa, F. 1985. Sliding distance of actin filament induced by a myosin cross-bridge during one ATP hydrolysis cycle. *Nature*, **316**, 366–369. {*239, 254*}

12. THE TOPOLOGY OF PHASE RESETTING AND THE ENTRAINMENT OF LIMIT CYCLES

Leon Glass

12.1   INTRODUCTION

Biological rhythms are ubiquitous. Their periods of oscillation range from fractions of a second to a year. Independent of the period of the oscillation and the precise mechanism underlying the generation of the oscillation, certain underlying mathematical concepts are broadly applicable. Appropriate stimuli delivered to the oscillators usually induce a resetting of the oscillation, so that the timing of the oscillation will be different from what it would have been without the stimulus. Occasionally, a stimulus delivered during an oscillation will terminate the oscillation, or lead to a different oscillation. Determining the response of an oscillator to perturbations administered at different phases of the cycle can give important information about the oscillator, and also may be useful in determining its behavior in the fluctuating environment. It is likely that in every subarea of biology in which oscillations are observed, there is a literature analyzing the oscillations from the idiosyncratic perspective of the particular discipline. Yet, from a mathematical perspective there is a commonality of ideas and approaches (Pavlidis 1973; Guckenheimer 1975; Kawato and Suzuki 1978; Kawato 1981; Winfree 1980; Guevara et al. 1981; Glass and Winfree 1984; Winfree 1987; Glass and Mackey 1988).

The main notions are:

- Resetting can be measured experimentally as a function of the phase of the stimulus.

- Continuity and topological properties of the dynamical systems generating the oscillations are probed by the resetting experiments.

- If one assumes that the effects of a single stimulus are known as a function of its phase, then provided the stimulus does not change the properties of the oscillator, and there is rapid relaxation back to the original oscillation, the effects of periodic stimulation at different frequencies can be computed.

My object in this chapter is to sketch out the basic mathematical concepts pointing out a number of open problems and difficulties. The actual applications of the mathematics in concrete biological situations are widespread. Applications stressed here involve cardiac electrophysiology, but the same methods are equally applicable to perturbation of neural rhythms with electrical shocks, modulation of respiration by lung inflation, resetting the circadian rhythms of plants and animals by light (e.g., Winfree (1980); Winfree (1987); Glass and Mackey (1988)). However, only rarely is there a critical testing of the theory by comparing experimental observations with theoretical computations. Thus, there is a need for combined theoretical and experimental studies testing the applicability of these methods in specific applications.

12.2   MATHEMATICAL BACKGROUND

12.2.1   Isochrons and the perturbation of biological oscillations by a single stimulus

Since biological oscillations often have "stable" periods and amplitudes (coefficient of variation of the order of 3%), it is usual to associate the oscillation with a stable limit cycle in some appropriate nonlinear theoretical model (Winfree 1980). Recall that a stable *limit cycle* is a periodic solution of a differential equation that is attracting in the limit of $t \to \infty$ for all points in the neighborhood of the limit cycle (see Appendix B). Say that the period of the oscillation is T_0. We will designate a particular event to be the fiducial event, designated as phase, $\phi = 0$. The *phase* at any subsequent time $t > 0$ is defined to be $\phi = t/T_0$ (mod 1). The phase here is defined to lie between 0 and 1; to convert it to radians, multiply it by 2π.

An illustration of the concept of a limit cycle in a concrete experimental setting is shown in Figure 12.1. A single stimulus delivered to a spontaneously beating aggregate of cells from embryonic chick heart leads to a rapid reestablishment of the original oscillation. This is experimental evidence that the rhythm is being generated by a stable-limit-cycle oscillation.

The set of all initial conditions that attract to the limit cycle in the limit $t \to \infty$ is called the *basin of attraction* of the limit cycle. Let $x(t)$ be on a limit cycle at time t and $y(t)$ be in the basin of attraction of the limit cycle. Denote the distance between a and b by $d[a, b]$. Let the phase of x at $t = 0$ be ϕ. Then, if in the limit $t \to \infty$,

$$d[x(t), y(t)] = 0,$$

the *latent* or *asymptotic phase* of $y(t)$ is also ϕ. We say that $y(t)$ is on the same *isochron* as $x(t)$.

The development of the concept of isochrons and the recognition of their significance is due to Winfree (1980). Many important mathematical results concerning isochrons were established by Guckenheimer (1975), who considered dynamical systems in n-dimensional Euclidean space. He proved the existence of isochrons and showed that every neighborhood of every point on the frontier of the basin of attraction of a limit cycle intersects every isochron. Moreover, the dimension of the frontier of the basin of attraction is $\geq n - 2$.

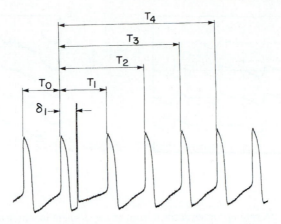

Figure 12.1. Resetting the intrinsic rhythm in a spontaneously beating aggregate of cells from embryonic chick heart. A single stimulus delivered at a phase $\phi = \delta_1/T_0$ leads to a resetting of the oscillation. The time from the action potential before the stimulus to the jth action potential after the stimulus is designated T_j. The reestablishment of an oscillation with the same amplitude and period as before the stimulus is evidence for a stable limit-cycle-oscillation in this preparation. (From Zeng et al. (1992).)

We now consider the effects of perturbations delivered to the biological oscillation. Assume that a perturbation delivered to an oscillation at phase ϕ shifts the oscillation to the latent phase $g(\phi)$. The function $g(\phi)$ is called the *phase transition curve*. The following *continuity rule* summarizes important aspects of the effects of perturbations on limit-cycle oscillations: **Provided the perturbation always leaves the system in the basin of attraction of the limit cycle, then the phase-transition curve $g(\phi)$ is a continuous circle map so that $g : S^1 \to S^1$.** The continuity rule seems obvious, but as far as I know, a proof of it has not yet been published.

Since $g(\phi)$ is a circle map, it is characterized by its *(topological) degree* or *winding number*. The degree of $g(\phi)$ measures the number of times $g(\phi)$ wraps around the unit circle as ϕ goes around the circle once. For example, if the perturbation is very weak, $g(\phi) \approx \phi$, and the degree is 1. In many situations, as Winfree (1980) discusses, the degree is 0 when the stimulation is strong. If the degree is 1 for weak stimuli and 0 for strong stimuli, there must be an intermediate stimulus (or stimuli) that will perturb the system outside of the basin of attraction of the limit cycle — though whether the limit cycle is eventually reestablished depends on whether the stimulus perturbs the system to the basin of attraction of another stable attractor.

These notions are directly related to experiment. The phase-transition curve can be measured experimentally. Assume once again that the marker event of an oscillation is defined as $t = 0$, $\phi = 0$. Assume that in response to a perturbation delivered at phase ϕ marker events recur at successive times $T_1(\phi), T_2(\phi), \ldots, T_n(\phi)$. Let us assume that for all j sufficiently large, the limit cycle is asymptotically approached, so that $T_j(\phi) - T_{j-1}(\phi) = T_0$, where T_0 is the control cycle length.

An experiment is illustrated in Figure 12.2. The panel on the left is typical of weak stimulation; the panel on the right, of strong stimulation.

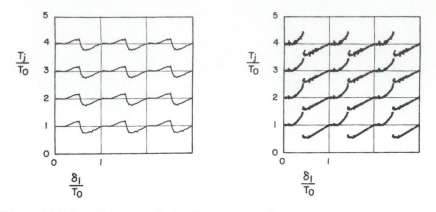

Figure 12.2. Resetting curves derived from an experiment in which a single stimulus is delivered to spontaneously beating heart-cell aggregates. The results are triple plotted. A stimulus of 13 nA gives weak resetting, and a stimulus of 26 nA gives strong resetting. The time from the action potential before the stimulus to the jth action potential after the stimulus is plotted as a function of the phase of the stimulus. (From Zeng et al. (1992).)

The phase-transition curve can be determined from the data given in Figure 12.2. It is given by

$$g(\phi) = \phi - \frac{T_j(\phi)}{T_0} \pmod 1 \tag{12.1}$$

Winfree (1980) gives many examples of resetting biological oscillators. The degree of the experimentally measured phase-transition curve is usually 1 or 0, though in some cases it was discontinuous, see, for example, pages 172, 318, 383, 438, 456 in Winfree (1980). Though most experimentalists are not much bothered by discontinuities in resetting experiments, understanding their origin is a challenge (Glass and Winfree 1984). I return to a discussion of the significance of discontinuities in resetting experiments in a later section.

12.2.2 Phase locking of limit cycles by periodic stimulation

The effects of periodic stimulation of a nonlinear oscillator can be modeled by a nonlinear map. Earliest studies involved the computation of the effects of periodic stimulation on a stretch receptor (Perkel et al. 1964) and circadian rhythms (Pavlidis 1973). The methods have been applied to a broad range of other problems subsequently. For an introduction to the basic theory and extensive references, see Glass and Mackey (1988).

Consider the effects of periodic stimulation with period t_s of a limit cycle with intrinsic period T_0. We assume that the stimulus instantaneously resets the oscillation, that the properties of the limit cycle are not affected by the stimulus, and that following the stimulus, the limit cycle is reestablished on a time scale that is fast compared to the period of stimulation t_s. Call ϕ_i the phase of stimulus i. Then, if the phase-transition curve is $g(\phi_i)$, the effects of periodic stimulation are given by

$$\phi_{i+1} = g(\phi_i) + \tau \pmod 1 \equiv f(\phi_i, \tau), \tag{12.2}$$

where $\tau = t_s/T_0$. Starting from an initial condition ϕ_0, we generate the sequence of points $\phi_1, \phi_2, \ldots, \phi_n$.

The sequence $\{\phi_i\}$ is well defined, provided no stimulus results in a resetting to a point outside the basin of attraction of the limit cycle. If $\phi_n = \phi_0$ and $\phi_i \neq \phi_0$ for $1 \leq i < n$, where i and n are positive integers, there is a periodic cycle of period n. A periodic cycle of period n is stable if

$$\left| \frac{\partial f^n(\phi_0)}{\partial \phi} \right| = \prod_{i=0}^{n-1} \left| \frac{\partial f}{\partial \phi} \right|_{\phi_i} < 1. \tag{12.3}$$

The *rotation number*, ρ, gives the average increase in ϕ per iteration. Calling

$$\Delta_{i+1} = g(\phi_i) + \tau - \phi_i, \tag{12.4}$$

we have

$$\rho = \limsup_{N \to \infty} \frac{1}{N} \sum_{i=1}^{N} \Delta_i. \tag{12.5}$$

Stable periodic orbits are associated with *phase locking*. In $n:m$ phase locking, there is a periodic orbit consisting of n stimuli and m cycles of the oscillator leading to a rotation number m/n. For periodically forced oscillators neither the periodicity nor the rotation number alone is adequate to characterize the dynamics.

12.3   THE POINCARÉ OSCILLATOR

We illustrate these concepts in a very simple ordinary differential equation that has been used extensively as a theoretical model in biology. Since this prototypical example of a nonlinear oscillation was first used by Poincaré as an example of stable oscillations, it has been called the Poincaré oscillator. However, it has been presented under a variety of other names in its 100-year history (Glass and Mackey 1988).

The Poincaré oscillator is probably the simplest differential equation that displays a stable-limit-cycle oscillation (Figure 12.3), and it has been considered many times as a model of biological oscillations (Winfree 1980; Guevara and Glass 1982; Hoppensteadt and Keener 1982; Keener and Glass 1984; Glass and Mackey 1988; Glass and Sun 1994). The model has uncanny similarities to experimental data and has been useful as a conceptual model to help one think about the effects of periodic stimulation of cardiac oscillators.

The Poincaré oscillator is most conveniently written in a polar coordinate system, where r is the distance from the origin and ϕ is the angular coordinate. The equations are written

$$\frac{dr}{dt} = kr(1 - r), \tag{12.6}$$

$$\frac{d\phi}{dt} = 2\pi, \tag{12.7}$$

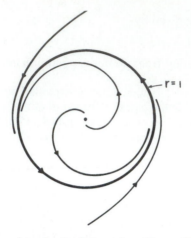

Figure 12.3. The phase plane for the Poincaré oscillator. (Reprinted from Glass and Mackey (1988).)

where k is a positive parameter. Starting at any value of r, except $r = 0$, the solution approaches the circle $r = 1$. The parameter k controls the relaxation rate. To make the connection with experiments in cardiac electrophysiology, such as the one shown in Figure 12.1, I assume that the phase, $\phi = 0$, corresponds to the upstroke of the action potential or the onset of the contraction.

Since the rate of change of ϕ is not a function of r, the isochrons are open sets lying along radii of the coordinate system. In this case the frontier of the basin of attraction of the limit cycle is the origin. The dimension of the frontier of the isochrons is 0, which is $\geq (n-2)$, in accord with Guckenheimer's theorem (see Figure 12.4(a)).

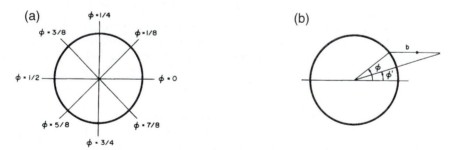

Figure 12.4. (a) Isochrons in the Poincaré oscillator. (b) A stimulus is assumed to induce a horizontal translation b. Poincaré oscillator. (Reprinted from Glass and Mackey (1988).)

We assume that perturbations are modeled by a horizontal translation to the right by a distance b (Figure 12.4 (b).) In the experimental setting shown in Figure 12.1, the perturbation is an electrical stimulus that depolarizes the membrane. A stimulus induces (after a delay) a new action potential if it is delivered in the latter part of the cycle.

This theoretical model facilitates analytical work because of its comparatively simple analytical form. The phase-transition curve, $g(\phi)$ is readily computed and is given by

$$g(\phi) = \frac{1}{2\pi} \arccos \frac{\cos 2\pi\phi + b}{(1 + b^2 + 2b\cos 2\pi\phi)^{1/2}} (\text{mod } 1). \qquad (12.8)$$

In computations using (12.8), in evaluating the arccosine function, take $0 < g(\phi) < 0.5$ for $0 < \phi < 0.5$, and $0.5 < g(\phi) < 1$ for $0.5 < \phi < 1$. In Figure 12.5, I plot the perturbed cycle length and the phase-transition curve for the Poincaré oscillator. The results in Figure 12.5 should be compared with the experimental data in Figure 12.2.

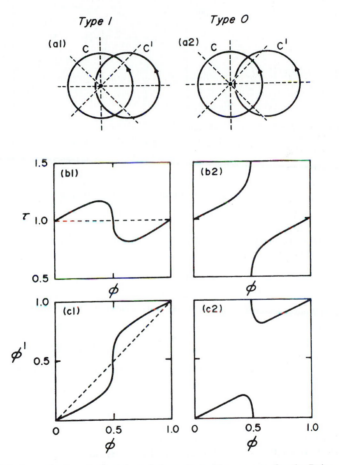

Figure 12.5. Perturbed cycle length and phase-transition curves for the Poincaré oscillator for weak and strong stimuli. (Reprinted from Glass and Winfree (1984).)

The effects of periodic stimulation can now be computed by application of (12.2) and (12.8). The geometry of the locking zones is very complicated; a partial representation is shown in Figure 12.6. Here I summarize several important properties. For further details the original references (Guevara and Glass 1982; Keener and Glass 1984; Glass and Sun 1994) should be consulted.

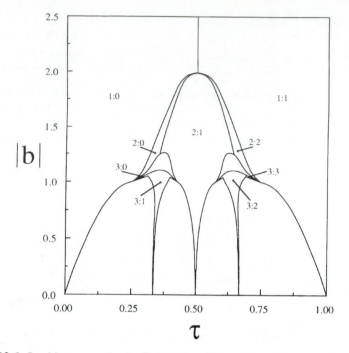

Figure 12.6. Locking zones for the Poincaré oscillator. (Based on Keener and Glass (1984).)

There are symmetries in the organization of the locking zones as originally derived in Guevara and Glass (1982). The symmetries are:

- *Symmetry 1.* Assume that there is a stable period-n cycle with fixed points $\phi_0, \phi_1, \ldots, \phi_{n-1}$ for $\tau = 0.5 - \delta$, $0 < \delta < 0.5$, associated with $n : m$ phase locking. Then, for $\tau = 0.5 + \delta$, there will be a stable cycle of period n associated with an $n : n - m$ phase-locking ratio. The n fixed points are $\psi_0, \psi_1, \ldots, \psi_{n-1}$, where $\psi_i = 1 - \phi_i$.

- *Symmetry 2.* Assume that there is a stable period-n cycle with fixed points $\phi_0, \phi_1, \ldots, \phi_{n-1}$ for $\tau = \delta$, $0 < \delta < 1.0$, associated with $n : m$ phase locking. Then for $\tau = \delta + k$, where k is a positive integer, there will be a stable cycle of period n associated with an $n : m + nk$ phase-locking ratio. The n fixed points are $\psi_0, \psi_1, \ldots, \psi_{n-1}$, where $\psi_i = 1 - \phi_i$.

Symmetry 1 is satisfied in Figure 12.6. Using the translational symmetry, symmetry 2, the zones in Figure 12.6 can be expanded to cover the region $\tau > 1$.

I now summarize the main features of the organization of the locking zones. The topology of $g(\phi)$ changes at $b = 1$, and this has profound effects on the organization of the locking zones.

$0 \le b < 1$. The map is an invertible differentiable map of the circle (Arnold 1983). An *Arnold tongue of rotation number m/n* is defined as the union of values in parameter space for which there is unique attracting $n : m$ phase locking for all initial conditions. For invertible diffeomorphisms of the circle of the form in (12.2), if there is $n : m$ phase locking for τ and $n' : m'$ phase locking for τ', then there exists a value $\tau < \tau* < \tau'$, leading to $n + n' : m + m'$ phase locking. Usually, the range of values of τ associated with a given Arnold tongue covers an open interval in parameter space. For a given set of parameters the rotation number is unique. If it is rational, there is phase locking, and if it is irrational there is *quasi-periodicity*. The organization of phase-locking zones for $0 \le b < 1$ shown in Figure 12.6 for $b < 1$ is typical and is called the *classic Arnold-tongue structure*. The periodic orbits lose stability via a tangent bifurcation.

1 < b. The map now has two local extrema. For any set of parameter values there is no longer necessarily a unique attractor. It is possible to have *bistability*, in which there exist two stable attractors for a given set of parameter values. The attractors are either periodic or chaotic. A *superstable cycle* is a cycle containing a local extremum. Such cycles are guaranteed to be stable. One way to get a good geometric picture of the structure of the zones is to plot the locus of the superstable cycles in the parameter space. The structure of bimodal interval maps and circle maps has been well studied and shows complex cascades of bifurcations in the two-dimensional parameter space. As b decreases in this zone, new phase-locking zones arise; however, almost all these zones disappear into the discontinuities of the circle map at $b = 1$. There are accumulation points of an infinite number of periodic orbits at the junction of the Arnold tongues with the line $b = 1$.

Analytic expressions for some of the bifurcations can be derived. For $0 < b < 1$ the stability is lost by a tangent bifurcation for which $\partial \phi_{i+1} / \partial \phi_i = 1$. This implies that at the boundary we have

$$b + \cos 2\pi \phi_0 = 0,$$

from which we compute

$$b = |\sin 2\pi \tau|. \tag{12.9}$$

The fixed point at the stability boundary is at

$$\phi_0 = \tau + \frac{1}{4}, \quad \text{for } 0 < \tau < \frac{1}{4}, \tag{12.10}$$

and

$$\phi_0 = \tau + \frac{3}{4}, \quad \text{for } \frac{3}{4} < \tau < 1. \tag{12.11}$$

For $1 < b < 2$, stability of the period-1 fixed point is lost by a period-doubling bifurcation for which $\partial \phi_{i+1} / \partial \phi_i = -1$. From this we compute that at the boundary we have

$$2 + b^2 + 3b \cos 2\pi \phi_0 = 0. \tag{12.12}$$

Carrying through the trigonometry, we find the stability boundary

$$b = \sqrt{4 - 3 \sin^2 2\pi \tau}. \tag{12.13}$$

The fixed point at the boundary is given by

$$\phi_0 = \tau + \frac{1}{2\pi} \sin^{-1} \sqrt{\frac{4 - b^2}{3b^2}}. \tag{12.14}$$

It is not generally appreciated that in this system there can be changes in the rotation number without a change in periodicity (Guevara and Glass 1982). For example, for $2 < b$, there is a change from 1:0 phase locking to 1:1 phase locking along the line $\tau = 0.5$.

12.4   PERIODIC STIMULATION OF A CARDIAC PACEMAKER

The computational machinery outlined above can be applied in practical situations. I will very briefly recount work from our group and give references to more complete descriptions.

Extensive studies of the effects of single and periodic stimulation on spontaneously beating aggregates of embryonic chick-heart cells have been carried out by Michael Guevara, Wanzhen Zeng, and Arkady Kunysz, working in Alvin Shrier's laboratory. The object has been to determine the phase resetting-behavior under single stimuli and to apply these results to compute the effects of periodic stimulation (Guevara et al. 1981; Guevara et al. 1986; Glass et al. 1983; Glass et al. 1984; Glass et al. 1987; Guevara et al. 1988; Zeng et al. 1990b; Kowtha et al. 1994). The results of these studies are shown in Figure 12.7, which summarizes computations of the different locking regions by numerical iteration of experimentally-determined resetting curves, using the methods described above, and shows examples of representative rhythms. The main findings of the experimental studies are:

- There are many different phase-locking regions. For low to moderate stimulation amplitudes: the largest zones which can be readily observed in every experiment are 1:1, 1:2, 3:2, 2:1, 3:1, 2:3. In addition, other zones corresponding to rational ratios $n : m$, where n and m are 4 or less, can usually be observed near the theoretically predicted region in Figure 12.7.

- For several different sets of stimulation amplitude and frequency there are aperiodic dynamics (Guevara et al. 1981). A particular zone, using moderate stimulation amplitude and frequencies slightly less than the intrinsic frequency, leads to period-doubling bifurcations and deterministic chaos. In this region, plots of ϕ_{i+1} as a function of ϕ_i based on the experimental data are approximately one dimensional with characteristic shape associated with one-dimensional maps that give chaotic dynamics (Figure 12.8.)

12.5   BEYOND THE ONE-DIMENSIONAL MAP

Modeling the effects of periodic stimulation of a biological oscillator by a one-dimensional map shows agreement with experiment over an impres-

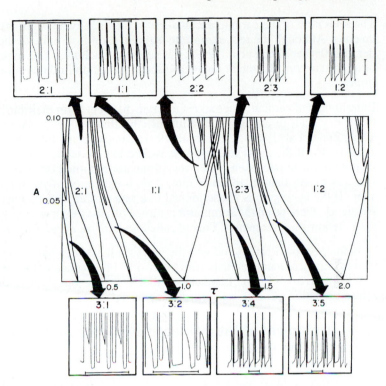

Figure 12.7. Locking zones for periodically stimulated heart-cell aggregates. The computations are based on experimentally measured resetting curves. The time bar is 1 sec. (Based on Glass et al. (1987).)

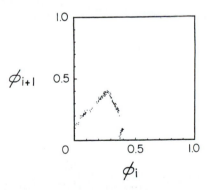

Figure 12.8. Return map for data obtained during aperiodic dynamics during periodic stimulation of spontaneously beating aggregates of chick-heart cells. The return map shows the phase of one stimulus plotted as a function of the phase of the preceding stimulus. This form for the map is similar to the quadratic map which is known to give chaotic dynamics. (Adapted from Glass et al. (1984).)

sive range of stimulation frequencies and amplitudes. Nevertheless, the situation is not perfect. In experimental work we have found situations in which two main assumptions of the one-dimensional map models break down.

- The relaxation of the cardiac pacemakers to the limit cycle is not instantaneous. Although the relaxation rate is rapid, estimated to be of the order of 100 msec (Zeng et al. 1992), it is nevertheless finite. This makes it of interest to consider the Poincaré oscillator in the finite-relaxation-time limit.

- Entrainment of cardiac pacemakers at a rate different from their intrinsic frequency leads to changes in the intrinsic beating frequency (Vassalle 1977; Zeng et al. 1990a; Zeng et al. 1991; Kunysz et al. 1995; Kunysz et al. 1996). If the aggregate is phase locked to a frequency more rapid than its intrinsic rate, then, when the stimulator is turned off, the oscillation is slower than its intrinsic rate. This is called *overdrive suppression*. If the aggregate is phase locked to a rate slower than its intrinsic frequency, then, when the stimulator is turned off, the frequency is higher than the intrinsic frequency. This is called *underdrive acceleration*.

In this section, I briefly recount the ways we have been incorporating these factors into theoretical and experimental work. I first consider the effects of finite relaxation times on the Poincaré oscillator. I then consider overdrive during the stimulation of heart-cell aggregates.

12.5.1   Entrainment of the Poincaré oscillator with finite relaxation times

As before, assume that a stimulus is schematically represented by a horizontal translation of magnitude b (Figure 12.4 (b)). The stimulus takes point (r, ϕ_i) to point (r_i', ϕ_i'), where

$$r_i' = (r_i^2 + b^2 + 2br_i \cos 2\pi \phi_i)^{1/2}, \tag{12.15}$$

$$\phi_i' = \frac{1}{2\pi} \arccos \frac{r_i \cos 2\pi \phi_i + b}{r_i'}. \tag{12.16}$$

Following the stimulus, the equations of motion take over, so that by direct integration, we find that immediately before stimulus $(i + 1)$ delivered at a time τ after the first stimulus, we have

$$r_{i+1} = \frac{r_i'}{(1 - r_i') \exp(-k\tau) + r_i'}, \tag{12.17}$$

$$\phi_{i+1} = \phi_i' + \tau \,(\text{mod } 1). \tag{12.18}$$

An important difference is present in the organization of locking zones — even for low stimulation amplitudes the classic Arnold-tongue structure described earlier does **NOT** apply. This fact does not seem to be widely appreciated. Even for low-amplitude stimulation, for any amplitude and frequency of stimulation there will always be a period-1 orbit for the map, i. e., a period τ orbit for the flow. In contrast, in the infinite relaxation limit, for $b < 1$, inside the Arnold tongues associated with locking of period $n \neq 1$, there is no period-1 cycle and all solutions in a tongue have the same rotation number. The existence of period-1 cycles follows immediately from an application of

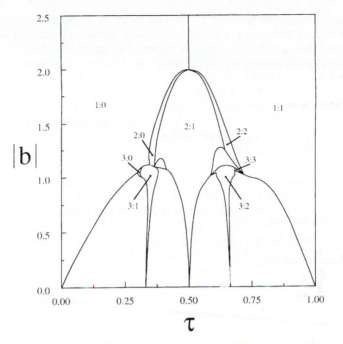

Figure 12.9. Locking zones for periodically stimulated Poincaré oscillator with finite relaxation times, $k = 10$. Based on Glass and Sun (1994).

the Brouwer fixed-point theorem. Consequently, the result is also applicable to a broad class of periodically stimulated oscillators and excitable systems, provided there is a sufficiently large contraction for large excursions from the limit cycle (Glass and Sun 1994). Of course, the period-1 cycle is not always stable, so that in experimental work it will often appear as though the classic Arnold tongue structure is being observed. After the publication of this result I found a similar result in Levinson (1944). The result deserves to be better known.

Finite relaxation to the limit cycle will also destroy the symmetries in the infinite-relaxation case. Moreover, the fine details of the locking zones change in subtle ways not yet well understood. For example, the points of accumulation of an infinite number of locking zones that occur at the intersection of the Arnold tongues with the line $b = 1$ have to "unfold" in some natural way. In Glass and Sun (1994) we observe that this unfolding appears to occur in a manner similar to that envisioned earlier by Arnold (1983); (see Fig. 153, p. 312 in Arnold (1983).)

12.5.2 Overdrive suppression

Phase locking of the heart-cell aggregates in a 1:1 fashion to periodic stimulation at a frequency faster than its intrinsic frequency leads to changes in the intrinsic frequency (Vassalle 1977; Zeng et al. 1990a; Zeng et al. 1991; Kunysz et al. 1995). When the stimulation is turned off, the intrinsic frequency will be slowed. The magnitude of the slowing depends both on the duration of the locking at the faster rate and on the rate itself. A simple theoretical model for overdrive suppression is that Na^+ builds up inside cells during the rapid stimulation. When the stimulation is turned off, the excess Na^+ is pumped out of the cells. The resulting (hyperpolarizing) current counteracts the pacemaker current slowing the beating rate until the normal Na^+ level is reestablished (Kunysz et al. 1995; Kunysz et al. 1996).

The presence of overdrive suppression raises a difficult question concerning the application of the theory to concrete situations. How should we determine the phase of a stimulus? In the earlier work it was assumed that the phase of a stimulus is simply the time interval from the start of an action to the start of the stimulus divided by the intrinsic cycle length. But if the intrinsic cycle length changes as a consequence of the stimulation, what cycle length should be used? Moreover, if the properties of the oscillator are changed as a consequence of the stimulation, does this also induce changes in the resetting curves?

Although the answers to these questions are not yet satisfactorily understood, some advances have been made. In computing the phase of a stimulus, it is necessary to take in the current value of the cycle length as it has been altered as a consequence of the periodic stimulation.

Consider what happens during stimulation at a rapid rate, but at a rate which still leads to a 1:1 locking rhythm. During the course of the stimulation, the intrinsic frequency of the oscillator will gradually decrease — it will beat at a slower rate. Consequently, the effective phase of the stimulus will fall earlier and earlier in the cycle. However, if the phase falls too early, it can no longer lead to an immediate action potential and there will be a missed beat. Such behavior is observed and has been analyzed by consideration of the points discussed here (Kunysz et al. 1996).

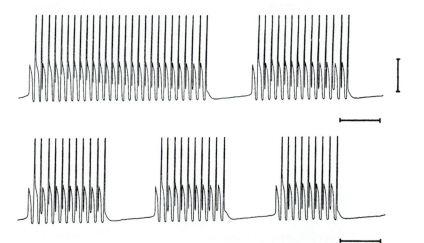

Figure 12.10. Bursting rhythms as a consequence of overdrive suppression during stimulation at a rate faster than the intrinsic rate. The time bar is 1 sec. (Based on Zeng et al. (1990a).)

To investigate these classes of effects more carefully, it is useful to consider the effects of stimulation at a fixed delay after the onset of an action potential. The original theory would assume that each stimulus has an identical effect. However, even for the Poincaré oscillator in the finite relaxation limit, stimulation at a fixed delay can lead to complex bifurcations. In work recently completed (Kunysz et al. 1996), complex dynamics were observed during fixed-delay stimulation — particularly for stimuli of moderate amplitude delivered at phases near the transition from shortening to lengthening in the plots of the perturbed cycle time. Development of a theory necessarily entails consideration of both the scaling of the cycle length and the resetting curves.

12.6   INFINITE-DIMENSIONAL SYSTEMS

Until recently, the preceding theory had been applied solely to finite-dimensional systems. However, some of the results might be equally applicable to infinite-dimensional systems. I consider two different problems: the resetting of a delay differential equation for negative feedback, and resetting of a reentrant rhythm in a ring of excitable media.

12.6.1   Resetting oscillations in delay differential equations modeling negative feedback

In physiology, time delays often are an essential feature of the control systems. It is well known that time delays in negative-feedback systems can lead to oscillations. A familiar example is the cycling on and off of a furnace during a cold Canadian winter. There is a long delay between the time the furnace turns on and the time the temperature at the thermostat is raised. Similarly, when the furnace turns off, there is a subsequent time lag until the temperature at the thermostat falls beneath the set point. Negative feedback in a delay differential equation was presented as a model for haematopoesis in Chapter 8 (see also Mackey and Glass (1977) and Glass and Mackey (1979)). In this context, let x be the level of circulating blood cells, and τ a delay. Assuming that the destruction of the blood cells takes place at a rate proportional to their concentration in the blood, but that the production is a nonlinear monotonically decreasing sigmoidal function $f(x_\tau)$ of the number of circulating cells x_τ at time $(t - \tau)$, we find

$$\frac{dx}{dt} = f(x_\tau) - x. \tag{12.19}$$

From a purely mathematical perspective, there are many open questions concerning the properties of this class of delay differential equations, see Hale and Lunel (1993). For computations, I assume

$$f(x_\tau) = \frac{1}{1 + x_\tau^6}. \tag{12.20}$$

For $\tau = 10$ there is a stable oscillation of length about 21.9.

The start of the cycle is arbitrarily defined as $x = 0.75$, $dx/dt > 0$. The stimulus is an instantaneous increase of x by 4. The stimulus is delivered at various phases of the cycle. The successive times of the start of the cycle are shown in Figure 12.11. Notice that there is a long transient — so that it takes approximately 10 cycles for the transients to dissipate. This is not surprising in view of the time delays in this system. However, the phase-transition curve, defined from (12.1), is type 0 — a strong resetting curve. I am aware of only one prior computation of resetting of limit cycles in a delay differential equation (Johnsson and Karlsson 1971). In view of the extremely long transients in this system, it is unlikely that one could accurately predict the effects of periodic stimulation by iteration of the phase-transition curve. Since delay differential equations are often appropriate models for blood disorders ((Mackey and Glass 1977; Glass and Mackey 1979); also see Chapter 8 in this volume for a detailed discussion and references), which are sometimes treated by periodic administration of drugs, additional study of the effects of periodic stimulation of delay differential systems would seem of interest.

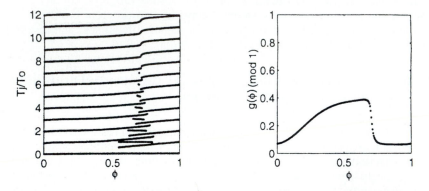

Figure 12.11. Resetting a limit cycle in a delay differential equation modeling negative feedback. The lefthand panel shows the time of successive starts of the cycle as a function of the phase of the stimulus. The righthand panel shows the associated phase-transition curve computed after all the transients are dissipated.

12.6.2   Resetting and entrainment of oscillations in partial differential equations

A second example of stable limit cycles in infinite-dimensional systems can be found in partial differential equations modeling excitable media. An excitable medium is one in which there is a large excursion from steady state in response to a small stimulus which is greater than some threshold. Nerve cells, cardiac tissue, and the Belousov-Zhabotinsky reaction are examples of excitable media and share many similar properties (Winfree 1980; Winfree 1987). A ring of excitable medium can support a circulating excitation — often called a reentrant wave. Reentrant waves have been demonstrated in a large number of experimental and theoretical systems (Quan and Rudy 1990; Rudy 1995; Courtemanche et al. 1993). They have a special importance to human health, since it is believed that many cardiac tachyarrhythmias (abnormally fast heart rhythms) are associated with reentrant mechanisms. There is a large cardiological literature that involves the resetting and entrainment of cardiac arrhythmias (Josephson et al. 1993; Gilmour 1990). However, it has not been recognized in cardiology that the methods developed to study resetting and phase locking might also be applicable to cardiology.

An initial attempt to apply this theory to a model system is given in two recent papers (Glass and Josephson 1995; Nomura and Glass 1996). I briefly summarize the main arguments and refer the reader to the original papers for full details.

The model system is the FitzHugh-Nagumo equations for excitable media (FitzHugh 1969):

$$
\begin{cases}
\dfrac{\partial v}{\partial t} = -v(v - 0.139)(v - 1) - w + I + D\dfrac{\partial^2 v}{\partial x^2}, \\[2mm]
\dfrac{\partial w}{\partial t} = 0.008(v - 2.54w).
\end{cases}
\tag{12.21}
$$

where D is a diffusion coefficient, I is a time- and space-dependent injected current, and the parameters are from Rinzel (1977). Parameters consistent with values appropriate for cardiac conduction are: the circumference is $L = 2 \times \sqrt{5}$ cm, $D = 1$ cm^2/sec, and cyclic boundary equations (Glass and Josephson 1995). The equations are integrated using the Euler method with $\Delta t = 0.1$ msec and $\Delta x = 0.005L$. These equations support a single circulating wave rotating around a ring with a period of about $T_0 = 356.1$ msec. The stimulation (injected current) is applied at a single grid point of the discretized equations with a magnitude I for 10 iteration steps (1 msec). Let $x_{\text{stim}} = 0.5$ be the locus where current is injected. This stimulus has one of three different effects, depending on the phase of the stimulus in the cycle.

- **No effect**. This will happen if the stimulus falls in the excited region. The time interval during which the stimulus has little effect is called the refractory period.

- **Initiation of two waves**. This will occur if the stimulus falls during the time the tissue is excitable. This time interval is sometimes called the excitable gap by cardiologists. The waves will propagate in both directions around the ring. Collision of one of these waves with the original wave will lead to an apparent resetting of the original oscillation.

- **Initiation of a single wave traveling in an opposite direction to the original wave**. This will occur when the stimulus falls in a narrow window just after the excitation. This window is sometimes called the vulnerable window by cardiologists (Starmer et al. 1993). Initiation of a single wave will lead to an annihilation of the original wave following collision of the two waves.

The resulting resetting curves and phase-transition curves for stimulation of an excitation in a ring of tissue are shown in Figure 12.12. Although the above results were found from computer simulation, a recent paper has argued that there must be a range of stimuli that lead to annihilation of the original wave based purely on the continuity principle discussed earlier and the arguments and the basic properties of reentrant excitation (Glass and Josephson 1995). Given the properties of excitable media, it is in general impossible to obtain continuous phase-transition curves. Therefore, there must be discontinuities associated with a range of phases that lead to annihilation. The effects

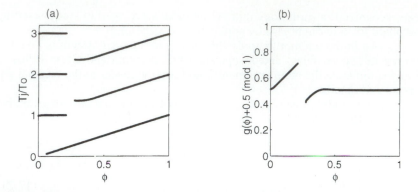

Figure 12.12. Resetting a limit cycle in a partial differential equation modeling reentrant excitation in a ring of tissue. The left-hand panel shows the time of successive starts of the cycle as a function of the phase of the stimulus. The right-hand panel shows the associated phase-transition curve computed after all the transients are dissipated. (From Nomura and Glass (1996).)

of periodic stimulation on a limit-cycle oscillation can be determined by iteration of the phase-transition curve just as was carried out in ordinary differential equations. Depending on the amplitude, the initial phase, the period of a train of stimuli, and the number of stimuli there can be either entrainment, resetting, or annihilation of a circulating wave. The dependence of these behaviors on the parameters in the model has a delicate structure that was not anticipated before the theory was carried out. The details are given in Nomura and Glass (1996). What is still missing is a clear demonstration that the theory really works in an experimental setting. I am confident that things will work out, but it is essential to carry out the experiments anyway.

One interesting aspect of this work is that resetting of reentrant excitations will always tend to be associated with discontinuous resetting curves. In view of the earlier observations that discontinuous resetting curves are often observed in experiments, it is possible that reentrant mechanisms might be a basic mechanism for a larger number of biological oscillators than is currently appreciated.

12.7   CONCLUSIONS

Single stimuli reset or annihilate stable nonlinear oscillations. Periodic stimuli delivered during the course of nonlinear oscillations can lead to a wide range of different behaviors, including quasi-periodicity, entrainment, chaos, and annihilation. The origin of all these different behaviors can be found in the iterations of low-dimensional maps. The number of examples that have been worked out carefully is still small.

In this review I have tried to cover a broad range, discussing the effects of single and periodic stimulation on nonlinear oscillations in biological systems. Although the theory has shown good agreement with experiment in a number of situations, there remain many additional experimental and theoretical problems that merit investigation.

- The stimulation can change the properties of the oscillator. For example, I have mentioned the phenomenon of overdrive suppression, in which the intrinsic period of the oscillation is increased as a consequence of rapid stimulation. Further work is needed to determine the best methods to include these effects in theoretical models.

- The relaxation time to the limit cycle may be long. Even in the Poincaré oscillator, when the relaxation time is finite, a host of theoretical problems have been barely analyzed. For example, multistability, chaos, and the bifurcations by which the stable periodic orbits lose stability have not yet been carefully investigated.

- There is little analysis of geometrical aspects of resetting and entrainment of limit-cycle oscillation in infinite-dimensional systems. For example, the structure of isochrons in infinite-dimensional systems has not been studied. Experiments are needed to focus the theory.

In order to understand the resetting and entrainment of limit-cycle oscillations in biological systems, there needs to be a mix of theory and experiment. Experimental studies of the effects of single and multiple stimuli delivered to biological oscillators often yield unambiguous data, in which the timing of key events can be measured over time. Since the rhythms that are observed depend on the parameters of the stimulation (amplitude of stimuli, frequency of stimuli, number of stimuli, initial phase of stimuli), systematic studies are essential. Since the detailed biochemical and physiological mechanisms generating rhythms are certainly different in different systems, a global geometric approach based on the mathematics of oscillations is essential.

12.8 ACKNOWLEDGMENTS

This chapter reviews work carried out with a number of colleagues. I have benefited greatly from these collaborations and give particular thanks to Michael Guevara, Alvin Shrier, Arthur T. Winfree, Zeng Wanzhen, Jiong Sun, Arkady Kunysz, and Taishin Nomura. This material is based on two series of lectures presented during the spring of 1996 — one during the Special Year for Mathematical Biology at the University of Utah, and the other at a Summer School in Nonlinear Dynamics in Biology and Medicine at McGill University. I appreciate the feedback from the students which helped shape this final summary. The work has been supported by the Natural Sciences and Engineering Council of Canada, the Canadian Heart and Stroke Foundation, and the Medical Research Council of Canada.

REFERENCES

Arnold, V. I. 1983. *Geometrical Methods in the Theory of Ordinary Differential Equations*. Grundlehren der mathematischen Wissenschaften, vol. 250. Berlin, Germany: Springer Verlag. ISBN 0-387-90681-9. Pages x + 334. Translated by J. Szucs; English translation edited by M. Levi. {*262, 267, 273*}

Courtemanche, M., Glass, L., and Keener, J. 1993. Instabilities of a propagating pulse in a ring of excitable media. *Physical Review Letters*, **70**(14), 2182–2185. {*270, 273*}

FitzHugh, R. 1969. Mathematical models of excitation and propagation in theoretical models of nerve. *Pages 1–85 of:* Schwan, Herman P., et al.(eds), *Biological Engineering*. Inter-university electronic series, vol. 9. New York, NY, USA: McGraw-Hill. {*271, 274*}

Gilmour, R. F. 1990. Phase resetting of circus movement reentry in cardiac tissue. *Pages 396–402 of:* Zipes, D. P. and Jalife, J. (eds), *Cardiac Electrophysiology: From Cell to Bedside*. Philadelphia, PA, USA: W. B. Saunders. {*270, 274*}

Glass, L. and Josephson, M. E. 1995. Resetting and annihilation of reentrant abnormally rapid heartbeat. *Physical Review Letters*, **75**(10), 2059–2063. {*271, 274*}

Glass, L. and Mackey, M. C. 1979. Pathological conditions resulting from instabilities in physiological control systems. *Annals of the New York Academy of Sciences*, **316**, 214–235. {*269, 270, 274*}

Glass, L. and Mackey, M. C. 1988. *From Clocks to Chaos: The Rhythms of Life*. Princeton, NJ, USA: Princeton University Press. ISBN 0-691-08495-5 (hardcover), 0-691-08496-3 (paperback). Pages xvii + 248. {*255, 256, 258–260, 274*}

Glass, L. and Sun, J. 1994. Periodic forcing of a limit cycle oscillator: Fixed points, Arnold tongues, and the global organization of bifurcations. *Physical Review E*, **50**, 5077–5084. {*259, 261, 267, 274*}

Glass, L. and Winfree, A. T. 1984. Discontinuities in phase-resetting experiments. *American Journal of Physiology*, **246**, R251–R258. {*255, 258, 261, 274*}

Glass, L., Guevara, M. R., Shrier, A., and Perez, R. 1983. Bifurcation and chaos in a periodically stimulated cardiac oscillator. *Physica*, **7D**, 89–101. {*264, 274*}

Glass, L., Guevara, M. R., Bélair, J., and Shrier, A. 1984. Global bifurcations of a periodically forced biological oscillator. *Physical Review A*, **29**, 1348–1357. {*264, 265, 274*}

Glass, L., Guevara, M. R., and Shrier, A. 1987. Universal bifurcations and the classification of cardiac arrhythmias. *Annals of the New York Academy of Sciences*, **504**, 168–178. {*264, 265, 274*}

Guckenheimer, J. 1975. Isochrons and phaseless sets. *Journal of Mathematical Biology*, **1**, 259–273. {*255, 256, 274*}

Guevara, M. R. and Glass, L. 1982. Phase locking, period doubling bifurcations, and chaos in a mathematical model of a periodically driven oscillator: A theory for the entrainment of biological oscillators and the generation of cardiac dysrhythmias. *Journal of Mathematical Biology*, **14**, 1–23. {*259, 261, 262, 264, 274*}

Guevara, M. R., Glass, L., and Shrier, A. 1981. Phase locking, period doubling bifurcations and irregular dynamics in periodically stimulated cardiac cells. *Science*, **214**, 1350–1353. {*255, 264, 274*}

Guevara, M. R., Shrier, A., and Glass, L. 1986. Phase resetting of spontaneously beating embryonic ventricular heart cell aggregates. *American Journal of Physiology*, **251**, H1298–H1305. {*264, 274*}

Guevara, M. R., Shrier, A., and Glass, L. 1988. Phase locked rhythms in periodically stimulated heart cell aggregates. *American Journal of Physiology*, **254**, H1–H10. Heart Circ. Physiol. 23. {*264, 274*}

Hale, J. K. and Lunel, S. M. V. 1993. *Introduction to Functional Differential Equations*. Applied mathematical sciences, vol. 99. Berlin, Germany: Springer Verlag. ISBN 0-387-94076-6 (New York), 3-540-94076-6 (Berlin). Pages x + 447. {*269, 274*}

Hoppensteadt, F. C. and Keener, J. 1982. Phase locking of biological clocks. *Journal of Mathematical Biology*, **15**, 339–349. {*259, 274*}

Johnsson, A. and Karlsson, H. G. 1971. Biological Rhythms: Singularities in Phase-Shift Experiments as Predicted from a Feedback Model. *Pages 263–267 of:* Broda, E., Locker, A., and Springer-Lederer, H. (eds), *First European Biophysics Congress, 14–17 Sept. 1971, Baden near Vienna, Austria*. Berlin, Germany: Springer Verlag. Six volumes. {*270, 274*}

Josephson, M. E., Callans, D., Almendral, J. M., Hook, B. G., and Kleiman, R. B. 1993. Resetting and entrainment of ventricular tachycardia associated with infarction: Clinical and experimental studies. *Pages 505–536 of:* Josephson, M. E. and Wellens, H. J. J. (eds), *Tachycardias: Mechanisms and Management*. The Bakken Research Center series, vol. 6. Mount Kisco, NY, USA: Futura. {*270, 274*}

Kawato, M. 1981. Transient and steady state phase response curves of limit cycle oscillators. *Journal of Mathematical Biology*, **12**, 13–30. {*255, 275*}

Kawato, M. and Suzuki, R. 1978. Biological oscillators can be stopped. Topological study of a phase response curve. *Biological Cybernetics*, **30**, 241–284. {*255, 275*}

Keener, J. P. and Glass, L. 1984. Global Bifurcations of a periodically forced nonlinear oscillator. *Journal of Mathematical Biology*, **21**, 175–190. {*259, 261, 262, 275*}

Kowtha, V. C., Kunysz, A., Clay, J. R., Glass, L., and Shrier, A. 1994. Ionic mechanisms and nonlinear dynamics of embryonic chick heart cell aggregates. *Progress in Biophysics and Molecular Biology*, **61**, 255–281. {*264, 275*}

Kunysz, A., Glass, L., and Shrier, A. 1995. Overdrive suppression of spontaneously beating chick heart cell aggregates: Experiment and theory. *American Journal of Physiology*, **269**, H1153–H1164. (Heart Circ. Physiol. 38). {*266–268, 275*}

Kunysz, A., Shrier, A., and Glass, L. 1996. Bursting behavior during fixed delay stimulation of spontaneously beating chick heart cell aggregates. *American Journal of Physiology*. Submitted. {*266, 268, 269, 275*}

Levinson, N. 1944. Transformation theory of non-linear differential equations of the second order. *Annals of Mathematics*, **45**, 723–737. {*267, 275*}

Mackey, M. C. and Glass, L. 1977. Oscillation and chaos in physiological control systems. *Science*, **197**, 287–289. {*191, 196, 269, 270, 275*}

Nomura, T. and Glass, L. 1996. Entrainment and termination of reentrant wave propagation in a periodically stimulated ring of excitable media. *Physical Review E*, **53**, 6353–6360. {*271, 272, 275*}

Pavlidis, T. 1973. *Biological Oscillators: Their Mathematical Analysis*. New York, USA: Academic Press. ISBN 0-12-547350-8. Pages xiii + 207. {*255, 258, 275*}

Perkel, D. H., Schulman, J. H., Bullock, T. H., Moore, G. P., and Segundo, J. P. 1964. Pacemaker neurons: Effects of Regularly spaced synaptic input. *Science*, **145**, 61–63. {*258, 275*}

Quan, W. and Rudy, Y. 1990. Unidirectional block and reentry of cardiac excitation: A model study. *Circulation Research*, **66**(2), 367–382. {*270, 275*}

Rinzel, J. 1977. Repetitive nerve impulse propagation: Numerical results and methods. *Pages 186–212 of:* Fitzgibbon, III, W. E. and Walker, H. F. (eds), *Nonlinear Diffusion*. Research notes in mathematics, vol. 14. London, UK: Pitman Publishing Ltd. {*271, 275*}

Rudy, Y. 1995. Reentry: Insights from theoretical simulations in a fixed pathway. *Journal of Cardiovascular Electrophysiology*, **6**(4), 294–312. {*270, 275*}

Starmer, C. F., Biktahev, V. N., Romashko, D. N., Stepanov, M. R., Makarova, O. N., and Krinsky, V. I. 1993. Vulnerability in an excitable medium: analytical and numerical study of initiating unidirectional propagation. *Biophysical Journal*, **65**(5), 1775–1787. {*271, 275*}

Vassalle, M. 1977. The relationship among cardiac pacemakers: Overdrive suppression. *Circulation Research*, **41**, 269–277. {*266, 267, 275*}

Winfree, A. T. 1980. *The Geometry of Biological Time*. Biomathematics, vol. 8. Berlin, Germany: Springer Verlag. ISBN 0-387-09373-7 (New York). 3-540-09373-7 (Berlin). Pages xiii + 530. {*188, 198, 255–259, 270, 275*}

Winfree, A. T. 1987. *When Time Breaks Down: The Three Dimensional Dynamics of Electrochemical Waves and Cardiac Arrhythmia*. Princeton, NJ, USA: Princeton University Press. ISBN 0-691-08443-2, 0-691-02402-2 (paperback). Pages xiv + 339. {*255, 256, 270, 275*}

Zeng, W., Morissette, J., Brochu, R., Glass, L., and Shrier, A. 1990a. Complex rhythms resulting from overdrive suppression in electrically stimulated heart cell aggregates. *Pacing and Clinical Electrophysiology: PACE*, **13**(12), 1678–1685. {*266–268, 275*}

Zeng, W., Courtemanche, M., Sehn, L., Shrier, A., and Glass, L. 1990b. Theoretical computation of phase locking in embryonic atrial heart cell aggregates. *Journal of Theoretical Biology*, **145**, 225–244. {*264, 276*}

Zeng, W., Glass, L., and Shrier, A. 1991. Evolution of rhythms during periodic stimulation of embryonic chick heart cell aggregates. *Circulation Research*, **69**(4), 1022–1033. {*266, 267, 276*}

Zeng, W., Glass, L., and Shrier, A. 1992. The topology of phase response curves induced by single and paired stimuli. *Journal of Biological Rhythms*, **7**(2), 89–104. {*257, 258, 266, 276*}

13. MODELING THE INTERACTION OF CARDIAC MUSCLE WITH STRONG ELECTRIC FIELDS

Wanda Krassowska

13.1 INTRODUCTION

The interaction of cardiac muscle with externally applied electric fields has been a subject of numerous studies, both experimental and theoretical. The importance of this topic stems from its connection to electrical induction and termination of cardiac arrhythmias, whose mechanisms are still not fully understood. This chapter will present mathematical models used in studying these phenomena.

Cardiac muscle (Figure 13.1), like other excitable tissues that consist of electrically-connected cells, exhibits an interesting duality. On one hand, it

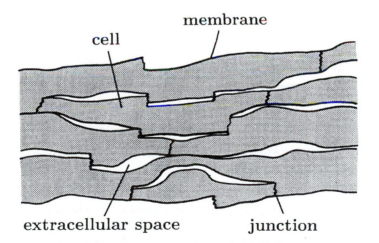

Figure 13.1. The anatomical structure of cardiac muscle.

is composed of individual cells, each independently capable of excitation and contraction. On the other hand, the presence of intercellular junctions enables such a tight coupling of their electrical and mechanical functions that the entire heart behaves in a highly synchronized manner. In the medical literature, this duality between a discrete microscopic structure and a synchronized macroscopic behavior is referred to as the syncytial nature of cardiac muscle.

The cytoplasm filling the interior of cells is usually modeled as a purely resistive, isotropic region. A typical value of its conductivity is 4mS/cm (Chapman and Fry 1978)[1]. The excitable membrane surrounding the cells is much more resistive. A typical value of membrane resistivity at rest is 10 $k\Omega$ cm^2 (Weidmann 1970). Assuming the membrane thickness of 7.5 nm,[2] the specific conductivity of the membrane is $7.5 \cdot 10^{-8}$ mS/cm, eight orders of magnitude smaller than the conductivity of the cytoplasm. This disparity of the conductive properties of the membrane and the cytoplasm has important consequences for the structure of electrical potentials in the myocardium. If cardiac muscle were placed in an external electric field and an ideal instrument were employed to measure the value of the potential at each point along the path traversing several cells, then the measured potentials would exhibit a complicated spatial pattern. The intracellular potential (Figure 13.2(a)) would consist of slow spatial changes inside the cells and of rapid jumps across the membrane between cells.

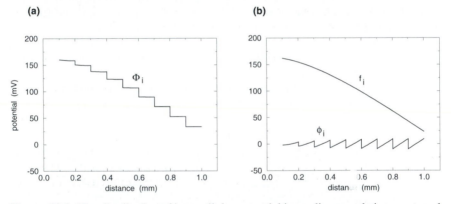

Figure 13.2. The distribution of intracellular potential in cardiac muscle in an external electric field. (a) Total intracellular potential Φ_i that actually exists in the tissue. (b) The components of Φ_i: macroscopic f_i and microscopic ("sawtooth") ϕ_i.

The presence of junctions would not disturb this pattern significantly because the effective junctional surface area is only about 2.3% of the cell surface (Hoyt et al. 1989). This "rice pools" structure is predominant when viewed on the level of a single cell. However, when several cells are traversed, one notices a macroscopic trend: a gradual transition from higher values on the left to lower values on the right. In many applications the knowledge of this macroscopic trend is sufficient to characterize the electrical behavior of cardiac muscle. Hence, there is a need for a continuum model that would disregard the cellular-level details of the potential distribution while capturing the most important features of cardiac muscle: its ability to form and propagate an action potential and to respond to external stimuli.

The concept of such a model was proposed by Schmitt (1969). The equations governing the macroscopic intra- and extracellular potentials were

---

[1]S denotes siemens, a unit of conductance. $S = \Omega^{-1}$, where Ω (ohm) is a unit of resistance.
[2]1 nm $= 10^{-9}$ meters.

developed in the late 1970s by Tung (1978) and Miller, III and Geselowitz (1978) and became known in the literature as the "bidomain" model of cardiac muscle.

The bidomain model has the form of two reaction-diffusion equations,

$$\nabla \cdot (\mathbf{T_i} \nabla f_i) = \beta \{C_m \frac{\partial}{\partial t} v_m + I_{\text{ion}}(v_m)\}$$

$$\nabla \cdot (\mathbf{T_e} \nabla f_e) = -\beta \{C_m \frac{\partial}{\partial t} v_m + I_{\text{ion}}(v_m)\},$$

(13.1)

The potentials of the bidomain model, intracellular f_i, extracellular f_e, and transmembrane $v_m \equiv f_i - f_e$, are understood as "coarse-grained" volume averages of potentials Φ_i, Φ_e, and Φ_m actually existing in the cardiac tissue. β is the surface-to-volume ratio of the myocardium, C_m is the membrane capacitance, ∇ is the gradient operator, and $\frac{\partial}{\partial t}$ is the derivative with respect to time. I_{ion} is the ionic current, which denotes in a succinct form the model of the membrane, passive or excitable, chosen by the modeler for a particular application. $\mathbf{T_i}$ and $\mathbf{T_e}$ are effective conductivity tensors for intra- and extracellular domains. If the coordinate system is aligned with the principal axes of the tissue, $\mathbf{T_i}$ and $\mathbf{T_e}$ have the diagonal form:

$$\mathbf{T_i} = \begin{bmatrix} g_{i1} & 0 & 0 \\ 0 & g_{i2} & 0 \\ 0 & 0 & g_{i3} \end{bmatrix} \quad \text{and} \quad \mathbf{T_e} = \begin{bmatrix} g_{e1} & 0 & 0 \\ 0 & g_{e2} & 0 \\ 0 & 0 & g_{e3} \end{bmatrix}.$$

(13.2)

Equation (13.1) can be written in one, two, or three space dimensions. In the one-dimensional case, (13.1) is equivalent to the core-conductor model which governs the behavior of a single fiber (Hodgkin and Rushton 1946). In some applications, the bidomain model is used in its simplified version of one reaction-diffusion equation for the transmembrane potential v_m (Shiba 1970; Bukauskas et al. 1974).

The bidomain model was not immediately accepted by the scientific community. Since the 1970s, however, it has been successfully used to reproduce results of a variety of experimental studies (Henriquez 1993), which led to its universal acceptance as a model of choice in almost all circumstances requiring the computations of electric potentials in cardiac muscle. In doing so, the bidomain is sometimes used in a manner that its original authors never intended. It may happen that in some cases the results of the bidomain model do not reflect the macroscopic trends of real potentials. However, since the bidomain model is now accepted without a question, the validity of the results is not questioned either.

The purpose of this chapter is to give a better understanding of the physical meaning of the bidomain model and a heightened awareness of its intrinsic limitations. To understand under what conditions the bidomain model is valid and under what conditions it breaks down, one must first understand the precise connection between the behavior of the myocardium at the cellular level and the quantities of the bidomain model. To this end, Section 13.2 presents the derivation of the bidomain equations from the cellular-level formulation. The main purpose is to identify all the assumptions that must be made in order to obtain (13.1). Section 13.3 outlines a similar derivation of the boundary conditions for the bidomain model. Section 13.4 summarizes the conditions under which the bidomain model is a correct representation of the macroscopic potentials in cardiac muscle and discusses physical reasons for its failure under strong electric fields. Sections 13.5 and 13.6 examine in more detail the two phenomena responsible for the breakdown of the bidomain model: electroporation of the membrane and polarization of individual cells. These sections also present the initial development of a homogenized syncytium model — a continuum model of cardiac muscle that is applicable under strong fields. Section 13.7 concludes the chapter by comparing the continuum models of cardiac muscle that are applicable under weak and strong electric fields.

13.2   DERIVATION OF THE BIDOMAIN EQUATIONS

13.2.1   Assumptions

The derivation of the bidomain equations (13.1) presented here uses a homogenization method based on two-scale expansions and follows closely the derivation given in (Krassowska 1996a). To proceed, the following assumptions must be made.

Representation of the myocardium

The cardiac tissue is modeled as the spatially periodic structure shown in Figure 13.3(a). A typical cell of a mammalian ventricle has length $l_c = 100 \mu m$ and diameter $d_c = 10 \ \mu m$ (Sommer and Scherer 1985). The interiors of these cells are all interconnected and form the intracellular space Ω_i. The narrow channels between cells contain interstitial fluid, collagen septa, blood vessels, and other extracellular structures. These channels form a similarly interconnected extracellular space Ω_e. Intra- and extracellular spaces are represented by purely resistive, homogeneous, and isotropic regions of typical conductivities $\sigma_i = 4$ mS/cm and $\sigma_e = 20$ mS/cm (Chapman and Fry 1978; Clerc 1976).

Because of the periodicity requirement, all cells in Figure 13.3(a) have the same shape and dimensions and the same pattern of connections between them. While this is a rather drastic simplification of the real cardiac muscle shown in Figure 13.1, it preserves the discreteness of the tissue structure. There are no restrictions on the geometry or on the pattern of cell-to-cell connections within a unit cell, which partially relaxes the restrictions imposed by the periodicity. The assumption of a periodic structure has been made in order to simplify the homogenization process: computations of volume and surface averages can be performed only on one unit cell (Figure 13.3(b)).

The membrane Γ surrounding the cells has a typical capacitance $C_m = 1 \ \mu F/cm^2$,[3] which enables it to maintain the difference of potentials between intra- and extracellular spaces. Transmembrane potential Φ_m changes from approximately -80 mV at rest to 20 mV during excitation, resulting in a typical amplitude of the cardiac action potential $\Delta V = 100$ mV. The ionic cur-

---

[3]F denotes farad, a unit of capacitance.

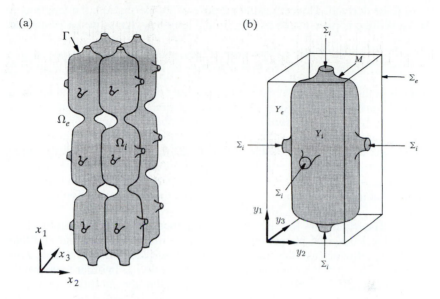

Figure 13.3. (a) Spatially periodic representation of cardiac tissue. Ω_i and Ω_e denote the interconnected intra- and extracellular spaces, separated by the membrane Γ. (b) Unit cell. Y_i and Y_e are intra- and extracellular portions of the unit cell, M denotes the part of the membrane that belongs to the unit cell, and Σ_i and Σ_e are the boundaries with the neighboring unit cells.

rent I_{ion} is determined by the kinetics of the membrane and consists of several nonlinear, dynamic currents carried by different ions. For the purpose of this derivation, the detailed description of the ionic currents is not important: the excitability of the membrane is represented by the total ionic current I_{ion} viewed as depending on the transmembrane potential Φ_m.

Choice of spatial and temporal scales

Cardiac muscle has several spatial and temporal scales. This derivation uses two spatial scales: the cellular length and the length constant. The cellular length is associated with phenomena occurring within a single cell and with cell-to-cell communication. It will be represented by the diameter of a cardiac cell, $d_c = 10 \ \mu$m. The length constant λ denotes the distance over which the intra- and extracellular fields equilibrate and is associated with the propagation of an action potential. Under subthreshold stimulation, λ measured in the principal direction x_1 (Figure 13.3(a)) is

$$\lambda = \sqrt{\frac{R_m}{\beta \left(\dfrac{1}{g_{i1}} + \dfrac{1}{g_{e1}} \right)}}, \tag{13.3}$$

where R_m is the membrane resistance at rest. The problem is that (13.3) contains the conductivities g_{i1} and g_{e1} coming from the effective conductivity tensors $\mathbf{T_i}$ and $\mathbf{T_e}$ (13.2), which at this stage are not available: the bidomain model is yet to be derived. Hence, for the purpose of the derivation, the length constant λ will be approximated by the liminal length l_b (Krassowska 1996a) defined as

$$l_b \equiv \sqrt{R_m \sigma_i d_c},   (13.4)$$

which is of the same order of magnitude as λ but contains only the fundamental material constants. For cardiac muscle, $l_b = 2$ mm and λ varies from approximately 0.3 mm when measured across cardiac fibers to 1 mm along fibers.

Cardiac muscle also has multiple temporal scales. The most important are the cellular time constant $\tau_c \equiv C_m d_c / \sigma_i$, associated with the direct polarization of individual cells by the external field, and the membrane time constant $\tau_m \equiv R_m C_m$, associated with phenomena such as excitation and propagation that involve the flow of current through the membrane. Typical values are $\tau_c = 0.25$ μs and $\tau_m = 10$ ms.

The equations that govern the behavior of the myocardium at the cellular-level describe phenomena associated with all these spatial and temporal scales. In contrast, the bidomain model describes only phenomena associated with the spatial scale l_b and the temporal scale τ_m. In order to derive the bidomain model from the cellular-level formulation, the short time and space scales must be eliminated. The first step is to recognize that the ratio of the length scales defines a small, dimensionless parameter ε,

$$\frac{d_c}{l_b} = \sqrt{\frac{d_c}{R_m \sigma_i}} \equiv \varepsilon = 5 \times 10^{-3}.   (13.5)$$

The square of the same dimensionless parameter ε is also equal to the ratio of the two temporal scales,

$$\frac{\tau_c}{\tau_m} = \frac{d_c}{R_m \sigma_i} = \varepsilon^2.   (13.6)$$

When the problem has multiple scales whose ratio forms a small parameter, two cases can occur. First, the solution may have a boundary-layer structure — a steep change next to the boundary — that occurs over a distance of the order of the shorter scale, and a more gradual change in the interior, occurring over a distance of the order of the longer scale (Bender and Orszag 1978). In the myocardium, the scale d_c gives rise to a boundary-layer on the longer scale l_b. A similar relationship exists between the two time constants, τ_c and τ_m. The polarization of individual cells by an external field, which takes 1 μs to complete, is an initial layer (i.e., a boundary layer in time) of the action potential that may follow and that would last for several milliseconds (Krassowska and Neu 1994b). The boundary layer phenomena can be eliminated by scaling: the position vector, and the time variable in the cellular-level formulation must be expressed in the units of the temporal and spatial scales we want to keep. In the case of the bidomain model, the position vector must be scaled with the liminal length l_b as a unit of distance, and the time variable must be scaled with the membrane time constant τ_m.

Second, the two spatial scales give rise to another phenomenon: the "rice-pools" structure of the intracellular potential shown in Figure 13.2(a). This obviously is not a boundary layer: the "rice pools," while associated with the microscopic scale d_c, are not confined to the close proximity of the boundary. They appear throughout the tissue riding on top of the macroscopic trend that becomes apparent over distances of the order of l_b. The equations governing this macroscopic trend can be extracted from the cellular level formulation by averaging the electric quantities over the unit cell (Bensoussan et al. 1978; Sanchez-Palencia 1980). This is the essence of the homogenization method used in this work.

Structure of potentials

The existence of the two spatial scales, d_c and l_b, has twofold implications for the structure of potentials. First, the chosen representation of potentials consists of two components (or two distinct groups of components), each associated with one spatial scale. Second, because the spatial scale d_c is an order of magnitude in ε smaller than l_b, two spatial variables are used, i.e., all functions depend upon the macroscopic position vector $\mathbf{x} \equiv (x_1, x_2, x_3)$, which measures distance in units of l_b, and upon the microscopic position vector $\mathbf{y} \equiv (y_1, y_2, y_3)$, which measures distance in units of d_c. The relationship between these two variables,

$$\mathbf{y} = \mathbf{x}/\varepsilon, \tag{13.7}$$

follows from (13.5). Hence, intracellular, extracellular, and transmembrane potentials are represented in the form of two-scale expansions:

$$
\begin{aligned}
\Phi_i^\varepsilon(\mathbf{x}, \mathbf{y}, t) &= f_i^\varepsilon(\mathbf{x}, t) + \varepsilon \phi_i^\varepsilon(\mathbf{x}, \mathbf{y}, t), \\
\Phi_e^\varepsilon(\mathbf{x}, \mathbf{y}, t) &= f_e^\varepsilon(\mathbf{x}, t) + \varepsilon \phi_e^\varepsilon(\mathbf{x}, \mathbf{y}, t), \\
\Phi_m^\varepsilon(\mathbf{x}, \mathbf{y}, t) &= v_m^\varepsilon(\mathbf{x}, t) + \varepsilon V_m^\varepsilon(\mathbf{x}, \mathbf{y}, t)..
\end{aligned}
\tag{13.8}
$$

The $O(1)$ terms of these expansions, f_i^ε and f_e^ε, are the macroscopic intra- and extracellular potentials, defined as volume averages of the total potentials Φ_i^ε and Φ_e^ε. Because of the assumption of spatial periodicity, the averaging can be conducted over one unit cell only:

$$
f_i^\varepsilon(\mathbf{x}, \mathbf{y}, t) \equiv\, < \Phi_i^\varepsilon > \;=\; \frac{1}{V_i} \int_{Y_i} \Phi_i^\varepsilon(\mathbf{x}, \mathbf{y}, t)\, d\mathbf{y},
$$

$$
f_e^\varepsilon(\mathbf{x}, \mathbf{y}, t) \equiv\, < \Phi_e^\varepsilon > \;=\; \frac{1}{V_e} \int_{Y_e} \Phi_e^\varepsilon(\mathbf{x}, \mathbf{y}, t)\, d\mathbf{y},
\tag{13.9}
$$

where Y_i and Y_e are the intra- and extracellular portions of the unit cell shown in Figure 13.3(b), and V_i and V_e are their volumes measured in units of d_c^3. The integrations in (13.9) eliminate the dependence upon the microscopic variable $\mathbf{y}$: the macroscopic intra- and extracellular potentials f_i^ε and f_e^ε depend only on the macroscopic spatial variable $\mathbf{x}$. Consequently, the macroscopic transmembrane potential, $v_m^\varepsilon \equiv f_i^\varepsilon - f_e^\varepsilon$, depends only on $\mathbf{x}$.

The $O(\varepsilon)$ terms, ϕ_i^ε, ϕ_e^ε, and V_m^ε, are the microscopic potentials that represent what is left of the total potentials Φ_i^ε, Φ_e^ε, and Φ_m^ε after the macroscopic trends have been subtracted. They represent changes in potentials arising due to the cellular structure of the tissue. Because the macroscopic potentials f_i^ε, f_e^ε, and v_m^ε do not always capture all of the $\mathbf{x}$ dependence, the microscopic potentials ϕ_i^ε and ϕ_e^ε must depend upon both spatial variables, $\mathbf{x}$ and $\mathbf{y}$.

The dependence upon **y** is periodic with a zero mean over a unit cell (Neu and Krassowska 1993).

Figure 13.2(a) shows an example of the total intracellular potential Φ_i^ε, and Figure 13.2(b) separates the two terms of Φ_i^ε. The $O(1)$ term f_i^ε is a smooth, monotonically decreasing function. The $O(\varepsilon)$ term ϕ_i^ε has a form of fast, sawtooth-like oscillations which represent the polarization of the individual cells. The period of this sawtooth is equal to the dimension of the cell.

13.2.2   Homogenization process

The starting point of the derivation is the equations that describe the electrical functioning of the myocardium on the level of physically existing cells and membranes. In accordance with the assumed representation of cardiac tissue (Figure 13.3), intra- and extracellular potentials, Φ_i and Φ_e, both satisfy Laplace's equation:

$$\Delta\Phi_i = 0 \quad \text{in } \Omega_i,$$

$$\Delta\Phi_e = 0 \quad \text{in } \Omega_e, \tag{13.10}$$

where Δ is the Laplacian operator. On the membrane Γ the potential is discontinuous, and its jump defines the transmembrane potential,

$$\Phi_m \equiv \Phi_i - \Phi_e. \tag{13.11}$$

The current through Γ is continuous and equal to the transmembrane current, i.e.,

$$-\hat{n} \cdot (\sigma_i \nabla\Phi_i) = -\hat{n} \cdot (\sigma_e \nabla\Phi_e) = C_m \frac{\partial}{\partial t}\Phi_m + I_{\text{ion}}(\Phi_m), \tag{13.12}$$

where $\hat{n}$ is the unit vector normal to the membrane and directed from intra- to extracellular space. As in (13.1), the transmembrane current is divided into two terms: the capacitive current proportional to the capacitance of the membrane C_m and the ionic current $I_{\text{ion}}(\Phi_m)$. Note that in (13.12) I_{ion} is computed using the transmembrane potential Φ_m: the *pointwise* difference between the real potentials Φ_i and Φ_e. In (13.1), I_{ion} is computed using the *averaged* macroscopic transmembrane potential v_m.

The homogenization process that converts (13.10)–(13.12) into bidomain equations (13.1) will be carried out explicitly only for the intracellular space. The derivation for the extracellular space is very similar (Krassowska 1996a). First, the cellular-level problem (13.10)–(13.12) is rewritten in a nondimensional form using the following system of units: time is measured in units of the membrane time constant τ_m, distance in units of the liminal length l_b, potentials in units of the action-potential amplitude ΔV, and conductivities in units of the cytoplasmic conductivity σ_i. The nondimensional equations are

$$\Delta\Phi_i = 0 \quad \text{in } \Omega_i$$

$$-\hat{n} \cdot \nabla\Phi_i = \varepsilon\{\frac{\partial}{\partial t}\Phi_m + I_{ion}(\Phi_m)\} \quad \text{on } \Gamma. \tag{13.13}$$

Next, the intracellular potential Φ_i is replaced by its two-scale expansion (13.8). In taking this step, one must recognize that the gradient must be computed with respect to both **x** and **y** variables,

$$\nabla \equiv \nabla_x + \frac{1}{\varepsilon}\nabla_y. \tag{13.14}$$

Laplace's equation (13.13) then takes the following form:

$$\nabla_y \cdot \{\nabla_y \phi_i^\varepsilon + 2\varepsilon\nabla_x \phi_i^\varepsilon\} + \Delta_x\{\varepsilon f_i^\varepsilon + \varepsilon^2 \phi_i^\varepsilon\} = 0 \qquad \text{in } \Omega_i \tag{13.15}$$

and the continuity of current condition becomes:

$$-\hat{n} \cdot \{\nabla_x f_i^\varepsilon + \nabla_y \phi_i^\varepsilon + \varepsilon\nabla_x \phi_i^\varepsilon\} = \varepsilon\{\frac{\partial}{\partial t}(v_m^\varepsilon + \varepsilon V_m^\varepsilon) + I_{\text{ion}}(v_m^\varepsilon + \varepsilon V_m^\varepsilon)\}$$

on Γ. The objective is to combine (13.15) and (13.16) into one equation that describes the *macroscopic* balance of currents and that depends only on *macroscopic* quantities. To this end, (13.15) is integrated over Y_i, the intracellular portion of the unit cell,

$$\int_{Y_i} \nabla_y \cdot \{\nabla_y \phi_i^\varepsilon + 2\varepsilon\nabla_x \phi_i^\varepsilon\}\,d\mathbf{y}$$

$$\tag{13.16}$$

$$+\varepsilon \int_{Y_i} \Delta_x f_i^\varepsilon\,d\mathbf{y} + \varepsilon^2 \int_{Y_i} \Delta_x \phi_i^\varepsilon\,d\mathbf{y} = 0.$$

The last term is zero, because ϕ_i^ε has zero mean. In the second term, the integrand does not depend on the microscopic variable **y** and can be taken outside the integral. The first term is transformed, using the divergence theorem, into

$$\int_{\partial Y_i} \hat{n} \cdot \{\nabla_y \phi_i^\varepsilon + 2\varepsilon\nabla_x \phi_i^\varepsilon\}\,da, \tag{13.17}$$

where da is measured in units of d_c^2. The boundary ∂Y_i consists of two parts: M, the membrane belonging to Y_i, and Σ_i, direct connections with adjacent cells (Figure 13.3(b)). The contributions from Σ_i cancel (Neu and Krassowska 1993) and the integral in (13.17) involves only the membrane M. Using boundary conditions (13.16), the integrand of (13.17) can be evaluated as

$$\hat{n} \cdot \{\nabla_y \phi_i^\varepsilon + 2\varepsilon\nabla_x \phi_i^\varepsilon\} = -\hat{n} \cdot \nabla_x f_i^\varepsilon + \varepsilon\hat{n} \cdot \nabla_x \phi_i^\varepsilon$$

$$-\varepsilon\{\frac{\partial}{\partial t}(v_m^\varepsilon + \varepsilon V_m^\varepsilon) + I_{\text{ion}}(v_m^\varepsilon + \varepsilon V_m^\varepsilon)\}. \tag{13.18}$$

The first term in the righthand side of (13.18), when integrated over M, is zero. After (13.18) is substituted into (13.17) and (13.17) into (13.16), the following integral identity is obtained:

$$\nabla_x \cdot \left\{V_i\,\nabla_x f_i^\varepsilon + \int_M \phi_i^\varepsilon \hat{n}\,da\right\}$$

$$\tag{13.19}$$

$$+\int_M \{\frac{\partial}{\partial t}(v_m^\varepsilon = \varepsilon V_m^\varepsilon) + I_{\text{ion}}(v_m^\varepsilon + \varepsilon V_m^\varepsilon)\}\,da,$$

which expresses the macroscopic balance of current in a unit cell.

Next, the potentials f_i^ε and ϕ_i^ε are sought in the form of asymptotic expansions in powers of ε:

$$
\begin{aligned}
f_i^\varepsilon &= f_i^0 + \varepsilon f_i^1 + \cdots \\
\phi_i^\varepsilon &= \phi_i^0 + \varepsilon \phi_i^1 + \cdots
\end{aligned}
$$

(13.20)

These expansions are introduced into the integral identity (13.19):

$$
\nabla_x \cdot \left\{ V_i \nabla_x (f_i^0 + \varepsilon f_i^1 + \cdots) + \int_M (\phi_i^0 + \varepsilon \phi_i^1 + \cdots) \hat{n} \, da \right\}
$$

$$
= \int_M \left\{ \frac{\partial}{\partial t} (v_m^0 + \varepsilon v_m^1 + \varepsilon V_m^0 + \varepsilon^2 V_m^1 + \cdots) \right.
$$

(13.21)

$$
\left. + I_{\text{ion}} (v_m^0 + \varepsilon v_m^1 + \varepsilon V_m^0 + \varepsilon^2 V_m^1 + \cdots) \right\} da.
$$

Taking the limit $\varepsilon \to 0$ allows us to retain only the leading-order terms. That must be done with caution. Note that terms multiplied by ε appear as an argument of the ionic current. I_{ion} is nonlinear and, potentially, an $O(\varepsilon)$ argument can result in an $O(1)$ current. For example, the question arises whether the sawtooth potential V_m^0, which is $O(\varepsilon)$, may induce an ionic current large enough to change the physiological state of the myocardium. Thus, by taking the limit $\varepsilon \to 0$, one makes an implicit assumption that the sawtooth or any other higher-order term *by itself* does not make I_{ion} large enough to influence the active processes of the membrane.

With this simplification, the transmembrane current no longer depends on the microscopic variable $\mathbf{y}$ and it can be taken outside the integral. Dividing both sides of (13.21) by V, the volume of the entire unit cell, we obtain

$$
\nabla_x \cdot \left\{ \frac{V_i}{V} \nabla_x f_i^0 + \frac{1}{V} \int_M \phi_i^0 \hat{n} \, da \right\} = \frac{S}{V} \left\{ \frac{\partial}{\partial t} v_m^0 + I_{\text{ion}}(v_m^0) \right\},
$$

(13.22)

where S is the surface area of the unit cell membrane M.

The last step is to eliminate from (13.22) the leading-order microscopic potential ϕ_i^0. To this end, expansions (13.20) are introduced into (13.15)–(13.16) and the $O(1)$ terms are collected. The resulting problem determines the leading-order microscopic potential ϕ_i^0:

$$
\begin{aligned}
\Delta_y \phi_i^0 &= 0 \qquad \text{in } Y_i \\
-\hat{n} \cdot \nabla_y \phi_i^0 &= \hat{n} \cdot \nabla_x f_i^0 \qquad \text{on } M.
\end{aligned}
$$

(13.23)

In addition, ϕ_i^0 must be periodic in $\mathbf{y}$ and have zero mean (Neu and Krassowska 1993). Problem (13.23) has a linear dependence on $\nabla_x f_i^0$, the macroscopic gradient of the leading-order intracellular potential. Thus, the microscopic potential ϕ_i^0 can be represented as an inner product of the intracellular potential gradient and a vector of weight functions $\mathbf{w_i}$,

$$
\phi_i^0 = -\nabla_x f_i^0 \cdot \mathbf{w_i}(\mathbf{y}).
$$

(13.24)

The weight functions $\mathbf{w_i}$ can be computed from

$$\Delta_y \mathbf{w_i} = 0 \quad \text{in } Y_i$$

$$(\hat{n} \cdot \nabla_y)\mathbf{w_i}(\mathbf{y}) = \hat{n} \quad \text{on } M,$$

$$(13.25)$$

supplemented by the periodicity and zero-mean requirements on $\mathbf{w_i}$. Equation (13.25) shows that weight functions $\mathbf{w_i}$ depend only on the geometry of a unit cell, not on the type or strength of an external stimulus. Hence, for a given type of tissue, $\mathbf{w_i}$ can be considered constant.

The form of (13.24) is useful in homogenization because it separates the influence of the macroscopic and microscopic variables: $\nabla_x f_i^0$ determines the magnitude and $\mathbf{w_i}(\mathbf{y})$ determines the shape of the microscopic potential ϕ_i^0. Using (13.24), the integral identity (13.22) becomes

$$\nabla_x \cdot \left\{ \left(\frac{V_i}{V}\mathbf{I} - \frac{1}{V}\int_M \hat{n}\mathbf{w_i}^T \, da \right) (\nabla_x f_i^0) \right\}$$

$$= \beta\left\{\frac{\partial}{\partial t}v_m^0 + I_{\text{ion}}(v_m^0)\right\},$$

$$(13.26)$$

where $\beta \equiv S/V$ is the microscopic surface-to-volume ratio of cardiac tissue. The expression in parentheses is a 3×3 tensor, whose terms are constant and represent the nondimensional form of $\mathbf{T_i}$, the effective, macroscopic conductivity of the intracellular domain. In dimensional form, $\mathbf{T_i}$ is

$$\mathbf{T_i} = \sigma_i \left(\frac{V_i}{V}\mathbf{I} - \frac{1}{V}\int_M \hat{n}\mathbf{w_i}^T \, da \right). \qquad (13.27)$$

Introducing (13.27) into (13.26) and converting to dimensional variables gives the final form of the differential equation governing the macroscopic intracellular potential:

$$\nabla_x \cdot (\mathbf{T_i}\nabla_x f_i^0) = \beta\left\{C_m \frac{\partial}{\partial t}v_m^0 + I_{\text{ion}}(v_m^0)\right\}. \qquad (13.28)$$

A similar homogenization procedure, applied to extracellular space, yields the following formula for the effective conductivity of the extracellular domain (Krassowska 1996a):

$$\mathbf{T_e} = \sigma_e \left\{ \frac{V_e}{V}\mathbf{I} + \frac{1}{V}\int_M \hat{n}\mathbf{w_e}^T \, da \right\}, \qquad (13.29)$$

where $\mathbf{w_e}$ is a vector of weight functions for the extracellular space, computed from a boundary-value problem analogous to (13.25). The equation governing the macroscopic extracellular potential is

$$\nabla_x \cdot (\mathbf{T_e}\nabla_x f_e^0) = -\beta\left\{C_m \frac{\partial}{\partial t}v_m^0 + I_{\text{ion}}(v_m^0)\right\}. \qquad (13.30)$$

The reversal of sign in the formula for effective conductivity (13.29) and the minus sign in the right-hand side of (13.30) result from the fact that the unit vector $\hat{n}$ normal to the membrane points *into* the extracellular space Y_e.

Equations (13.28) and (13.30) are identical in form with the equations of the bidomain model (13.1). However, the derivation presented here gives a more precise meaning to the quantities appearing in (13.1). The macroscopic potentials of the bidomain model, f_i and f_e, are the leading order volume averaged potentials, $f_i^0 \equiv \lim_{\varepsilon \to 0} < \Phi_i^\varepsilon >$ and $f_e^0 \equiv \lim_{\varepsilon \to 0} < \Phi_e^\varepsilon >$. The gradient operators in the bidomain equations must be understood as taken with respect to the macroscopic variable $\mathbf{x}$, which measures distance in units of l_b. The right-hand sides of (13.1) and (13.28), (13.30) contain the standard, pointwise ionic current — the same ionic current that governs the spatially clamped membrane. However, here the ionic current operates on the leading order of the volume-averaged transmembrane potential v_m^0 and not on the pointwise transmembrane potential Φ_m that exists in the tissue.

13.3   DERIVATION OF THE BOUNDARY CONDITIONS FOR THE BIDOMAIN MODEL

The previous section presented the derivation of the bidomain equations that govern the macroscopic potentials in cardiac muscle. This section concentrates on the derivation of the conditions that should be satisfied by these potentials and by their derivatives on the interface between cardiac muscle and the surrounding medium. These boundary conditions must be related to the potentials existing on the interface in the same way the governing equations of the bidomain model are related to the potentials actually existing deep in the tissue.

The microscopic model of the interface is shown in Figure 13.4. The

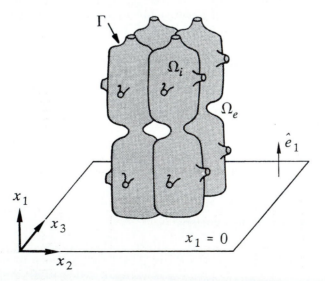

Figure 13.4. The microscopic model of the interface. Surface $x_1 = 0$ denotes the macroscopic boundary between cardiac muscle (above) and the outside medium (below), $\hat{e}_1$ is a unit vector normal to this interface.

surface $x_1 = 0$ acts as a macroscopic boundary between cardiac muscle and the outside medium. The cellular-level behavior of the potentials and currents at the interface follows two physiologically motivated assumptions. First, the extracellular space is in a direct contact with the outside medium, allowing free flow of charge and diffusion of molecules. Second, since no cell can be dissected and still remain viable, it is assumed that all cells on the interface are intact. Consequently, there is no direct contact between intracellular space and the outside medium.

These two assumptions imply that on the microscopic level, there is a continuity of potentials and current densities between extracellular space and the outside medium. It is generally accepted that the continuity of potentials holds also on the macroscopic level,

$$f_e^0 = f^0 \qquad \text{on } x_1 = 0, \tag{13.31}$$

i.e., the leading-order extracellular potential, f_e^0, is equal to the leading-order macroscopic potential in the outside medium, f^0. However, the microscopic continuity of currents between extracellular space and the outside region does not have to be automatically satisfied on the macroscopic level. Considering the cellular-level model of the interface (Figure 13.4), one recognizes that the membrane has some finite, albeit small conductivity and that a small current will seep through the membrane on the interface and enter intracellular space. The macroscopic boundary conditions may need to recognize that the intracellular domain receives this small transmembrane current, and, subsequently, the current received by the extracellular domain is somewhat diminished.

The problem with this reasoning is that the potentials of the bidomain model constitute only one of the terms, $O(1)$, of the expansions (13.8). Currents have similar expansions, and only their leading-order components belong to the bidomain model. The small transmembrane current seeping through the membrane on the interface may or may not belong to the bidomain model: it depends on its position in the asymptotic expansion for currents. If on the macroscopic boundary $x_1 = 0$ the macroscopic intra- and extracellular currents are both $O(1)$, then the outside current must indeed be divided between intra- and extracellular domains. However, if the intracellular current is $O(\varepsilon)$ while the extracellular current is $O(1)$, then the intracellular domain must be considered insulated, and there will be a continuity of current between extracellular domain and the outside region.

This question was resolved by deriving the macroscopic boundary conditions for the bidomain model using the same homogenization process as used to derive the governing equations (Krassowska and Neu 1994a). Because of the similarity of these derivations, the homogenization process is not repeated here. Instead, this section briefly summarizes the assumptions and the final result.

The cardiac tissue is again assumed spatially periodic, and it terminates at the macroscopic interface $x_1 = 0$. The derivation uses the same spatial and temporal scales and imposes the same conditions on the magnitude of electric potentials and fields as in the derivation of the bidomain equations. However, the potentials around the interface have a boundary-layer structure and are represented by two types of expansions. Inner expansions,

$$\Phi_i^\varepsilon(\mathbf{x}', \mathbf{y}, t) = F_i^\varepsilon(\mathbf{x}', t) + \varepsilon\phi_i^\varepsilon(\mathbf{x}', \mathbf{y}, t),$$

$$\Phi_e^\varepsilon(\mathbf{x}', \mathbf{y}, t) = F_e^\varepsilon(\mathbf{x}', t) + \varepsilon\phi_e^\varepsilon(\mathbf{x}', \mathbf{y}, t),$$

(13.32)

describe intra- and extracellular potentials very close to the interface $x_1 = 0$. Note that these potentials depend on the position vector $\mathbf{x}' \equiv (x_2, x_3)$ that includes the macroscopic variables x_2 and x_3, which are tangential to the interface, but not on x_1, which is normal to the interface. Outer expansions are the same as expansions (13.8) used to derive the bidomain model. Here, they describe the intra- and extracellular potentials deep in the tissue and away from the boundary. In addition, outer expansions similar to (13.8) are introduced to represent the potential f^ε in the outside medium.

Next, homogenization is used to actually compute what fraction of the current, delivered from the outside, flows out of the boundary layer through the intracellular domain. This current is initially expressed in terms of inner expansions, i.e., in terms of the microscopic electrical quantities. In this form it cannot be used as a boundary condition for the macroscopic bidomain currents. Hence, asymptotic matching is used to express this current in terms of outer expansions, i.e., in terms of the macroscopic potentials and the effective conductivities of the bidomain model.

The homogenization process described in detail in Krassowska and Neu (1994a) shows that if the outside current is $O(1)$, then the current that flows out of the boundary layer via the intracellular domain is only $O(\varepsilon)$. Thus, in the bidomain model, the intracellular domain must be treated as insulated, and all outside current must enter the extracellular domain,

$$\hat{e}_1 \cdot (\mathbf{T_i}\nabla_x f_i^0) = 0$$

$$\hat{e}_1 \cdot (\mathbf{T_e}\nabla_x f_e^0) = \hat{e}_1 \cdot (\sigma_o\nabla_x f^0) \qquad \text{on } x_1 = 0.$$

(13.33)

Here $\hat{e}_1$ is the unit vector normal to the macroscopic interface $x_1 = 0$ and σ_o is the conductivity of the outside region. This form of the boundary conditions was proposed originally by Tung (1978) and is the most widely used with the bidomain equations.

13.4 VALIDITY OF THE BIDOMAIN MODEL AND ITS BREAKDOWN UNDER STRONG FIELDS

The derivation presented in the two previous sections demonstrated that the processes responsible for the change of the physiological state of the myocardium are adequately described by the macroscopic, volume-averaged potentials. Hence, bidomain equations (13.1) constitute a correct and, to a leading order, complete model of the macroscopic electrical behavior of the myocardium. The derivation confirmed that the original intuitive understanding of the bidomain potentials as "coarse-grained volume averages" of the potentials actually existing in the real tissue is correct. What has not been understood previously is that the bidomain potentials are also asymptotic limits as a small parameter ε approaches zero. That means that the bidomain potentials capture only the leading-order macroscopic phenomena, and not all the macroscopic phenomena that may take place in the tissue. Whether or not these neglected higher-order terms are of consequence depends, among other factors, on the smallness of the nondimensional parameter ε. In the derivation, ε was defined as d_c/l_b and had a reasonably small value of $5 \cdot 10^{-3}$. For real cardiac muscle, a better measure is the ratio of the cell width (length) to the actual length constant λ measured across (along) cardiac fibers. Typical values of this ratio are larger than ε: 0.03 across and 0.1 along fibers. Hence, under certain conditions, e.g., if cells become uncoupled or the membrane too leaky, this ratio would not be sufficiently small and the bidomain equations would not constitute a complete model: the change of the physiological state of the myocardium may be caused by phenomena not described by the bidomain.

It must be stressed that the bidomain model describes only phenomena associated with certain temporal and spatial scales: the membrane time constant τ_m and the tissue-length constant λ (in the derivation approximated by the liminal length l_b). The bidomain is not an adequate tool to study the phenomena associated with the cellular time constant τ_c or with the cellular length scale d_c.

Finally, the expansions in (13.8) assume that macroscopic potentials change no faster than ΔV, the amplitude of an action potential, over a distance of l_b. Hence, macroscopic electric fields, e.g., $\nabla_x f_i^0$, must be of the order of $\Delta V/l_b$, which for cardiac muscle is approximately 0.5 V/cm. This assumption is readily satisfied during the propagation of an action potential or during pacing with small stimuli. The situation changes drastically when the macroscopic electric fields become larger (Krassowska 1996b). For example, fields established in the myocardium by defibrillation shock range from 5 to 120 V/cm (Wharton et al. 1992; Zhou et al. 1993). Under fields of such strengths, the response of the myocardium may be determined, in addition to the bidomain, by the phenomena associated with other temporal and spatial scales. In consequence, the bidomain model would no longer faithfully reproduce the macroscopic behavior of the heart.

To illustrate the phenomena evoked by strong fields, consider an idealized strand of cardiac muscle that consists of cells connected by gap junctions (Figure 13.5).

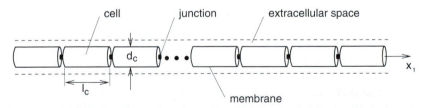

Figure 13.5. A one-dimensional model of a discrete cardiac strand.

To investigate the effects of the electric field without interference from the active processes of the membrane, the membrane of this strand is passive, represented in the model by a constant resistance R_m and capacitance C_m. This strand is exposed to an external electric field of 20 V/cm, i.e., 40 times stronger than the 0.5 V/cm field of the bidomain model. Figure 13.6 shows the distribution of the transmembrane potential $\Phi_m^0 \equiv v_m^0 + \varepsilon V_m^0$ along this strand 20 ms after the onset of the field.

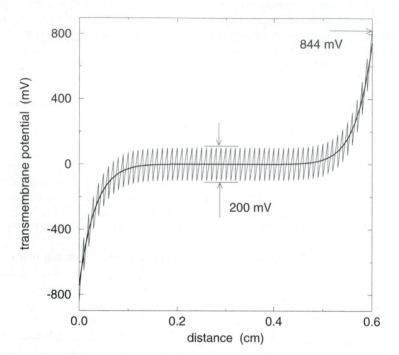

Figure 13.6. The distribution of the transmembrane potential computed using the model of a discrete cardiac strand (Figure 13.5) with passive membrane. The strand is exposed to an electric field of 20 V/cm, and the distribution shown in this figure corresponds to the time instant 20 ms after the onset of the field. The two plots are the macroscopic transmembrane potential v_m^0 (smooth line) and the total transmembrane potential Φ_m^0, which consists of v_m^0 and of the sawtooth potential V_m^0.

First, at the ends of the strand, the value of Φ_m^0 was computed to be 844 mV. However, biological membranes cannot support potentials of such a magnitude. In a real strand, the membrane at the ends of the strand would undergo electroporation: in the presence of strong transmembrane potential, the lipid bilayer matrix of the membrane would destabilize and develop small, conductive pores. These pores would short-circuit part of the external stimulating current directly to the intracellular domain, and consequently, the distribution of Φ_m^0 would change. Second, the amplitude of the sawtooth potential V_m^0 is 200 mV. This violates the assumption made while deriving the bidomain model that the sawtooth potential must be small and, by itself, unable to change the physiological state of the membrane. Therefore, the model that is to be used under strong fields must account for the effects of electroporation and for the contribution of the sawtooth potential to the ionic current.

The next two sections investigate the influence of these two phenomena on the macroscopic behavior of the heart and examine why they limit the applicability of the standard bidomain model. These sections also discuss how these phenomena should be represented in the model and present the initial development of the homogenized syncytium model, an extension of the bidomain model that is applicable under strong electric fields.

13.5 ELECTROPORATION OF THE MEMBRANE

The physical basis of electroporation can be briefly described as follows. The membrane consists of the lipid bilayer matrix and the proteins imbedded in it. Lipid molecules are tightly packed and form a nonconducting barrier that separates the interior of the cell from the outside medium. Because of thermal fluctuations, lipid molecules often separate, forming so-called hydrophobic pores (Figure 13.7(a)).

(a)   **(b)**

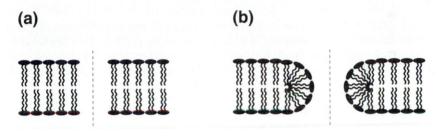

Figure 13.7. The schematic representation of hydrophobic and hydrophilic pores.

These pores are nonconductive and have a very short lifetime of approximately 10 ps (Glaser et al. 1988)[4]. Occasionally, a hydrophobic pore attains enough energy to change its configuration and to form an inverted or hydrophilic pore (Figure 13.7(b)). These pores are conducting and have a lifetime of the order of seconds (Glaser et al. 1988). Under normal conditions the membrane contains only very few hydrophilic pores; they account for the very small background conductivity of the lipid bilayer. The presence of a large transmembrane potential facilitates the inversion process, increasing the number of hydrophilic pores and causing the expansion of their radius. In consequence, the membrane partially loses its barrier function and allows a relatively easy flow of ions between intra- and extracellular space. When the transmembrane potential is removed, the pores shrink and eventually reseal. This process is called reversible electrical breakdown. Stimuli that last longer or have higher magnitudes may cause irreversible electrical breakdown and the mechanical rupture of the membrane.

The physics of electroporation is complex, and its faithful representation requires fairly sophisticated models (Pastushenko et al. 1979; Powell et al. 1986; Freeman et al. 1994). The use of these models in a spatially distributed problem such as the strand of Figure 13.5 would be impractical: at each point in space a partial differential equation describing the evolution of pores would have to be solved. However, such a detailed representation is not necessary: to study the effect of electroporation on the potentials arising in cardiac muscle, it is sufficient to trace an overall change of the conductivity of the membrane. Hence, a simplified model of electroporating membrane was proposed

---

[4] 1 ps $= 10^{-12}$ seconds

(Glaser 1986; Krassowska 1995a; Krassowska 1995b), in which the total transmembrane current I_m consists of the capacitive and ionic currents of the intact membrane (13.12) supplemented with the electroporation current:

$$I_m = C_m \frac{\partial}{\partial t} \Phi_m + I_{\text{ion}}(\Phi_m) + G(\Phi_m, t)\, \Phi_m \quad \text{on } \Gamma.$$

$$(13.34)$$

Here, G is the dynamically changing conductance that represents the increase in membrane conductivity owing to the creation of pores. The dynamics of G is governed by an ordinary differential equation,

$$\frac{dG}{dt} = \alpha\, e^{\beta(\Phi_m)^2} \left(1 - e^{-\gamma(\Phi_m)^2}\right).$$

$$(13.35)$$

The constants $\alpha = 4 \cdot 10^{-3}\text{mS/(cm}^2\text{ms)}$, $\beta = 2.5 \cdot 10^{-5}\text{mV}^{-2}$, and $\gamma = 1^{-4}\text{mV}^{-2}$ were chosen based on the voltage-clamp experiments of Chernomordik and Chizmadzhev (1989) and Zhou et al. (1996). This model contains many simplifications; for example, it does not represent the resealing of pores nor the fact that the conductivity of a pore is nonohmic. Thus, equation (13.35) cannot be expected to capture all the details of the electroporation process. Rather, it is a "cartoon" describing only the essential features of the response of the membrane to very strong shocks.

To illustrate a typical response of the electroporating membrane, we show the time course of the transmembrane potential Φ_m when a strong constant-current pulse is applied to a space-clamped membrane patch (see Figure (13.8(a))).

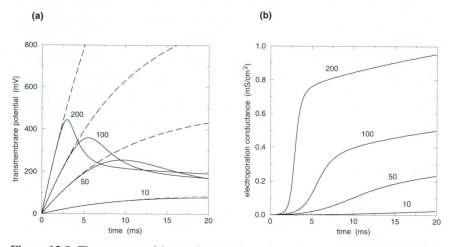

Figure 13.8. The response of the passive membrane model to constant-current pulses of 10, 50, 100 and 200 $\mu\text{A/cm}^2$. (a) Transmembrane potential Φ_m of electroporating (solid line) and nonelectroporating (dashed line) membrane. (b) Electroporation conductance G.

Were the membrane represented by a linear circuit, consisting of R_m and C_m, the transmembrane potential would have increased monotonically to a steady state. With the addition of the electroporation conductance G the transmembrane potential initially increases, and, after it exceeds a certain critical voltage, starts decreasing. The steady-state value of Φ_m in an electroporating membrane is much lower than in the intact membrane: to prevent further development of pores, it must fall below the critical voltage. The precise shape of the time course of the transmembrane potential depends on the strength of the stimulus. Small pulses, e.g., $10\ \mu A/cm^2$, do not cause any appreciable degree of electroporation, and one observes only a monotonic increase in the transmembrane potential. For stronger stimuli, the time course of the transmembrane potential is characteristic for electroporation: an initial rise, peak, and then decrease below the critical voltage. The time and the amplitude of the peak depend upon the strength of the pulse and coincide in time with the increase of the electroporation conductance G (Figure 13.8(b)). These features of the time course of the transmembrane potential, computed from (13.34) and (13.35), agree qualitatively with experimental results (Benz and Zimmermann 1980).

Figure 13.8(a) demonstrates that the potential of the electroporating membrane is considerably lower than that of the nonelectroporating membrane. This "saturation effect" has been observed experimentally by Zhou et al. (Zhou et al. 1996; Zhou et al. 1995b), who exposed a 4-mm strand of a guinea pig papillary muscle to electric fields 1.8–18 V/cm and measured the shock-induced changes in the transmembrane potential. The measurements were performed during the plateau of an action potential with a double-barreled microelectrode impaled 1 mm away from the end of the strand. Figure 13.9 shows that the magnitude of the shock-induced Φ_m saturates above approximately 7 V/cm (squares with error bars).

This saturation was reproduced using the model of a strand with electroporating membrane (solid line) (DeBruin and Krassowska 1996). In contrast, the model of a strand with nonelectroporating membrane (dashed line) does not exhibit the saturation effect. These experiments confirm that under strong fields, the bidomain model (13.1) significantly overestimates the macroscopic transmembrane potential v_m^0. This overestimation is not limited to regions where v_m^0 is the largest, but it occurs at all sites throughout the tissue.

The saturation of the transmembrane potential, caused by electroporation, has important consequences on the structure of the potentials in cardiac muscle. Under strong electric fields, expansions (13.8) should be modified by recognizing the fact that the macroscopic intra- and extracellular potentials grow with the increase in shock strength, yet the magnitude of the macroscopic transmembrane potential remains bounded. To this end, the expansions applicable under strong electric fields assume that intra- and extracellular potentials have *two* macroscopic components, not one. One of these components saturates with the increase of the electric field while the other grows without bounds. Hence, the structure of potentials is represented by the following expansions:

$$\Phi_i^\varepsilon = \frac{1}{\varepsilon}\Psi^\varepsilon(\mathbf{x},t) + f_i^\varepsilon(\mathbf{x},t) + \phi_i^\varepsilon(\mathbf{x},\mathbf{y},t),$$

$$\Phi_e^\varepsilon = \frac{1}{\varepsilon}\Psi^\varepsilon(\mathbf{x},t) + f_e^\varepsilon(\mathbf{x},t) + \phi_e^\varepsilon(\mathbf{x},\mathbf{y},t), \qquad (13.36)$$

$$\Phi_m^\varepsilon = v_m^\varepsilon(\mathbf{x},t) + V_m^\varepsilon(\mathbf{x},\mathbf{y},t).$$

In (13.36), the macroscopic potentials of the bidomain model, intra-

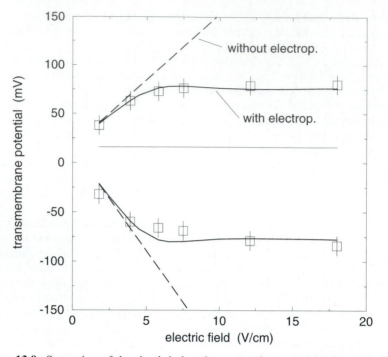

Figure 13.9. Saturation of the shock-induced transmembrane potential measured experimentally (squares with error bars) (Zhou et al. 1996) and reconstructed using a model of a one-dimensional passive strand with electroporating membrane (solid line). The strand with nonelectroporating membrane (dashed line) does not exhibit the saturation effect. The horizontal line represents the transmembrane potential immediately before the shock was delivered.

cellular f_i^ε and extracellular f_e^ε, are limited in size to $O(1)$, so that they do not give rise to an unphysiologically high macroscopic transmembrane potential v_m^ε. The magnitude of the other macroscopic potential Ψ^ε grows without limits with the increasing shock strength; to underscore this, in expansions (13.36), Ψ^ε is assumed to be $O(1/\varepsilon)$. However, since Ψ^ε is the same in the intra- and extracellular domains, it does not contribute to the macroscopic transmembrane potential v_m^0.

   The unresolved problem is the exact form of the differential equations governing f_i^ε, f_e^ε, and Ψ^ε. As will be shown in the next section, in a special case of a uniform macroscopic electric field, the leading-order macroscopic potentials f_i^0 and f_e^0 are governed by the reaction-diffusion equations similar to the bidomain. In a general case, when the macroscopic electric field is not uniform, potentials f_i^ε, f_e^ε, and Ψ^ε may not have such a straightforward interpretation and a more complicated picture may emerge, with electroporation affecting not only the magnitude but also the spatial distribution of the shock-induced transmembrane potential (Trayanova and Krassowska 1995).

13.6 POLARIZATION OF INDIVIDUAL CELLS

13.6.1 Derivation of a continuum model applicable under strong fields

In the previous section we demonstrated that the structure of the potentials will change under strong fields. In particular, expansions (13.36) now include an $O(1/\varepsilon)$ potential Ψ^ε. Consequently, the leading-order microscopic potentials, ϕ_i^0, ϕ_e^0, and V_m^0, are now determined by the gradient of Ψ^0, which is $O(1/\varepsilon)$, rather than by the gradient of f_i^0 or f_e^0, which are $O(1)$. Hence, in (13.36), the sawtooth potential V_m^0 is $O(1)$, not $O(\varepsilon)$ as it was in expansions (13.8) of the standard bidomain. Accordingly, in Figure 13.6, the sawtooth is ± 100 mV; i.e., its magnitude is comparable to the amplitude of the action potential ΔV. Such a sawtooth potential is in a position to compete with the macroscopic transmembrane potential v_m^0 in changing the physiological state of the tissue. This indicates that the bidomain equations (13.1) should be replaced by a new continuum model that would correctly account for the effect of the sawtooth on the macroscopic electrical behavior of the heart.

To assure the consistency with the actual behavior of cardiac muscle, the model applicable under strong electric fields should be derived from the cellular-level formulation (13.10)–(13.12), using the same homogenization process that was used in Section 13.2 to derive the bidomain equations. Such a derivation was carried out for a special case, when the external electric field **E** is uniform (Neu and Krassowska 1993). Since it is similar to the derivation of the bidomain equations, only the critical differences will be presented here. In particular, the potentials were assumed to have the following structure:

$$\Phi_i^\varepsilon = -\frac{1}{\varepsilon}\mathbf{E} \cdot \mathbf{x} + f_i^\varepsilon(\mathbf{x}, t) + \phi_i^\varepsilon(\mathbf{x}, \mathbf{y}, t),$$

$$\Phi_e^\varepsilon = -\frac{1}{\varepsilon}\mathbf{E} \cdot \mathbf{x} + f_e^\varepsilon(\mathbf{x}, t) + \phi_e^\varepsilon(\mathbf{x}, \mathbf{y}, t), \qquad (13.37)$$

$$\Phi_m^\varepsilon = v_m^\varepsilon(\mathbf{x}, t) + V_m^\varepsilon(\mathbf{x}, \mathbf{y}, t).$$

This expansion is a special case of (13.36), in which the potential Ψ^ε is a linear function of the macroscopic position vector **x**. With (13.37), the homogenization process yields the following integral identity for the intracellular space:

$$\nabla_x \cdot \left\{ V_i \nabla_x (f_i^0 + \varepsilon f_i^1 + \cdots) + \frac{1}{\varepsilon} \int_M (\phi_i^0 + \varepsilon \phi_i^1 + \cdots) \hat{n}\, da \right\}$$

$$= \int_M \left\{ \frac{\partial}{\partial t}(v_m^0 + \varepsilon v_m^1 + V_m^0 + \varepsilon V_m^1 + \cdots) \right. \qquad (13.38)$$

$$+ \left. I_{\text{ion}}(v_m^0 + \varepsilon v_m^1 + V_m^0 + \varepsilon V_m^1 + \cdots) \right\} da.$$

(13.38), just like its counterpart (13.21) in Section 13.2, represents the macroscopic balance of current in a unit cell. Comparing (13.21) and (13.38) shows two differences. First, the integral over M appearing in the lefthand side is now $O(1/\varepsilon)$ rather than $O(1)$, but this difference is of no consequence for the final result. Second, the expansions of the transmembrane potential in the right-hand side now contain two $O(1)$ terms: v_m^0 and V_m^0, the macroscopic and the sawtooth components of the transmembrane potential. Previously, V_m^0 was $O(\varepsilon)$, so it disappeared in the limit $\varepsilon \to 0$ and its effect on I_{ion} could be ignored. In (13.38), v_m^0 and V_m^0 are of the same order of magnitude, and both must be accounted for in the model. To do so, V_m^0 is expressed in terms of $\mathbf{E}$, the external electric field, and $\mathbf{w_i}$ and $\mathbf{w_e}$, the weight functions for intra- and extracellular space introduced in Section 13.2:

$$V_m^0 = \mathbf{E} \cdot (\mathbf{w_i} - \mathbf{w_e}). \tag{13.39}$$

For a given type of tissue, $\mathbf{w_i}$ and $\mathbf{w_e}$ are independent of the external stimulus. Hence, the magnitude of the sawtooth is determined by the external field $\mathbf{E}$.

Equation (13.39), when substituted into (13.38), allows the following simplification. On the time scale τ_m, $\mathbf{w_i}$ and $\mathbf{w_e}$ do not depend on time. Hence, $\partial \mathbf{E}/\partial t$ can be taken outside the integral. Since $\mathbf{w_i}$ and $\mathbf{w_e}$ have zero mean, $\partial V_m^0/\partial t$ is zero and only $\partial v_m^0/\partial t$ remains. Taking the limit $\varepsilon \to 0$ and converting to the dimensional form leads to the equations governing f_i^0 and f_e^0,

$$\nabla_x \cdot (\mathbf{T_i} \nabla_x f_i^0) = \beta \{ C_m \frac{\partial}{\partial t} v_m^0 + i_{ion}(v_m^0, \mathbf{E}) \} \tag{13.40}$$

$$\nabla_x \cdot (\mathbf{T_e} \nabla_x f_e^0) = -\beta \{ C_m \frac{\partial}{\partial t} v_m^0 + i_{ion}(v_m^0, \mathbf{E}) \}.$$

These equations are called the homogenized syncytium model (Neu and Krassowska 1993), and they allow us to predict the macroscopic electrical behavior of cardiac muscle under strong electric fields. Superficially, (13.40) looks like the equations of the bidomain model (13.1). The difference is that the right-hand side now contains the *macroscopic* ionic current, defined as a surface average of I_{ion} over the membrane of the unit cell,

$$i_{ion}(v_m^0, \mathbf{E}) \equiv \frac{1}{S} \int_M I_{ion}(v_m^0 + \mathbf{E} \cdot (\mathbf{w_i} - \mathbf{w_e})) \, da. \tag{13.41}$$

The argument of the pointwise ionic current I_{ion} contains not only v_m^0, but also the sawtooth term $\mathbf{E} \cdot (\mathbf{w_i} - \mathbf{w_e})$, indicating that the change in the macroscopic potentials f_i^0 and f_e^0 can now be caused by the macroscopic transmembrane potential, by the sawtooth potential, or by the combination of both. To underscore this fact, i_{ion} has two arguments, v_m^0 and $\mathbf{E}$, the external electric field that via (13.39) determines the magnitude of the sawtooth component. In some applications the argument of I_{ion} should also involve an $O(\varepsilon)$ component of the sawtooth potential, V_m^1. The magnitude of V_m^1 is determined by the gradients of the macroscopic intra- and extracellular potentials (Neu and Krassowska 1993):

$$V_m^1 = -(\nabla_x f_i^0 \cdot \mathbf{w_i} - \nabla_x f_e^0 \cdot \mathbf{w_e}). \tag{13.42}$$

The inclusion of this term into the computation of i_{ion} is important for shocks of intermediate strengths, when gradients of f_i^0 and f_e^0 are comparable in magnitude to $\mathbf{E}$, or under conditions when the ratio of the cellular dimension to the length constant is not sufficiently small.

13.6.2   Effects of the sawtooth on the dynamics of cardiac muscle

The similarities and differences between the response of a myo-cardium to weak versus strong electric fields can be demonstrated most clearly when the membrane dynamics are described by the FitzHugh-Nagumo model (FitzHugh 1961; Nagumo et al. 1962), in which the excitability of the membrane is represented by a simple cubic nonlinearity. For the space-clamped membrane, the governing equations are:

$$\frac{\partial}{\partial t} v = -v(a - v)(1 - v) - u$$

$$\frac{\partial}{\partial t} u = b(cv - u),$$

(13.43)

where v denotes the fast-activation variable (e.g., transmembrane potential), u denotes the slow-inactivation variable, and a, b, and c are constants.

Under weak electric fields, the leading-order potentials in the myo-cardium are described by the standard bidomain equations (13.1). Ionic current appearing in the right-hand side of (13.1) can be obtained from (13.43) by substituting the macroscopic transmembrane potential v_m^0 in place of v,

$$I_{\text{ion}}(v_m^0) = v_m^0(a - v_m^0)(1 - v_m^0) + u^0.$$

(13.44)

Here, u^0 is the macroscopic inactivation variable defined as

$$u^0(\mathbf{x}, t) \equiv \frac{1}{S} \int_S u(\mathbf{x}, \mathbf{y}, t)\, da$$

(13.45)

i.e., as a surface average of the pointwise inactivation variable u which actually exists in the tissue and which, as a function of $\mathbf{y}$, may vary with the position within a unit cell. The governing equation for u^0 has the same form as for the space-clamped membrane,

$$\frac{\partial}{\partial t} u^0 = b(cv_m^0 - u^0),$$

(13.46)

confirming that under weak fields the dynamics of the bidomain model is the same as the dynamics of the space-clamped membrane but depends upon averaged state variables.

Under strong electric fields, the leading-order potentials in the myo-cardium are described by the homogenized syncytium equations (13.40), which use the macroscopic ionic current i_{ion}. For the FitzHugh-Nagumo dynamics, i_{ion} can be computed analytically using (13.41) and (13.43). At this point, in order to compute weight functions $\mathbf{w_i}$ and $\mathbf{w_e}$, one must choose the geometry of the unit cell. The simplest case results when the unit cell has a shape of a thin cylinder of diameter 1 and length l_c (both nondimensional, measured in units of d_c) and when the electric field $\mathbf{E}$ points in the direction y_1, parallel to the long axis of the cell (Figure 13.10(a)).

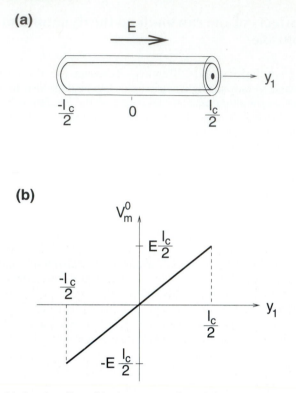

Figure 13.10. (a) A unit cell used in the computation of the macroscopic ionic current. The myocyte itself has a cylindrical shape, and the black dot on its face indicates the location of the low-resistance connection with the neighboring cell. The extracellular space forms a cylindrical layer surrounding the myocyte. (b) The sawtooth potential V_m^0 arising in the unit cell of panel (a) when the external field $\mathbf{E}$ is aligned with the cell axis.

If the connections with neighboring cells do not significantly alter the distribution of potentials inside and outside of the cell, then $\mathbf{w_i} - \mathbf{w_e}$ has a simple form

$$\mathbf{w_i} - \mathbf{w_e} = y_1 \hat{e}_1, \qquad -l_c/2 \leq y_1 \leq l_c/2, \qquad (13.47)$$

where $\hat{e}_1$ is a unit vector pointing in the direction y_1. Hence, the sawtooth potential,

$$\mathbf{E} \cdot (\mathbf{w_i} - \mathbf{w_e}) = E y_1 \qquad -l_c/2 \leq y_1 \leq l_c/2, \qquad (13.48)$$

changes linearly within the cell, and its slope is proportional to E, the strength of the electric field $\mathbf{E}$ (Figure 13.10(b)).

With (13.48), i_{ion} can be computed as

$$i_{\text{ion}}(v_m^0, E) = \frac{1}{2\pi l_c} \int_0^{2\pi} \int_{-l_c/2}^{l_c/2} \left[(v_m^0 + Ey_1)\right.$$

$$\cdot \left. (a - v_m^0 - Ey_1)(1 - v_m^0 - Ey_1) + u\right] dy_1 d\theta \qquad (13.49)$$

$$= v_m^0(a - v_m^0)(1 - v_m^0) + u^0 + \frac{3v_m^0 - a - 1}{3}\left(\frac{El_c}{2}\right)^2.$$

Using (13.44), (13.49) can be written as

$$i_{\text{ion}}(v_m^0, E) = I_{\text{ion}}(v_m^0) + \frac{3v_m^0 - a - 1}{3}\left(\frac{El_c}{2}\right)^2. \qquad (13.50)$$

An equivalent expression for i_{ion} was obtained by Keener (1996) for cardiac muscle stimulated with external current.

    The form of (13.50) shows that under strong electric fields, the ionic current comes from two sources. The first term is identical to the ionic current of the standard bidomain model: spaced-clamped I_{ion} depending on averaged state variables. The second term describes the contribution of the sawtooth. To see this, recognize that $El_c/2$ is the value attained by the sawtooth potential at the ends of the cell (Figure 13.10(b)). Hence, the contribution of the sawtooth is proportional to the square of its magnitude and, to a leading order, independent of the polarity of the field. However, the sawtooth potential can have different effects on i_{ion}, depending on the initial state of the tissue. To see this, analyze expression $(3v_m^0 - a - 1)/3$ which combines the influence of the cell geometry with the membrane dynamics. If the electric field is applied to the resting tissue, i.e., when $v_m^0 = 0$, the ionic current caused by the sawtooth is negative (inward) and moves the tissue toward depolarization. If the electric field is applied during the plateau, i.e., when $v_m^0 \approx 1$, the ionic current is positive (outward) and moves the tissue toward hyperpolarization.

    Expression (13.50) for the macroscopic ionic current is valid only for the FitzHugh-Nagumo dynamics. For a physiologically motivated dynamics, such as the Luo-Rudy model (Luo and Rudy 1991), an explicit expression for i_{ion} may not exist because of the difficulty of the analytical evaluation of integral (13.41). Nevertheless, the homogenized syncytium equations (13.40) are general and apply to membrane dynamics described by any model and to stimulation in any phase of an action potential.

    To demonstrate that the predictions of the FitzHugh-Nagumo model are general, Figure 13.11 illustrates the effects of the sawtooth potential on the macroscopic response of a discrete cardiac strand with the Luo-Rudy membrane dynamics (Luo and Rudy 1991). Panel (a) shows the field stimulation of the strand by a 5.7 V/cm external field applied for 3 ms. As predicted by (13.50), owing to the contribution of the sawtooth potential to the total ionic current, the field activates directly almost the entire strand. When the ionic current does not contain the contribution from the sawtooth (panel (b)), the field excites only the right end of the strand and initiates the propagating action potential. Panel (c) compares the time courses of the transmembrane potential computed at the center of the strand for the two cases shown in panels (a) and

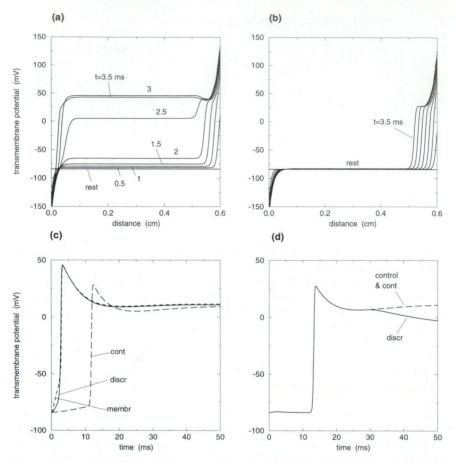

Figure 13.11. The effects of the sawtooth potential on the macroscopic response of the one-dimensional cardiac strand to strong electric fields. In this figure, the distribution of the transmembrane potential was computed using the model of a discrete cardiac strand (Fig. 13.5) with the membrane dynamics represented by the Luo-Rudy model. (a) Response to the field stimulation of 5.7 V/cm applied for 3 ms. The distribution of the macroscopic transmembrane potential v_m^0 is shown at rest and at time instants 0.5, 1, 1.5, ... , 3.5 ms after the onset of the field. (b) Stimulation with the same 5.7 V/cm, 3 ms field but computed without the contribution of the sawtooth to the macroscopic ionic current. (c) Time courses of v_m^0 at the center of the strand. Solid line represents field stimulation of panel (a), dashed line, stimulation without the sawtooth contribution of panel (b), and short-dashed line, an action potential of the space-clamped membrane. (d) The effect of the sawtooth on the response to the shock delivered during the plateau. The action potential was initiated at the right end of the strand by a 2 V/cm, 3 ms stimulus and propagated to the left. At 30 ms, a 20 V/cm shock was applied. Solid line shows v_m^0 in the center and computed with the contribution of the sawtooth. Dashed line shows v_m^0 computed without the sawtooth and the control case (no second shock).

(b). The action potential elicited by the direct field stimulation (solid line) appears very shortly after the shock and is very similar to the action potential of the space-clamped membrane (short-dashed line). If the sawtooth is not taken into account (dashed line), the action potential appears with the delay needed to propagate from the right end of the strand. Panel (d) shows the effect of the sawtooth on the response to the shock delivered during the plateau. The action potential was initiated at the right end of the strand by a small, 2-V/cm, 3-ms stimulus. This stimulus was too low to cause field stimulation, so the action potential appeared in the center with an approximately 12 ms delay. At 30 ms, a 20-V/cm shock was applied. When the ionic current includes the contribution of the sawtooth, the shock causes v_m^0 to decrease, as predicted by equation (13.50). When the sawtooth contribution is not included, the shock has no effect and the time course of v_m^0 is the same as in the control case (no shock).

The direct effect of the electric field on cardiac muscle via polarization of individual cells is currently being subject to numerous theoretical and experimental studies. The sawtooth potential is invoked as a mechanism of field stimulation and termination of arrhythmias by defibrillation shocks (Plonsey and Barr 1986; Krassowska et al. 1987; Chernysh et al. 1988; Dillon 1991; Trayanova and Pilkington 1993; Pumir and Krinsky 1996; Keener 1996), and as an explanation of why biphasic shocks require less energy for successful defibrillation (Sobie et al. 1995). What makes this hypothesis so attractive is that the magnitude of the sawtooth is proportional to the electric field (13.39) and thus provides an explanation of the experimental finding that the success or failure of stimulating and defibrillating shocks depends on the strength of the electric field (Wharton et al. 1992; Zhou et al. 1993; Frazier et al. 1988). In addition, experimental (Lepeschkin et al. 1980; Chen et al. 1986; Tang et al. 1992) and modeling (Rush et al. 1969; Claydon, III et al. 1988; Kothiyal et al. 1988; Sepulveda et al. 1990) studies have shown that even though the electric field decays with the distance from the electrodes, it retains an appreciable magnitude throughout most of the heart. Hence, the sawtooth provides the mechanism for the shock to reach the distant myocardium. Unfortunately, to date, the sawtooth has been observed only in single cells (Knisley et al. 1993; Windish et al. 1995), not in multicellular tissues (Zhou et al. 1995a). Hence, the evidence supporting the sawtooth is indirect and based on qualitative similarities between theoretical and experimental results. Quantitatively, the theory and experiments disagree: for example, thresholds for field stimulation measured experimentally are 5 to 10 times lower than predicted from the model. This disagreement suggests that structural units of the heart larger than cells may also play a role in the mechanisms of field stimulation and defibrillation (Plonsey and Barr 1986; Krassowska et al. 1990; Krassowska and Kumar 1996).

13.7   CONCLUSION

The bidomain model, originally proposed on basis of an intuitive physical argument, has been derived directly from the tissue cellular structure and basic physical principles. This derivation has shown that under weak electric fields, the bidomain model (13.1) is a correct and, to a leading order, complete representation of the macroscopic electrical behavior of cardiac muscle. However, under strong fields, the bidomain model loses its predictive power. This happens because the bidomain model is an asymptotic limit of the cellular-level formulation (13.10)–(13.12), taken under certain spatial and temporal scales and under certain electrical conditions. Thus, the bidomain model describes some, but not all, of the phenomena that may occur in a myocardium. There exist other, complementary phenomena, but under weak fields they have no leading-order consequences (Krassowska 1996a).

Strong electric fields awaken some of these complementary phenomena (Krassowska 1996b). For example, under strong fields, cardiac membrane electroporates and causes the saturation of the magnitude of the macroscopic transmembrane potential v_m^0 with the increase of the external field (Figure 13.9). The other phenomenon that gains importance under strong fields is the polarization of the individual cells of the tissue. Under weak fields, the magnitude of this polarization is so small that its existence can be ignored without any leading-order consequences on the macroscopic potentials. Under strong fields, the sawtooth is an order of magnitude (in ε) larger and changes the dynamics of the myocardium. For example, the presence of the sawtooth enables the direct excitation of the tissue by the external field (Figure 13.11 (a)–(c)), the result not predicted by the standard bidomain model.

The homogenized syncytium model, whose initial development was presented in this chapter, has additional mechanisms dealing with phenomena specific to strong shocks. Even though not all details of the model have yet been worked out, it is clear that there is an important qualitative difference between the models applicable under weak versus strong fields. Under weak fields, potentials actually existing in the tissue, Φ_i, Φ_e, and Φ_m, are well approximated by the leading-order macroscopic potentials, f_i^0, f_e^0, and v_m^0. Thus, it suffices to use the bidomain equations (13.1) to predict the electrical behavior of the tissue. The bidomain model is in this case a self-contained representation of the tissue: it takes the external stimulus and uses (13.1) to compute f_i^0, f_e^0 and v_m^0. Equations (13.1) deal only with macroscopic quantities: macroscopic, volume-averaged potentials, effective conductivities, and the membrane dynamics depending on the macroscopic state variables.

Under strong fields, potentials actually existing in the tissue, Φ_i, Φ_e, and Φ_m, contain, in addition to the bidomain potentials f_i^0, f_e^0 and v_m^0, also a volume conductor-type potential Ψ^0 caused by electroporation and the microscopic potentials ϕ_i^0, ϕ_e^0, and V_m^0 caused by the polarization of individual cells. To account for the possible influence of these potentials on the macroscopic behavior of the heart, the model applicable under strong fields is built around the homogenized syncytium equations (13.40). Just like the bidomain equations (13.1), the homogenized syncytium equations (13.40) predict the macroscopic potentials f_i^0, f_e^0 and v_m^0 and, consequently, the physiological state of the tissue. In contrast to (13.1), equations (13.40) are not self-contained: in order to compute f_i^0, f_e^0 and v_m^0, (13.40) must have information regarding the sawtooth potential V_m^0. Since the magnitude of V_m^0 is determined by the macroscopic electric field, (13.40) receives, indirectly, also information regarding the volume conductor potential Ψ^0. In addition, the outcome of (13.40) depends on the electroporation process that occurs on the boundary of the tissue. Therefore, the homogenized syncytium model (13.40) acts as a focal point that brings together all phenomena contributing to the macroscopic behavior of cardiac muscle.

In conclusion, this chapter demonstrated the existence of qualitative differences in the way cardiac muscle responds to weak and to strong electric fields. These differences are reflected by two different continuum models, the bidomain model (13.1) and the homogenized syncytium model (13.40). It is therefore important to understand under what conditions each model is valid, so that in a particular application the model that describes the dominant phenomena correctly and at a lowest computational cost is used.

REFERENCES

Bender, C. M. and Orszag, S. A. 1978. *Advanced Mathematical Methods for Scientists and Engineers*. New York, NY, USA: McGraw-Hill. ISBN 0-07-004452-X. Pages xiv + 593. {*282, 305*}

Bensoussan, Alain, Lions, Jacques-Louis, and Papanicolaou, George. 1978. *Asymptotic analysis for periodic structures*. Amsterdam, The Netherlands: North-Holland Publishing Co. ISBN 0-444-85172-0. Pages xxiv + 700. {*283, 305*}

Benz, R. and Zimmermann, U. 1980. Relaxation studies on cell membranes and lipid bilayers in the high electric field range. *Bioelectrochemistry and Bioenergetics*, **7**, 723–739. {*295, 305*}

Bukauskas, F. F., Kukushkin , N. I., and Sakson, M. Y. 1974. Model of a two-dimensional anisotropic syncytium. Communication I. *Biofizika*, **19**, 712–716. {*279, 305*}

Chapman, R. A. and Fry, C. H. 1978. An analysis of cable properties of frog ventricular myocardium. *Journal of Physiology*, **283**, 263–282. {*278, 280, 305*}

Chen, P.-S., Wolf, P. D., Claydon, III, F. J., Dixon, E. G., Vidaillet, Jr., H. J., Danieley, N. D., Pilkington, T. C., and Ideker, R. E. 1986. The potential gradient field created by epicardial defibrillation electrodes in dogs. *Circulation*, **74**, 626–636. {*303, 305*}

Chernomordik, L. V. and Chizmadzhev, Y. A. 1989. Electrical breakdown of lipid bilayer membranes. Phenomenology and mechanism. *Pages 83–95 of:* Neumann, E., Sowers, A. E., and Jordan, C. A. (eds), *Electroporation and Electrofusion in Cell Biology*. New York, NY, USA; London, UK: Plenum Press. {*294, 305*}

Chernysh, A. M., Tabak, V. Y., and Bogushevich, M. S. 1988. Mechanisms of electrical defibrillation of the heart. *Resuscitation*, **16**, 169–178. {*303, 305*}

Claydon, III, F. J., Pilkington, T. C., Tang, A. S. L., Morrow, M. N., and Ideker, R. E. 1988. A volume conductor model of the thorax for the study of defibrillation fields. *IEEE Transactions on Biomedical Engineering*, **BME-35**, 331–341. {*303, 305*}

Clerc, L. 1976. Directional differences of impulse spread in trabecular muscle from mammalian heart. *Journal of Physiology*, **255**, 335–346. {*280, 305*}

DeBruin, K. A. and Krassowska, W. 1996. Electroporation as a mechanism of the saturation of transmembrane potential induced by large electric fields. *In: 18th Annual Conference of the IEEE Engineering in Medicine and Biology Society*. 1109 Spring Street, Suite 300, Silver Spring, MD 20910, USA: IEEE Computer Society Press. Accepted. {*295, 305*}

Dillon, S. M. 1991. Optical recordings in the rabbit heart show that defibrillation strength shocks prolong the duration of depolarization and the refractory period. *Circulation research*, **69**, 842–856. {*303, 305*}

FitzHugh, R. 1961. Impulses and physiological states in theoretical models of nerve membrane. *Biophysical Journal*, **1**, 445–466. {*98, 299, 305*}

Frazier, D. W., Krassowska, W., Chen, P.-S., Wolf, P. D., Dixon, E. G., Smith, W. M., and Ideker, R. E. 1988. Extracellular field required for excitation in three-dimensional anisotropic canine myocardium. *Circulation research*, **63**, 147–164. {*303, 305*}

Freeman, S. A., Wang, M. A., and Weaver, J. C. 1994. Theory of electroporation of planar bilayer membranes: Predictions of the aqueous area, change in capacitance, and pore-pore separation. *Biophysical Journal*, **67**, 42–56. {*293, 305*}

Glaser, R. W. 1986. Appearance of a "critical voltage" in reversible electric breakdown. *Studia Biophysica*, **16**, 77–86. {*294, 305*}

Glaser, R. W., Leikin, S. L., Chernomordik, L. V., Pastushenko, V. F., and Sokirko, A. I. 1988. Reversible electrical breakdown of lipid bilayers: formation and evolution of pores. *Biochimica et Biophysica Acta*, **940**, 275–287. {*293, 305*}

Henriquez, C. S. 1993. Simulating the electrical behavior of cardiac tissue using the bidomain model. *CRC critical reviews in biomedical engineering*, **21**, 1–77. {*279, 306*}

Hodgkin, A. L. and Rushton, W. A. H. 1946. The electrical constants of a crustacean nerve fibre. *Proceedings of the Royal Society of London. Series B. Biological sciences*, **133**, 444–479. {*279, 306*}

Hoyt, R. H., Cohen, M. L., and Saffitz, J. E. 1989. Distribution and three-dimensional structure of intercellular junctions in canine myocardium. *Circulation research*, **64**, 563–574. {*278, 306*}

IEEE (ed). 1995. *Proceedings of the 17th Annual International Conference of the IEEE Engineering in Medicine and Biology Society (EMBS): Montreal, Canada, September 20–23, 1995*. 1109 Spring Street, Suite 300, Silver Spring, MD 20910, USA: IEEE Computer Society Press. {*306, 307*}

Keener, J. P. 1996. Direct activation and defibrillation of cardiac tissue. *Journal of Theoretical Biology*, **178**(313–324). {*301, 303, 306*}

Knisley, S. B., Blitchington, T. F., Hill, B. C., Grant, A. O., Smith, W. M., Pilkington, T. C., and Ideker, R. E. 1993. Optical measurements of transmembrane potential changes during electric field stimulation of ventricular cells. *Circulation research*, **72**(2), 255–270. {*303, 306*}

Kothiyal, K. P., Shankar, B., Fogelson, L. J., and Thakor, N. V. 1988. Three-dimensional computer model of electric fields in internal defibrillation. *Proceedings of the IEEE*, **76**, 720–730. {*303, 306*}

Krassowska, W. 1995a. Effects of electroporation on transmembrane potential induced by defibrillation shocks. *Pacing and Clinical Electrophysiology: PACE*, **18**, 1644–1660. {*294, 306*}

Krassowska, W. 1995b. Macroscopic model of electroporating membrane. *In:* (IEEE 1995), pages 253–254. {*294, 306*}

Krassowska, W. 1996a. The bidomain model. Part I: Derivation. *IEEE Transactions on Biomedical Engineering*. Submitted. {*280, 282, 284, 287, 304, 306*}

Krassowska, W. 1996b. The bidomain model. Part II: Limitations. *IEEE Transactions on Biomedical Engineering*. Submitted. {*291, 304, 306*}

Krassowska, W. and Kumar, M. S. 1996. The role of spatial interactions in creating the dispersion of transmembrane potential by premature electric shocks. *Annals of biomedical engineering*. Submitted. {*303, 306*}

Krassowska, W. and Neu, J. C. 1994a. Effective boundary conditions for syncytial tissues. *IEEE Transactions on Biomedical Engineering*, **41**, 143–150. {*289, 290, 306*}

Krassowska, W. and Neu, J. C. 1994b. Response of a single cell to an external electric field. *Biophysical Journal*, **66**, 1768–1776. {*282, 306*}

Krassowska, W., Pilkington, T. C., and Ideker, R. E. 1987. The closed form solution to the periodic core-conductor model using asymptotic analysis. *IEEE Transactions on Biomedical Engineering*, **34**(519–531). {*303, 306*}

Krassowska, W., Frazier, D. W., Pilkington, T. C., and Ideker, R. E. 1990. Potential distribution in three-dimensional periodic myocardium: Part II. Application to extracellular stimulation. *IEEE Transactions on Biomedical Engineering*, **BME-37** (3), 267–284. {*303, 306*}

Lepeschkin, E., Herrlich, H. C., Rush, S., Jones, J. L., and Jones, R. E. 1980. Cardiac potential gradients between defibrillation electrodes. *Medical Instrumentation*, **14**, 57. {*303, 306*}

Luo, C-H. and Rudy, Y. 1991. A model of the ventricular cardiac action potential: Depolarization, repolarization, and their interaction. *Circulation research*, **68**, 1501–1526. {*301, 306*}

Miller, III, W. T. and Geselowitz, D. B. 1978. Simulation studies of the electrocardiogram. I. The normal heart. *Circulation research*, **43**, 301–315. {*279, 307*}

Nagumo, J., Arimoto, S., and Yoshizawa, S. 1962. An active pulse transmission line simulating nerve axon. Proc. IRE,. *Proceedings of the Institute of Radio Engineers*, **50**, 2061–2070. {*98, 299, 307*}

Neu, J. C. and Krassowska, W. 1993. Homogenization of syncytial tissues. *CRC critical reviews in biomedical engineering*, **21**, 137–199. {*284–286, 297, 298, 307*}

Pastushenko, V. F., Chizmadzhev, Y. A., and Arakelyan, V. B. 1979. Electric breakdown of bilayer lipid membranes: II. Calculations of the membrane lifetime in the steady-state diffusion approximation. *Bioelectrochemistry and Bioenergetics*, **6**, 53–62. {*293, 307*}

Plonsey, R. and Barr, R. C. 1986. Effect of microscopic and macroscopic discontinuities on the response of cardiac tissue to defibrillating (stimulating) currents. *Medical & Biological Engineering & Computing*, **24**, 130–136. {*303, 307*}

Powell, K. T., Derrick, E. G., and Weaver, J. C. 1986. A quantitative theory of reversible electrical breakdown of bilayer membranes. *Bioelectrochemistry and Bioenergetics*, **15**, 243–255. {*293, 307*}

Pumir, A. and Krinsky, V. I. 1996. How does an electric field defibrillate cardiac muscle? *Physica D*, **91**, 205–219. {*303, 307*}

Rush, S., Lepeschkin, E., and Gregoritsch, A. 1969. Current distribution from defibrillation electrodes in a homogeneous torso model. *Journal of electrocardiology*, **2**, 331–341. {*303, 307*}

Sanchez-Palencia, E. 1980. Non-homogeneous media and vibration theory. *Pages ix + 398 of: Lecture Notes in Physics*. Berlin, Germany / Heidelberg, Germany / London, UK / etc.: Springer Verlag. {*283, 307*}

Schmitt, O. H. 1969. Biological information processing using the concept of interpenetrating domains. *Pages 325–331 of:* Leibovic, K. N. (ed), *Information processing in the nervous system; proceedings of a symposium held at the State University of New York at Buffalo, 21st–24th October, 1968*. Berlin, Germany / Heidelberg, Germany / London, UK / etc.: Springer Verlag. {*278, 307*}

Sepulveda, N. G., Wikswo, Jr., J. P., and Echt, D. S. 1990. Finite element analysis of cardiac defibrillation current distributions. *IEEE Transactions on Biomedical Engineering*, **BME-37**(4), 354–365. {*303, 307*}

Shiba, H. 1970. An electric model for flat epithelial cells with low resistive junctional membranes. A mathematical supplement. *Japanese Journal of Applied Physics*, **9**, 1405–1409. {*279, 307*}

Sobie, E. A., Fishler, M. G., and Tung, L. 1995. Biphasic shocks induce more uniform repolarization than monophasic shocks in a cardiac cellular field excitation model. *In:* (IEEE 1995), pages 257–258. {*303, 307*}

Sommer, J. R. and Scherer, B. 1985. Geometry of cell and bundle appositions in cardiac muscle: light microscopy. *American Journal of Physiology*, **248**, H792–H803. {*280, 307*}

Tang, A. S. L., Wolf, P. D., Afework, Y., Smith, W. M., and Ideker, R. E. 1992. Three-dimensional potential gradient fields generated by intracardiac catheter and cutaneous patch electrodes. *Circulation*, **85**(5), 1857–1864. {*303, 307*}

Trayanova, N. and Krassowska, W. 1995. Virtual electrode effects in a bidomain model with electroporating membrane. *In:* (IEEE 1995), pages 255–256. {*296, 307*}

Trayanova, N. A. and Pilkington, T. C. 1993. A bidomain model with periodic intracellular junctions. One-dimensional analysis. *IEEE Transactions on Biomedical Engineering*, **40**, 424–433. {*303, 307*}

Tung, L. 1978. *A bi-domain model for describing ischemic myocardial D-C potentials*. Ph.D. thesis, MIT, Cambridge, MA. {*279, 290, 307*}

Weidmann, S. 1970. Electrical constants of trabecular muscle from mammalian heart. *Journal of Physiology*, **210**, 1041–1054. {*278, 307*}

Wharton, J. M., Wolf, P. D., Smith, W. M., Chen, P-S., Frazier, D. W., Yabe, S.,

Danieley, N., and Ideker, R. E. 1992. Cardiac potential and potential gradient fields generated by single, combined, and sequential shocks during ventricular defibrillation. *Circulation*, **85**, 1510–1523. {*291, 303, 307*}

Windish, H., Ahammer, H., Schaffer, P., Müller, W., and Platzer, D. 1995. Optical multisite monitoring of cell excitation phenomena in isolated cardiomyocyte. *Pflugers Archiv: European Journal of Physiology*, **430**(4), 508–518. {*303, 308*}

Zhou, X., Daubert, P., Wolf, P. D., Smith, W. M., and Ideker, R. E. 1993. Epicardial mapping of ventricular defibrillation with monophasic and biphasic shocks in dogs. *Circulation research*, **72**, 145–160. {*291, 303, 308*}

Zhou, X., Ideker, R. E., Blitchington, T. F., Smith, W. M., and Knisley, S. B. 1995a. Optical transmembrane potential measurements during defibrillation-strength shocks in perfused rabbit hearts. *Circulation research*, **77**(3), 593–602. {*303, 308*}

Zhou, X., Rollins, D. L., Smith, W. M., and Ideker, R. E. 1995b. Responses of the transmembrane potential of myocardial cells during a shock. *Journal of Cardiovascular Electrophysiology*, **6**, 252–263. {*295, 308*}

Zhou, X., Smith, W. M., Ideker, R. E., and Rollins, D. L. 1996. Transmembrane potential changes caused by shocks in guinea pig papillary muscle. *American Journal of Physiology*. to appear. {*294–296, 308*}

14. FLUID DYNAMICS OF THE HEART AND ITS VALVES

Charles S. Peskin and David M. McQueen

14.1 INTRODUCTION

Blood is a viscous incompressible fluid which is propelled through the arteries, capillaries, and veins of the circulation by a collection of elastic and contractile fibers known as the heart. The left side of the heart receives bright red oxygenated blood from the lungs, and it pumps this blood into the aorta through which it is distributed to all of the tissues of the body, including the heart muscle (via the coronary arteries). As it flows through these various tissues, part of the oxygen is removed, and the color changes from bright red to bluish red. At the same time, carbon dioxide that has been generated by tissue metabolism is picked up by the blood. The deoxygenated blood returns to the right side of the heart, which pumps it through the pulmonary arterial tree to the lungs, where the carbon dioxide is removed and the blood becomes once more saturated with oxygen.

Each side of the heart has two chambers, an atrium and a ventricle. Each ventricle has an inflow (atrioventricular) and an outflow (arterial) valve. The valves are primarily passive structures that move in response to the flow of blood, although the atrioventricular valves are supported by muscles that prevent prolapse when those valves are closed. Both types of valves are constructed in such a way that they open freely to allow forward flow but close to prevent backflow. When the ventricles are relaxed (diastole), their inflow valves are open and their outflow valves are closed: during this time period the ventricular pressures are low, and each ventricle fills with blood from the corresponding atrium. When the ventricles contract (systole), the inflow valves close first, and then the outflow valves open as the ventricular pressures rise. Once the outflow valves are open, each ventricle ejects blood into its corresponding artery.

The familiar heart sounds are associated with the closure of the valves and the subsequent vibration of the cardiac and arterial chambers. The lower pitch "Lub" is associated with the nearly synchronous closure of the two atrioventricular valves, and the higher pitched "Dup" is associated with the nearly synchronous closure of the two arterial valves. Valve opening is normally silent but may produce sounds in disease states. Heart murmurs are associated with turbulence that may be generated by jets of fluid that form when valves fail to open fully or fail to close properly.

For further detail concerning the physiology of the heart, see Guyton (1991). The anatomy of the heart will be described in Section 14.6 of the present chapter.

Our concern here is with the mathematical formulation and computer solution of the coupled equations of motion of the blood, the muscular heart walls, and the flexible heart valve leaflets. The goal of the work described herein is to provide a realistic computer model of the heart which can be used in applied studies concerning normal cardiac physiology, diseases affecting the mechanical function of the heart and its valves, and the computer-assisted design of prosthetic cardiac valves. For examples of such studies conducted in an earlier, two-dimensional model of the left heart only, see McQueen et al. (1982); McQueen and Peskin (1983); McQueen and Peskin (1985); Meisner et al. (1985). The present chapter, however, is concerned with a three-dimensional whole-heart model (Peskin and McQueen 1992; Peskin and McQueen 1993; Peskin and McQueen 1995).

14.2   IMMERSED ELASTIC FIBERS IN A VISCOUS INCOMPRESSIBLE FLUID

In this section we formulate the fiber-fluid problem of cardiac dynamics. The formulation that we give involves both fluid dynamics and elasticity theory, coupled together in an unusual way. For background in these two fields, see Chorin and Marsden (1993); Green and Adkins (1970).

Consider an idealized composite material made of elastic fibers embedded in a viscous incompressible fluid. The fibers occupy zero volume fraction, and they have no mass, yet they are so finely divided that a continuum description of the material may be used. The fibers stick to the fluid and move at the local fluid velocity $\mathbf{u}$. At each point of the fiber-fluid composite there is a well-defined fiber direction given by the unit vector τ.

Let (q, r, s) be curvilinear coordinates chosen in such a way that a fixed value of the triple (q, r, s) denotes a material point, and also in such a way that a fixed value of the pair (q, r) designates a fiber. Let

$$\mathbf{x} = \mathbf{X}(q, r, s, t) \tag{14.1}$$

be the position at time t of the material point that carries the label (q, r, s). Then the unit tangent to the fibers is given by

$$\tau = \frac{\partial \mathbf{X}/\partial s}{|\partial \mathbf{X}/\partial s|}. \tag{14.2}$$

Let $T(q, r, s, t)$ be the fiber tension, in the sense that $T \, dq dr$ is the force transmitted by the fibers corresponding to the patch $dq dr$ of the (q, r) parameter plane. This force points in the fiber direction $\pm \tau$.

Since the fibers are elastic, the fiber tension is related to the fiber strain, which is determined by $|\partial \mathbf{X}/\partial s|$. We use a generalized Hooke's law of the form

$$T = \sigma(|\partial \mathbf{X}/\partial s|; q, r, s, t). \tag{14.3}$$

Note in particular the explicit time dependence of the stress-strain relation (14.3). It is this explicit time dependence that makes it possible for the heart to act as a pump and to do net work on the blood over a cardiac cycle.

We now determine the force applied by the fibers to the fluid in which they are immersed. Consider the collection $\mathcal{S}$ of fiber segments defined by $(q, r) \in \Omega$, $s_1 \le s \le s_2$, where Ω is an arbitrary region of the (q, r) parameter plane, and where s_1 and s_2 (with $s_1 < s_2$) are arbitrary values of the parameter s. Let $\mathcal{F}$ denote the total force applied by the fiber segments in $\mathcal{S}$ to the fluid. Then $-\mathcal{F}$ is the force of the fluid on $\mathcal{S}$. The only other forces acting on $\mathcal{S}$ are the fiber forces transmitted across the surfaces $s = s_1$, and $s = s_2$.

Note that the total force acting on $\mathcal{S}$ must be zero, since the fibers are massless. (As described above, all of the mass is carried by the fluid, which permeates the same space that is occupied by the fibers.) This yields the equation

$$0 = -\mathcal{F} + \int_\Omega (T\tau) \, dq \, dr \bigg|_{s=s_1}^{s=s_2} \tag{14.4}$$

or, by the fundamental theorem of calculus,

$$\mathcal{F} = \int_{s_1}^{s_2} \int_\Omega \frac{\partial}{\partial s} (T\tau) \, dq \, dr \, ds. \tag{14.5}$$

Let

$$\mathbf{f} = \frac{\partial}{\partial s} (T\tau). \tag{14.6}$$

Since s_1, s_2 and Ω are arbitrary, (14.5) shows that $\mathbf{f}$ is the density (with respect to the measure $dq \, dr \, ds$) of the force applied by the fibers to the fluid in which they are immersed.

For the sake of interpretation, one may expand the derivative in (14.6) to obtain

$$\mathbf{f} = \frac{\partial T}{\partial s} \tau + T \frac{\partial \tau}{\partial s}. \tag{14.7}$$

Note that these two terms are orthogonal: τ is the unit tangent to the fibers, and $\partial \tau / \partial s$ is the vector whose direction defines the principal normal to the fibers. There is no component of $\mathbf{f}$ in the binormal direction.

In the following, we shall need an expression for the force density in Cartesian coordinates. That is, we require a vector field $\mathbf{F}(\mathbf{x}, t)$ such that

$$\int_V \mathbf{F}(\mathbf{x}, t) \, d\mathbf{x} = \int_{\mathbf{X}^{-1}(V, t)} \mathbf{f}(q, r, s, t) \, dq \, dr \, ds, \tag{14.8}$$

where V is an arbitrary region in the physical space, and where

$$\mathbf{X}^{-1}(V, t) = \{(q, r, s) : \mathbf{X}(q, r, s, t) \in V\} \tag{14.9}$$

This can be achieved by setting

$$\mathbf{F}(\mathbf{x}, t) = \int \mathbf{f}(q, r, s, t) \delta(\mathbf{x} - \mathbf{X}(q, r, s, t)) \, dq \, dr \, ds, \tag{14.10}$$

in which the integral extends over the entire (q, r, s) parameter space, and in which $\delta(\mathbf{x})$ is the three-dimensional Dirac delta function. To recover (14.8) from (14.10), integrate over the volume V, interchange the order of integration, and recall that

$$\int_V \delta(\mathbf{x} - \mathbf{X}) \, d\mathbf{x} = \begin{cases} 1, \mathbf{X} \in V, \\ 0, \mathbf{X} \notin V \end{cases} \qquad (14.11)$$

Note that $\mathbf{F}(\mathbf{x}, t) = 0$ unless the point $\mathbf{x}$ happens to be within the region that is occupied by the fibers at time t.

Equation (14.10) is not the only way to write the relationship between $\mathbf{F}$ and $\mathbf{f}$. Another possibility, which follows directly from (14.8), is

$$\mathbf{F}(\mathbf{X}(q, r, s, t), t) J(q, r, s) = \mathbf{f}(q, r, s, t), \qquad (14.12)$$

where

$$J(q, r, s) = \left(\frac{\partial \mathbf{X}}{\partial q} \times \frac{\partial \mathbf{X}}{\partial r} \right) \cdot \frac{\partial \mathbf{X}}{\partial s} \qquad (14.13)$$

is the Jacobian of the map $(q, r, s) \to \mathbf{X}(q, r, s, t)$. J is independent of t because the fluid is incompressible. The advantages of using the Dirac delta function as in (14.10) are that it avoids the introduction of J, that it gives an explicit formula for $\mathbf{F}(\mathbf{x}, t)$, and that it generalizes easily to the situation in which the fibers are confined to a (moving) surface embedded in the three-dimensional space, as in the case of a fiber-reinforced heart-valve leaflet. To obtain this generalization, we simply drop one of the two parameters (q or r) that label the fibers. Then the surface in question is given at any particular time t by an equation of the form $\mathbf{x} = \mathbf{X}(r, s, t)$, and we have

$$\mathbf{F}(\mathbf{x}, t) = \int \mathbf{f}(r, s) \delta(\mathbf{x} - \mathbf{X}(r, s, t)) \, dr \, ds. \qquad (14.14)$$

In this expression there are only two integrals, but the delta function is still three-dimensional, so $\mathbf{F}(\mathbf{x}, t)$ is singular like a one-dimensional delta function. Despite this important difference between the relations given by (14.10), wherein $\mathbf{F}$ is finite, and (14.14), the use of the Dirac delta function provides a unified framework for modeling the thick heart walls and the thin heart-valve leaflets.

14.3   EQUATIONS OF MOTION

Our purpose here is to state the equations of motion of the fiber-fluid system described in the previous section. These equations are as follows

$$\rho \left(\frac{\partial \mathbf{u}}{\partial t} + \mathbf{u} \cdot \nabla \mathbf{u} \right) = -\nabla p + \mu \nabla^2 \mathbf{u} + \mathbf{F}, \qquad (14.15)$$

$$\nabla \cdot \mathbf{u} = 0, \qquad (14.16)$$

$$\mathbf{F}(\mathbf{x}, t) \;=\; \int \mathbf{f}(q, r, s, t)\delta(\mathbf{x} - \mathbf{X}(q, r, s, t))\, dq\, dr\, ds, \quad (14.17)$$

$$\frac{\partial \mathbf{X}}{\partial t}(q, r, s, t) \;=\; \mathbf{u}(\mathbf{X}(q, r, s, t), t) \qquad\qquad (14.18)$$

$$=\; \int \mathbf{u}(\mathbf{x}, t)\delta(\mathbf{x} - \mathbf{X}(q, r, s, t))\, d\mathbf{x}, \qquad (14.19)$$

$$\mathbf{f} \;=\; \frac{\partial}{\partial s}(T\tau), \qquad\qquad (14.20)$$

$$T \;=\; \sigma\left(\left|\frac{\partial \mathbf{X}}{\partial s}\right|; q, r, s, t\right), \qquad (14.21)$$

$$\tau \;=\; \frac{\partial \mathbf{X}/\partial s}{|\partial \mathbf{X}/\partial s|}. \qquad\qquad (14.22)$$

Equations (14.15)–(14.16) are the *fluid equations*, (14.20)–(14.22) are the *fiber equations*, and (14.17)–(14.19) are the *interaction equations* of the fiber-fluid system.

Note that the fluid equations are in *Eulerian* form: they involve several unknown functions of $(\mathbf{x}, t)$, where $\mathbf{x} = (x_1, x_2, x_3)$ are fixed Cartesian coordinates and t is the time. These unknown functions are the fluid velocity, $\mathbf{u}(\mathbf{x}, t)$, the fluid pressure, $p(\mathbf{x}, t)$, and the Eulerian fiber force density $\mathbf{F}(\mathbf{x}, t)$. The constants ρ and μ are the density and viscosity of the fluid. Equations (14.15)–(14.16) are the familiar Navier-Stokes equations of a viscous, incompressible fluid. The only novel feature is the use of the applied-force density $\mathbf{F}(\mathbf{x}, t)$ for the representation of the fiber force.

The fiber equations derived in Section 14.2 are in *Lagrangian* form: they involve several unknown functions of (q, r, s, t), where (q, r, s) are moving curvilinear coordinates attached to the material points of the fibers. These unknown functions are the fiber configuration $\mathbf{X}(q, r, s, t)$, the unit tangent to the fibers $\tau(q, r, s, t)$, the fiber tension $T(q, r, s, t)$ and the Lagrangian form of the fiber force density $\mathbf{f}(q, r, s, t)$. The fiber equations may be used to determine the fiber force density at any given time t from the fiber configuration at that same time t. This is done by substituting (14.21)–(14.22) into (14.20). The fact that the fiber forces can be determined from the fiber configuration is an expression of the elasticity of the fibers.

The interaction equations connect the Lagrangian and Eulerian variables. Note that both of the interaction equations take the form of integral transformations in which the kernel is $\delta(\mathbf{x}-\mathbf{X}(q, r, s, t))$. In (14.17) the integration $(dqdrds)$ is over the fiber parameter space, but in (14.19) the integration $(d\mathbf{x})$ is over the physical space. The latter may be thought of as an integration over the fluid, since the volume fraction occupied by the fibers is zero (see Section 14.2).

The first of the interaction equations (14.17) has already been discussed in Section 14.2. The second interaction equation (14.18)–(14.19) is the no-slip condition of a viscous fluid. In the present context, it states that the fibers move at the local fluid velocity, and it serves as an equation of motion for the fibers, not as a constraint on the fluid motion, since the motion of the fibers is unknown.

The structure of (14.15)–(14.22) is as follows. At any given time t, the state of the system is determined by the fiber configuration $\mathbf{X}(\ ,\ ,\ ,t)$ and the fluid velocity $\mathbf{u}(\ ,t)$. Given these quantities, $\partial \mathbf{X}/\partial t$ may be found directly from (14.18)–(14.19), and $\partial \mathbf{u}/\partial t$ may be determined as follows. First use (14.21) and (14.22) to find T and τ, substitute these results into (14.20) to find $\mathbf{f}$, and then use (14.17) to find $\mathbf{F}$. With $\mathbf{F}$ known, the Navier-Stokes equations, (14.15)–(14.16) may be used to find $\partial \mathbf{u}/\partial t$. (As is well known, this requires the elimination of p through the solution of a Poisson equation.) In effect, then, (14.15)–(14.22) are a first-order system in the state variables $\mathbf{X}$ and $\mathbf{u}$. The numerical solution of this system will be discussed in the following section.

14.4   THE IMMERSED-BOUNDARY METHOD

Equations (14.15)–(14.22) are solved by the immersed-boundary method, see Peskin and McQueen (1992); Peskin and McQueen (1993); Peskin and McQueen (1995), and references therein.

The philosophy underlying the immersed-boundary method is that the fluid equations (14.15)–(14.16), which are in Eulerian form, should be discretized on a fixed cubic lattice, whereas the fiber equations (14.20)–(14.22), which are in Lagrangian form, should be discretized on a moving collection of points that need not coincide with the lattice points of the fluid equations. This immediately raises the question of how to handle the fiber-fluid interaction (14.17)–(14.19), a question which will be answered below through the introduction of a smoothed approximation to the Dirac delta function.

Let time proceed in discrete steps of duration Δt, and use a superscript as the time step index: $\mathbf{X}^n(q, r, s) = \mathbf{X}(q, r, s, n\Delta t)$. Let a discrete collection of fibers be chosen, e.g., $(q, r) = (k\Delta q, \ell\Delta r)$, where k and ℓ are integers, and let each fiber be represented by a discrete collection of points: $s = m\Delta s$, where m is an integer. It is convenient to define the fiber tension and the unit tangent τ at the "half-integer" points given by $s = (m + \frac{1}{2})\Delta s$.

This is done as follows. For any function $\phi(s)$, let

$$(D_s\phi)(s) = \frac{\phi(s + \dfrac{\Delta s}{2}) - \phi(s - \dfrac{\Delta s}{2})}{\Delta s}. \tag{14.23}$$

Then make the definitions

$$T^n = \sigma(|D_s\mathbf{X}^n|; q, r, s, n\Delta t), \tag{14.24}$$

$$\tau^n = \frac{D_s\mathbf{X}^n}{|D_s\mathbf{X}^n|}, \tag{14.25}$$

both of which hold for $s = (m + \frac{1}{2})\Delta s$. Finally, we can use T^n and τ^n to define $\mathbf{f}^n$ at the points $s = m\Delta s$:

$$\mathbf{f}^n = D_s(T^n\tau^n) \tag{14.26}$$

Note that $\mathbf{f}^n$ is defined at the same values of s as $\mathbf{X}^n$. Equations (14.24)–(14.26) are discrete approximations to (14.21), (14.22), and (14.20), respectively.

Let the fluid velocity and pressure be defined on the cubic lattice of points $\mathbf{x} = \mathbf{j}h$, where h is the meshwidth and $\mathbf{j} = (j_1, j_2, j_3)$ is a vector with integer components. Our next task is to construct a force-field $\mathbf{F}$, defined on this cubic lattice, to represent the effect of the fibers in the fluid-dynamics computation. This done using (14.17) as a guide. The formula for $\mathbf{F}$ is as follows:

$$\mathbf{F}^n(\mathbf{x}) = \sum_{q,r,s} \mathbf{f}^n(q, r, s)\delta_h(\mathbf{x} - \mathbf{X}^n(q, r, s))\Delta q\,\Delta r\,\Delta s, \tag{14.27}$$

The notation $\sum_{q,r,s}$ is here understood to mean the sum over the discrete collection of points of the form $(q, r, s) = (k\Delta q, \ell\Delta r, m\Delta s)$, where k, ℓ, m are integers. The function δ_h is a smoothed approximation to the three-dimensional Dirac delta-function. The choice of δ_h will be discussed in the following section.

With $\mathbf{F}^n$ defined, we are ready to solve the Navier-Stokes equations (14.15)–(14.16) on the cubic lattice of meshwidth h introduced above. Note that the Navier-Stokes solver does not need to know anything about the complicated time-dependent geometry of the cardiac fibers, since the influence of those fibers on the fluid is completely described by the forcefield $\mathbf{F}$, which is defined on the regular cubic lattice of the fluid computation. This is the central idea of the immersed-boundary method.

There are many different schemes that could be used to integrate the Navier-Stokes equations. The one that we currently employ is implicitly defined as follows:

$$\rho\left(\frac{\mathbf{u}^{n+1}-\mathbf{u}^n}{\Delta t} + \sum_{\alpha=1}^{3} u_\alpha^n D_\alpha^\pm \mathbf{u}^n\right) - \mathbf{D}^0 p^{n+1}$$

$$= \mu \sum_{\alpha=1}^{3} D_\alpha^+ D_\alpha^- \mathbf{u}^{n+1} + \mathbf{F}^n, \tag{14.28}$$

$$\mathbf{D}^0 \cdot \mathbf{u}^{n+1} = 0. \tag{14.29}$$

In (14.28)–(14.29), $\mathbf{D}^0$ is the central-difference approximation to ∇, defined by

$$\mathbf{D}^0 = (D_1^0, D_2^0, D_3^0) \tag{14.30}$$

where

$$(D_\alpha^0\phi)(\mathbf{x}) = \frac{\phi(\mathbf{x} + h\mathbf{e}_\alpha) - \phi(\mathbf{x} - h\mathbf{e}_\alpha)}{2h}. \tag{14.31}$$

and where $\{\mathbf{e}_1, \mathbf{e}_2, \mathbf{e}_3\}$ is the standard basis of $\mathbb{R}^3$. Thus $\mathbf{D}^0 p$ approximates ∇p, and $\mathbf{D}^0 \cdot \mathbf{u}$ approximates $\nabla \cdot \mathbf{u}$. The operators $D_\alpha^\pm$ are forward and backward difference approximations to $\partial/\partial x_\alpha$, defined as follows:

$$(D_\alpha^+\phi)(\mathbf{x}) = \frac{\phi(\mathbf{x} + h\mathbf{e}_\alpha) - \phi(\mathbf{x})}{h}, \tag{14.32}$$

$$(D_\alpha^-\phi)(\mathbf{x}) = \frac{\phi(\mathbf{x}) - \phi(\mathbf{x} - h\mathbf{e}_\alpha)}{h}. \tag{14.33}$$

Thus, the expression $\sum_{\alpha=1}^{3} D_\alpha^+ D_\alpha^-$, which appears in the viscous term of (14.28), is a difference approximation to the Laplace operator. Similarly, the expression

$\sum\limits_{\alpha=1}^{3} u_\alpha D_\alpha^\pm$, which appears in the convection term, is a difference approximation
to $\mathbf{u} \cdot \nabla$. In this latter case, the plus or minus sign is chosen to yield upwind differencing:

$$u_\alpha D_\alpha^\pm = \begin{cases} u_\alpha D_\alpha^+, & u_\alpha < 0, \\ u_\alpha D_\alpha^-, & u_\alpha > 0. \end{cases} \tag{14.34}$$

The motivation for this choice of sign can be understood by considering the stability of the difference equation

$$\frac{\phi^{n+1} - \phi^n}{\Delta t} + \sum_{\alpha=1}^{3} u_\alpha^n D_\alpha^\pm \phi^n = 0, \tag{14.35}$$

which is a difference approximation to the transport (or advection) equation

$$\frac{\partial \phi}{\partial t} + \mathbf{u} \cdot \nabla \phi = 0. \tag{14.36}$$

With the upwind choice of sign, (14.35) takes the form

$$\phi^{n+1}(\mathbf{x}) = \left(1 - \frac{\Delta t}{h} \sum_{\alpha=1}^{3} |u_\alpha^n(\mathbf{x})| \right) \phi^n(\mathbf{x})$$

$$+ \frac{\Delta t}{h} \sum_{\alpha=1}^{3} |u_\alpha^n(\mathbf{x})| \phi^n(\mathbf{x} \pm h\mathbf{e}_\alpha). \tag{14.37}$$

Provided that

$$\Delta t \sum_{\alpha=1}^{3} |u_\alpha^n(\mathbf{x})| \le h \tag{14.38}$$

for all $n, \mathbf{x}$, the righthand side of (14.37) is a weighted average. Because of this, we can derive the inequality

$$\max_{\mathbf{x}} |\phi^{n+1}(\mathbf{x})| \le \max_{\mathbf{x}} |\phi^n(\mathbf{x})| \tag{14.39}$$

which says that (14.35) is a *stable* difference scheme. The condition (14.38) is known as the Courant-Friedrichs-Lewy or CFL condition.

It should be mentioned that the upwind difference approximation to $\mathbf{u} \cdot \nabla \mathbf{u}$ used above is only first-order accurate and is just the simplest of a variety of available upwind schemes. For a second-order-accurate upwind scheme, see Bell et al. (1989).

It is important to note that (14.28)–(14.29) provide only an implicit definition of $(\mathbf{u}^{n+1}, p^{n+1})$ given $\mathbf{u}^n$ and $\mathbf{F}^n$. Surprisingly, these equations are linear (with constant coefficients) in the unknowns $\mathbf{u}^{n+1}$ and p^{n+1}; the non-linear terms of the Navier-Stokes equations are present but have been entirely expressed in terms of $\mathbf{u}^n$. It is therefore natural to use Fourier methods, implemented by the Fast Fourier Transform (FFT) algorithm, to solve for $(\mathbf{u}^{n+1}, p^{n+1})$, see Press et al. (1986).

Since Fourier methods are most easily applied in the context of periodic functions, we choose a periodic cube (3–torus) with period L, as the domain that is occupied by the fluid. The meshwidth h is chosen to be of the form $h = L/N$, where N is an integer, and the computational lattice of points $\mathbf{x_j} = \mathbf{j}h = (j_1 h, j_2 h, j_3 h)$ is chosen to be the set of points given by $j_\alpha \in \{0, 1, \ldots, N-1\}$, for $\alpha = 1, 2, 3$. Arithmetic involving the j_α is understood to be performed modulo N. To accommodate the simplest form of the FFT algorithm, N is chosen to be a power of 2.

Thus, the problem we actually consider is that of a heart immersed in fluid, the fluid being contained in a cubic box with periodic boundary conditions imposed. The fluid external to the model heart may be regarded as representing (in a simplified way) the tissues of the thorax that are adjacent to the heart. These tissues must move when the heart walls move, and they therefore have a (modest) influence on cardiac dynamics. As for the periodic boundary conditions, it should be noted that these are far less restrictive than rigid walls, since they do allow fluid to flow freely through any face of the cube, provided only that such fluid must return instantaneously through the opposite face. Since all points of a 3-torus are equivalent, we do not have to worry about peculiar behavior near the faces of the cube, as we would if rigid-wall boundary conditions had been used. The periodic box does, however, enforce a constant-volume constraint, which is unwanted in the present context, since the volume of the heart changes during the cardiac cycle. This is taken care of through the provision of sources and sinks (including an external source/sink). For ease of exposition, however, we omit the sources and sinks from the present discussion.

The Fourier solution of (14.28) and (14.29) is accomplished as follows. First we rewrite these equations in the form

$$\left(I - \frac{\mu \Delta t}{\rho} \sum_{\alpha=1}^{3} D_\alpha^+ D_\alpha^- \right) \mathbf{u}^{n+1} + \frac{\Delta t}{\rho} \mathbf{D}^0 p^{n+1} = \mathbf{v}^n, \qquad (14.40)$$

$$\mathbf{D}^0 \cdot \mathbf{u}^{n+1} = 0, \qquad (14.41)$$

where

$$\mathbf{v}^n = \mathbf{u}^n - \frac{\Delta t}{\rho} \sum_{\alpha=1}^{3} u_\alpha^n D_\alpha^\pm \mathbf{u}^n + \frac{\Delta t}{\rho} \mathbf{F}^n. \qquad (14.42)$$

The three-dimensional discrete Fourier transform is defined as follows. Let $\phi(\mathbf{x})$ be a periodic function with period L in all three space directions. Only the value of ϕ at mesh points will be used in the following definition. Let

$$\begin{aligned} \hat{\phi}(\mathbf{k}) &= \frac{1}{L^3} \sum_{\mathbf{x}} \phi(\mathbf{x}) \exp\left(-i 2\pi \frac{\mathbf{k} \cdot \mathbf{x}}{L} \right) h^3 \\ &= \frac{1}{N^3} \sum_{\mathbf{j}} \phi(\mathbf{j}h) \exp\left(-i \frac{2\pi}{N} \mathbf{k} \cdot \mathbf{j} \right). \end{aligned} \qquad (14.43)$$

The sums in (14.43) are over the N^3 points given by $j_\alpha \in \{0, 1, \ldots, N-1\}$ for $\alpha = 1, 2, 3$. Here $\mathbf{k} = (k_1, k_2, k_3)$ is a vector with integer components.

Note that $\hat{\phi}(\mathbf{k})$ is periodic in all the components of $\mathbf{k}$ with period N. In the following, we shall restrict k_α to be chosen from the same set as j_α, namely $\{0, 1, \cdots, N-1\}$.

The inversion formula, which is exact at mesh points, is given by

$$\phi(\mathbf{x}) = \sum_{\mathbf{k}} \hat{\phi}(\mathbf{k}) \exp\left(\frac{2\pi i \mathbf{k} \cdot \mathbf{x}}{L}\right) \tag{14.44}$$

or

$$\phi(\mathbf{j}h) = \sum_{\mathbf{k}} \hat{\phi}(\mathbf{k}) \exp\left(\frac{2\pi i}{N} \mathbf{k} \cdot \mathbf{j}\right). \tag{14.45}$$

By considering displacements in $\mathbf{x}$, it is easy to find the Fourier transforms of the difference operators appearing in (14.40)–(14.41). They are

$$\hat{D}_\alpha^+(\mathbf{k})\hat{D}_\alpha^-(\mathbf{k}) = -\frac{4}{h^2}\sin\frac{\pi k_\alpha}{N}, \tag{14.46}$$

$$\hat{\mathbf{D}}^0 = \frac{i}{h}\sin\left(\frac{2\pi}{N}\mathbf{k}\right). \tag{14.47}$$

In the latter formula, the expression $\sin(\frac{2\pi}{N}\mathbf{k})$ denotes the vector with components $\sin(\frac{2\pi}{N}k_\alpha)$, $\alpha = 1, 2, 3$. Using these results, we may write the discrete Fourier transformation of (14.40)–(14.41) as follows:

$$A(\mathbf{k})\hat{\mathbf{u}}^{n+1}(\mathbf{k}) + \frac{i\Delta t}{\rho h}\sin\left(\frac{2\pi}{N}\mathbf{k}\right)\hat{p}^{n+1}(\mathbf{k}) = \hat{\mathbf{v}}^n(\mathbf{k}), \tag{14.48}$$

$$\frac{i}{h}\left(\sin\left(\frac{2\pi}{N}\mathbf{k}\right)\right) \cdot \hat{\mathbf{u}}^{n+1}(\mathbf{k}), = 0 \tag{14.49}$$

where

$$A(\mathbf{k}) = 1 + \frac{4\mu\Delta t}{\rho h^2}\sum_{\alpha=1}^{3}\left(\sin\frac{\pi k_\alpha}{N}\right)^2. \tag{14.50}$$

Note that the different values of $\mathbf{k}$ are now uncoupled. This is the (usual) benefit of applying a Fourier transformation to a translation-invariant problem.

For each $\mathbf{k}$, the system (14.48)–(14.49) can be solved for the four unknowns $(\hat{\mathbf{u}}^{n+1}(\mathbf{k}), \hat{p}^{n+1}(\mathbf{k}))$. The solution may be expressed as follows:

$$\hat{\mathbf{u}}^{n+1}(\mathbf{k}) = \frac{\hat{\mathbf{v}}^n(\mathbf{k}) - \frac{i\Delta t}{\rho h}(\sin\frac{2\pi}{N}\mathbf{k})\hat{p}^{n+1}(\mathbf{k})}{A(\mathbf{k})}, \tag{14.51}$$

where

$$\hat{p}^{n+1}(\mathbf{k}) = \frac{\frac{i}{h}(\sin\frac{2\pi}{N}\mathbf{k}) \cdot \hat{\mathbf{v}}^n(\mathbf{k})}{-\frac{\Delta t}{\rho h^2}(\sin\frac{2\pi}{N}\mathbf{k}) \cdot (\sin\frac{2\pi}{N}\mathbf{k})} \tag{14.52}$$

and where $A(\mathbf{k})$ is given by (14.50)

The values of $\mathbf{k}$ at which $\sin(\frac{2\pi}{N}\mathbf{k}) = (0, 0, 0)$ require special consideration. There are eight such points in the $\mathbf{k}$-lattice; they are given by $k_\alpha = 0$ or $N/2$, for $\alpha = 1, 2, 3$. At these special values of the wave vector $\mathbf{k}$, (14.49) is automatically satisfied, and (14.48) reduces to

$$A(\mathbf{k})\hat{\mathbf{u}}^{n+1}(\mathbf{k}) = \hat{\mathbf{v}}^n(\mathbf{k}), \tag{14.53}$$

where $k_\alpha \in \{0, N/2\}$, $\alpha = 1, 2, 3$. There is no danger of $A(\mathbf{k})$ being zero, since $A(\mathbf{k}) \geq 1$ for all $\mathbf{k}$, see (14.50). Thus, (14.53) determines $\hat{\mathbf{u}}^{n+1}(\mathbf{k})$ at the eight special wave vectors $\mathbf{k}$ that are under consideration here. As for $\hat{p}^{n+1}(\mathbf{k})$, its values at these eight wave vectors are completely arbitrary, so we may set $\hat{p}^{n+1}$ equal to zero at these eight wave vectors in order to have a well-defined pressure.

Once $\mathbf{u}^{n+1}(\mathbf{x})$ has been determined (by applying the inverse FFT algorithm to $\hat{\mathbf{u}}^{n+1}(\mathbf{k})$), the fiber points are moved at the local fluid velocity in this new velocity field. This done according to the following interpolation scheme:

$$\frac{\mathbf{X}^{n+1}(q, r, s) - \mathbf{X}^n(q, r, s)}{\Delta t} = \sum_{\mathbf{x}} \mathbf{u}^{n+1}(\mathbf{x})\delta_h(\mathbf{x} - \mathbf{X}^n(q, r, s))h^3, \tag{14.54}$$

which is a discretization of (14.20). Here $\sum_{\mathbf{x}}$ denotes the sum over the computational lattice $\mathbf{x} = \mathbf{j}h$, $j_\alpha \in \{0, 1, \ldots, N-1\}$, $\alpha = 1, 2, 3$. Note that the same δ-function weights which are here used for interpolation were previously employed in the application of the fiber force to the computational lattice of the fluid, see (14.27). The construction of the function δ_h will be described in the next section.

In summary, the immersed-boundary method proceeds as follows. Given the fiber configuration $\mathbf{X}^n(q, r, s)$ and the fluid velocity $\mathbf{u}^n(\mathbf{x})$, begin by evaluating the fiber forces $\mathbf{f}^n(q, r, s)$:

$$T^n = \sigma(|D_s\mathbf{X}^n|; q, r, s, n\Delta t), \tag{14.55}$$

$$\tau^n = \frac{D_s\mathbf{X}^n}{|D_s\mathbf{X}^n|}, \tag{14.56}$$

$$\mathbf{f}^n = D_s(T^n\tau^n). \tag{14.57}$$

Next apply the fiber forces to the computational lattice of the fluid:

$$\mathbf{F}^n(\mathbf{x}) = \sum_{q,r,s} \mathbf{f}^n(q, r, s)\delta_h(\mathbf{x} - \mathbf{X}^n(q, r, s))\Delta q \Delta r \Delta s \tag{14.58}$$

Then update the fluid velocity by solving the following system of equations for $(\mathbf{u}^{n+1}, p^{n+1})$:

$$\rho\left(\frac{\mathbf{u}^{n+1} - \mathbf{u}^n}{\Delta t} + \sum_{\alpha=1}^{3} u_\alpha^n D_\alpha^\pm \mathbf{u}^n\right) + \mathbf{D}^0 p^{n+1}$$

$$= \mu \sum_{\alpha=1}^{3} D_\alpha^+ D_\alpha^- \mathbf{u}^{n+1} + \mathbf{F}^n, \tag{14.59}$$

$$\mathbf{D}^0 \cdot \mathbf{u}^{n+1} = 0. \tag{14.60}$$

Finally, interpolate the new velocity field and move the fibers

$$\mathbf{X}^{n+1}(q, r, s) = \mathbf{X}^n(q, r, s) + (\Delta t) \sum_{\mathbf{X}} \mathbf{u}^{n+1}(\mathbf{x}) \delta_h(\mathbf{x} - \mathbf{X}^n(q, r, s)) h^3 \tag{14.61}$$

Since $\mathbf{X}^{n+1}$ and $\mathbf{u}^{n+1}$ have been found, the time step is complete.

14.5 CONSTRUCTION OF A SMOOTHED δ-FUNCTION

The function $\delta_h(\mathbf{x})$, which plays a prominent role in the immersed-boundary method, is a smoothed approximation to the three-dimensional Dirac δ-function. Let

$$\delta_h(\mathbf{x}) = h^{-3} \phi\left(\frac{x_1}{h}\right) \phi\left(\frac{x_2}{h}\right) \phi\left(\frac{x_3}{h}\right), \tag{14.62}$$

where $\mathbf{x} = (x_1, x_2, x_3)$, and where ϕ has the following properties:

(i) ϕ is a continuous function.

(ii) $\phi(r) = 0$ for $|r| \geq 2$.

(iii) For all r,

$$\sum_{j \text{ even}} \phi(r - j) = \sum_{j \text{ odd}} \phi(r - j) = \frac{1}{2}. \tag{14.63}$$

(iv) For all r,

$$\sum_j (r - j) \phi(r - j) = 0. \tag{14.64}$$

(v) For all r,

$$\sum_j (\phi(r - j))^2 = C, \tag{14.65}$$

where C is independent of r.

We shall show presently that these five postulates uniquely determine the function ϕ, including the specific numerical value of the constant C, and hence (via (14.62)) that they uniquely determine the function δ_h. Before doing so, however, we give an informal discussion of the motivation for proposing these five postulates in the first place.

Continuity of ϕ implies continuity of δ_h, and this avoids sudden changes as the fiber points move in space. An alternative strategy (for example) would be to let each fiber point interact only with the nearest lattice point, but this would introduce spurious discontinuities as the fiber points cross the planes that are equidistant between lattice points.

As a practical matter, bounded support of the function ϕ is essential for the efficient operation of the immersed-boundary method. In the interaction steps, where the function δ_h is used, each fiber point $\mathbf{X}$ interacts with all lattice points $\mathbf{x}$ that lie within the support of $\delta_h(\mathbf{x}-\mathbf{X})$. With the choice $\phi(r) = 0$ for $|r| \geq 2$, each fiber point interacts with exactly $4^3 = 64$ lattice points, already a substantial number. From this standpoint, the support of ϕ should be as small as possible. In postulate (ii), we have made the smallest choice that is consistent with the other postulates.

The third postulate (14.63) immediately implies that, for all r

$$\sum_j \phi(r - j) = 1. \tag{14.66}$$

When δ_h is used for interpolation, this means that the interpolation of constant functions is exact; when it is used for spreading the force to the fluid lattice, this implies conservation of momentum: that the total force is not altered by the spreading process. These considerations do not, however, explain why we need to impose the stronger conditions given by (14.63).

The motivation for these stronger conditions comes from a peculiar property of the central difference operator $\mathbf{D}^0$ that is used in this work as an approximation to the vector differential operator ∇. Of course, $\nabla\phi = 0 \Rightarrow \phi =$ constant, but the analogous statement is not true for $\mathbf{D}^0$. As an example, suppose $\phi(\mathbf{j}h) = (-1)^{j_1}$, independent of j_2 and j_3. This function is certainly not constant, it oscillates in the x_1 direction from one lattice point to the next, but $\mathbf{D}^0\phi = 0$ anyway. The effect of this is that spurious long-range oscillations can be introduced into the computed solution by the action of localized forces unless the conditions (14.63) are imposed. These conditions guarantee that the force field $\mathbf{F}$ will have a discrete Fourier transform $\hat{\mathbf{F}}(\mathbf{k})$ that vanishes for any wave vector $\mathbf{k}$ having any component k_α equal to $N/2$.

The fourth postulate, (14.64), when combined with (14.66), guarantees that the interpolation scheme based on δ_h will be exact for any linear function. When δ_h is used for spreading the force to the fluid lattice, these equations guarantee conservation of angular momentum: that the correct total torque is applied to the fluid by each element of the fiber force.

The fifth postulate, (14.65), is motivated an inequality which follows from it by an application of the Schwarz inequality:

$$\sum_j \phi(r_1 - j)\phi(r_2 - j) \leq C \tag{14.67}$$

for all r_1, r_2. Expressions like the one on the lefthand side of this inequality arise when a fiber quantity is applied to the fluid lattice and then interpolated back to the fiber points According to (14.66) and its corollary (14.67), the influence of a fiber point on itself during such an operation is constant, independent of the location of that fiber point with regard to the fluid lattice, and the influence of one fiber point on another is no bigger than the influence of a fiber point on itself.

We now turn to the actual determination of the function ϕ. First note that (14.64) can be simplified through the use of (14.63). The result is

$$\sum_j j\phi(r - j) = r. \tag{14.68}$$

For $r \in [0, 1]$, (14.63), (14.68), and (14.65) read as follows:

$$\phi(r - 2) + \phi(r) = \frac{1}{2}, \quad (14.69)$$

$$\phi(r - 1) + \phi(r + 1) = \frac{1}{2}, \quad (14.70)$$

$$2\phi(r - 2) + \phi(r - 1) - \phi(r + 1) = r, \quad (14.71)$$

$$(\phi(r - 2))^2 + (\phi(r - 1))^2 + (\phi(r))^2 + (\phi(r + 1))^2 = C, \quad (14.72)$$

where we have made use of the postulate that $\phi(r') = 0$ for $|r'| \geq 2$. The constant C can now be determined by setting $r = 0$, and noticing that $\phi(-2) = 0$. With $r = 0$, (14.69) yields $\phi(0) = \frac{1}{2}$, (14.70)–(14.71) imply $\phi(-1) = \phi(1) = \frac{1}{4}$, and (14.72) therefore states that $C = \frac{3}{8}$.

With C known, we can solve for the function ϕ on $[0, 1]$. Equations (14.69)–(14.71) can be used to express $\phi(r - 2)$, $\phi(r - 1)$, and $\phi(r + 1)$ in terms of $\phi(r)$. These results can be substituted with (14.72) to yield a quadratic equation for $\phi(r)$. This equation has two solutions, and it might seem that the choice could be made arbitrarily and separately for each $r \in [0, 1]$, but the postulate of continuity together with the requirement that $\phi(0) = \frac{1}{2}$ (derived above) determines uniquely which solution must be chosen. Once ϕ is known on $[0, 1]$, it can be extended to $[-2, 2]$ by using the abovementioned expressions for $\phi(r - 2)$, $\phi(r - 1)$, and $\phi(r + 1)$ in terms of $\phi(r)$. The result is as follows:

$$\phi(r) = \begin{cases} \dfrac{3 - 2|r| + \sqrt{1 + 4|r| - 4r^2}}{8}, & |r| \leq 1, \\[3mm] \dfrac{5 - 2|r| - \sqrt{-7 + 12|r| - 4r^2}}{8}, & 1 \leq |r| \leq 2, \\[3mm] 0, & 2 \leq |r|. \end{cases} \quad (14.73)$$

Note that ϕ is an even function, and that not only ϕ but also its derivative is continuous. This last fact is a pleasant surprise, since the continuity of the derivative of ϕ was not one of our postulates.

This completes the construction of ϕ and hence of δ_h. The function ϕ is plotted in Figure 14.1.

14.6   THREE-DIMENSIONAL HEART MODEL

The foregoing sections have described a general numerical method for solving the equations of motion of a viscous incompressible fluid containing an immersed system of elastic or contractile fibers. Our purpose here is to apply that method to the heart. To do so, we must first create an arrangement of fibers in space that mimics the actual layout of muscle fibers in the heart wall and of collagen fibers in the heart-valve leaflets. Elastic parameters, time dependent in the case of the muscle fibers, must also be specified.

We shall not attempt a complete technical description of the heart model here, but the general principles on which it is based will be broadly described. Many of these principles were laid down in the pioneering anatomical

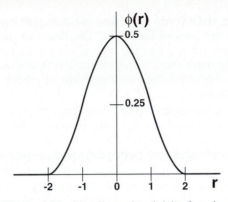

Figure 14.1. One-dimensional delta function.

research of Carolyn Thomas (Thomas 1957), who has given a global qualitative description of the layout of muscle fibers in the left and right ventricles of the mammalian heart. We have also been influenced by Streeter and his colleagues (Streeter, Jr. et al. 1969; Streeter, Jr. et al. 1978) who have made detailed quantitative measurements concerning the distribution of fiber angles in the left ventricular wall.

The two ventricles of the heart together form a somewhat conical structure, with a *base* and an *apex*. In the body, the axis of this cone slopes downward, forward, and leftward as it passes from base to apex. When an anatomical discussion is confined to the heart itself, however, it is customary to think of this axis as though it were vertical, with the base at the top and the apex below. In this viewpoint, which is the one that we shall adopt, the base is a horizontal plane at the top of the venticles. All four valves lie (more or less) within the plane of the base. Above the base one finds the structures to which the ventricles are attached: the left and right atria, the ascending aorta, and the main pulmonary artery.

The conical description of the ventricles that we have given would seem to imply that the ventricles are widest at the base, but this is not quite correct. In fact, there is a plane parallel to the base and slightly below it in which the ventricles achieve their maximum diameter. This is known as the equatorial plane of the heart.

A horizontal cross section of the heart cutting through the left and right ventricles reveals striking asymmetry. The left ventricle is thick walled and roughly circular, whereas the right ventricle is thin walled and crescent shaped. These differences are related to the fact that the pressure developed by the left ventricle is considerably greater (by a factor of about 6) than that of the right ventricle, though the two sides of the heart pump equal volumes of blood per unit time.

According to Thomas (1957), the muscle fibers of the cardiac ventricles begin and end at the valve rings, in the plane of the base of the heart. She describes the fibers as being organized in layers. Each layer has two sheets: one on which the fibers spiral away from the base and another on which they return. The following is a somewhat simplified description of the layers that Thomas found.

Outer/inner layer: The outer sheet of this layer surrounds both ventricles, and the inner sheet forms the innermost lining of the left ventricle. The two sheets meet at the apex of the heart. The fibers of the outer/inner layer have very little swirl. Almost like the rays of a cone, they run nearly directly to the apex along the outer sheet, penetrate the left ventricular wall at the apex (where the wall is very thin), and run almost directly back to the base of the heart along the inner sheet.

Right-inner/left-outer layer: This layer is atypical in that its two sheets lie side by side instead of one inside the other. One sheet forms the inner lining of the right ventricle and the other surrounds the left ventricular wall. The two sheets coincide along the right-ventricular surface of the interventricular septum. On the right-ventricular sheet of this layer, fibers leave the base and spiral clockwise (viewed from above). In the septum, they make a smooth transition to the left-ventricular sheet, after which they spiral counterclockwise (viewed from above) around the left ventricle back to the base of the heart.

Internal left-ventricular layers: These layers make up the bulk of the left-ventricular wall. They are nested, one inside another. As before, each layer has two sheets. Fibers spiral away from the base along the outer sheet, make a smooth transition to the inner sheet where the two sheets meet, and spiral back to the base along the inner sheet.

Although Thomas gives a qualitative discussion of how the fibers run (as summarized above), she does not enunciate a mathematical principle that would determine the fiber paths once the fiber surfaces are known. Such a principle does, however, emerge from the work of Streeter, Jr. et al. (1978), who made detailed measurements of fiber angle in the left ventricular wall.

According to Streeter and his colleagues, the fibers follow geodesic curves on the fiber surfaces. Since the left ventricle has (approximate) axial symmetry, the term "fiber surface" has a well-defined meaning: it is the surface of revolution that one gets by rotating a fiber about the axis of symmetry. Whether the fibers are geodesics on these surfaces or not can be tested by using Clairaut's theorem, that a geodesic on a surface of revolution has $r \cos \phi = $ constant, where r is the radial coordinate (in cylindrical coordinates) and ϕ is the angle between the tangent direction to the geodesic and the circumferential ($\hat{\theta}$) direction. What Streeter and his colleagues actually did was to determine loci of constant $r \cos \phi$, and then to check that these loci were tangent to the fibers. Satisfactory agreement was found, confirming the principle that the fibers are geodesics on the fiber surfaces.

The general method that we use to construct the initial configuration of the model ventricles may now be described. First, we define double-sheeted surfaces on which the fibers lie (see below). In typical cases, the two sheets meet along a common boundary curve somewhere in the interior of the heart wall. The surfaces are generally not tangent where they meet, so the only way that a fiber can make a smooth transition from the sheet to the other is by being locally tangent to the common boundary curve of the two-sheeted surface. This defines an initial direction for each fiber at the common boundary curve, and the fiber can then be continued as a geodesic in both directions along each of the two sheets, until it encounters the valve rings at the base of the heart.

An exception the above description occurs in the case of the right-inner/left-outer layer, where the two sheets of the layer lie side by side and actually share a common surface along the right-ventricular face of the interventricular septum. We think of the transition between the two sheets as being made at the vertical midline of the common surface, and we assume that the fibers are horizontal there (i.e., perpendicular to the midline). This defines an initial direction, and the fibers can then be contained in both directions as geodesics along the two sheets until they encounter the valve rings at the base of the heart as before.

Below the equator, of the heart, we take all of the surfaces to be portions of cones, although not necessarily cones with circular cross section. In the case of the outer/inner layer and of the right-inner/left-outer layer these conical surfaces extend all the way to the apex of the heart, which is also the common apex of the two cones forming the two sheets of the layer. For the interior layers of the left-ventricular wall, however, we use truncated cones, with the two sheets of the layer meeting along the common curve where this truncation occurs. (This is also the curve along which the fibers make a smooth transition from one sheet to the other.)

All of these conical surfaces must somehow be continued above the equator of the heart, so that they can connect with valve rings in the plane of the base. This process generally involves a bifunction, since a given sheet of a given ventricular layer has a cross section which is a single closed curve in the equatorial plane but which becomes a pair of closed curves in the plane of the valve rings (each ventricle has both an inflow and an outflow valve).

To deal with this situation, we have introduced a fairly general method for constructing a surface whose cross-section in the plane $z = z_1$, consists of n_1 given rings (with nonoverlapping interiors) and whose cross section in the plane $z = z_2$ consists of n_2 given rings (with non-overlapping interiors). By "ring" we just mean a simple closed curve. Note that n_1, and n_2 may be different. If they are, the topology of the cross section necessarily changes as we pass from z_1 to z_2,

Let the n_p rings in the plane $z = z_p$, with $p = 1, 2$, be given in the form $g_{pi}(x, y) = 0$, $i = 1 \cdots n_p$, with $g_{pi}(x, y) < 0$ denoting the interior of the ring and with $g_{pi}(x, y) > 0$ denoting its exterior. Since the interiors are nonoverlapping, any point in the plane belongs to at most one of them. It follows that the function

$$G_p(x, y) = \prod_{i=1}^{n_p} g_{pi}(x, y)$$

has the property that $G_p(x, y) < 0$ on the union of the interiors of the various rings in plane p, whereas $G_p(x, y) > 0$ on the intersection of the exteriors in that plane.

Now consider the surface

$$0 = G(x, y, z) = G_1(x, y)\frac{z_2 - z}{z_2 - z_1} + G_2(x, y)\frac{z - z_1}{z_2 - z_1}$$

This has the required cross section in the planes $z = z_1$ and $z = z_2$, and we can hope that it defines a plausible means of interpolating between these cross sections. Unfortunately, this is not guaranteed: sometimes the surface $G(x, y, z) = 0$ has two disconnected components and makes no connection at all between $z = z_1$ and $z = z_2$. We have managed to avoid this possibility in the construction of the heart, and we have therefore been able to use this interpolation scheme to connect the ventricular fiber surfaces to the valve rings.

The surfaces $G(x, y, z) = 0$ do not join smoothly with the conical surfaces that we have constructed below the equator of the heart (the surface normal is discontinuous at the equator), but there is no difficulty continuing a geodesic in a unique way across such an edge, and this lack of smoothness tends to disappear in any case as the heart is pressurized.

The construction that we have just described for determining a surface connecting specified rings in two given planes is also used to build the atria of the model heart. Here, the rings in question are those of the atrioventricular valves and those at which the various veins connect to the atria. Finally the arteries and veins that connect to the heart are initially defined as cylinders with hemispherical caps, on which geodesic fibers are wrapped.

Since the model vessels have blind ends, it is necessary to provide sources and sinks in the hemispherical caps, to simulate the connection of the heart to the rest of the circulation. At present, each such source and sink is connected to a pressure reservoir through a fixed hydraulic resistance. These provide appropriate pressure loads for the model heart. An external source/sink allows for changes in volume as the heart fills and ejects.

The valves of the model heart are also constructed out of fibers. In the case of the atrioventricular valves this is done in much the same style as has been previously described: by defining rather arbitrary initial surfaces and then wrapping fibers as geodesics on these surfaces. For the arterial valves, however, we have found a less arbitrary procedure in which the both the fibers and the surfaces are simultaneously defined by solving a system of partial differential equations. These equations describe the mechanical equilibrium of a one-parameter family of fibers under tension, and their numerical solution determines the closed configuration of the arterial valves (Peskin and McQueen 1994). A future goal is to formulate and solve partial differential equations for the fiber architecture of the heart as a whole. For preliminary results in this direction, see (Peskin 1989).

Finally, we have to consider the elasticity of the model fibers. Let R be the length of a short fiber segment and let T be the tension in that segment. We use a nonlinear length-tension relationship of the form

$$T = \begin{cases} S_0 \left(\dfrac{R - R_0}{R_0} \right)^2, & R \geq R_0, \\ 0, & R \leq R_0, \end{cases}$$

where R_0 is the rest length of the segment and S_0 is a stiffness parameter. Both R_0 and S_0 may be time dependent to simulate active, contractile muscle.

The time dependence R_0 and S_0 can be programmed differently in different parts of the heart, in such a way as to simulate the waves of activation and deactivation that propagate through the cardiac tissue and coordinate the heartbeat. At present, however, we do this only in a rudimentary way: First the atrial muscle contracts synchronously, and then, after a realistic delay, the ventricular muscle contracts synchronously. This procedure ignores the delays that occur within the atria or within the ventricles.

The construction of the model heart is shown in Figures 14.2 through 14.11, and its dynamics are depicted in Figures 14.12 through 14.19.

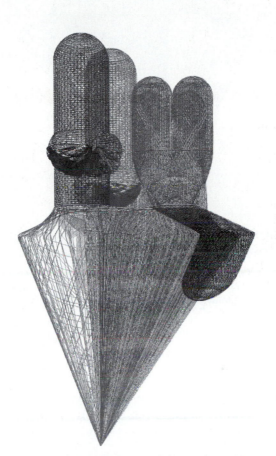

Figure 14.2. Computer model of the heart and the nearby great vessels. The heart is viewed from the front, so the left ventricle is at the lower right, and the thin-walled right ventricle is at the lower left. The vertical vessels at the top of the figure from left to right are the pulmonary artery (in front), the aorta, and two of the four pulmonary veins. The pulmonic and aortic valves (outflow valves of the right and left ventricles, respectively) can be seen within the pulmonary artery and the aorta, and the tricuspid valve (inflow valve of the right ventricle) can just barely be seen at the top of the right ventricle. The mitral valve (inflow valve of the left ventricle) is obscured by the dense left-ventricular wall. The left atrium is visible below the pulmonary veins, and the left-atrial appendage (auricle) is prominently seen hanging over the left-ventricular wall at the right side of the figure. The right atrium is behind the aorta and cannot be seen in this view. The model veins and arteries have blind ends, but sources and sinks are provided, with properties chosen to establish realistic pressure loads on the model heart.

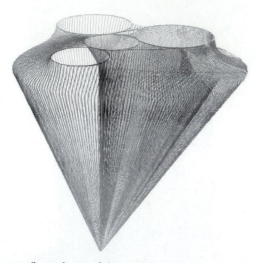

Figure 14.3. The outer/inner layer of the model ventricles. Again, the heart is viewed from the front, so the left ventricle is on the right side of the figure. The four valve rings at the top of the figure form the base of the heart. The larger rings are the locations at which the atria join the ventricles, and the smaller rings are the locations at which the arteries are attached. The fibers of the outer/inner layer form the outermost layer of the ventricles as well as the innermost layer of the left ventricle only.

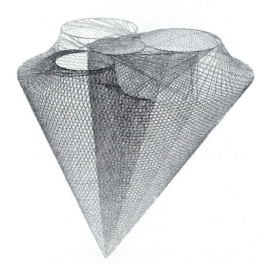

Figure 14.4. The right-inner/left-outer layer of the model ventricles. The fibers of this layer form the inner lining of the right ventricle and the outer lining of the left ventricle. Viewed from the base, these fibers follow a figure-eight trajectory, spiraling clockwise around the right ventricle and then counterclockwise around the left ventricle. The transition is made on the right-ventricular face of the interventricular septum.

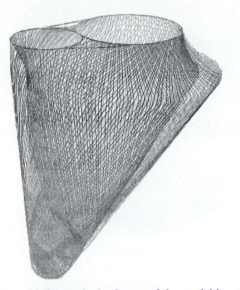

Figure 14.5. The internal left-ventricular layers of the model heart are shown in this and the following three figures. The smaller of the two rings at the top of each figure is the aortic-valve ring and the larger is the mitral-valve ring. Each layer is composed of two sheets of fibers which meet at the two valve rings and also at a third ring that forms the lower boundary of that layer. The fibers are tangent to this lower ring and therefore make a smooth transition there from one sheet to the other. The layers are nested, so that both sheets of the smallest layer lie between the two sheets of the next larger layer and so on. Both sheets of the largest of these internal left-ventricular layers lie between the left-ventricular parts of the outer/inner layer and the right-inner/left-outer layer.

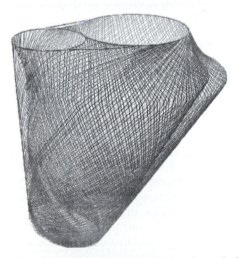

Figure 14.6. An internal left-ventricular layer (see legend of Figure 14.5).

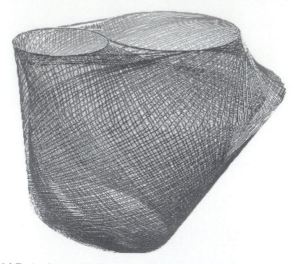

Figure 14.7. An internal left-ventricular layer (see legend of Figure 14.5).

Figure 14.8. An internal left-ventricular layer (see legend of Figure 14.5).

14.7   SUMMARY AND CONCLUSION

This chapter has described the mathematical formulation of the fiber-fluid problem of cardiac mechanics, a technique known as the immersed-boundary method which is suitable for the numerical solution of that problem, and a detailed three-dimensional model of the heart for use in conjunction with the immersed-boundary method.

Both the mathematical formulation and the numerical method described in this chapter are based on the notion that the blood, valves, and muscular heart walls can all be modeled within a common framework. Thus, we think of the cardiac tissues as though they were a part of the fluid, in which, however, there are additional tissue stresses besides those that would normally be present in a fluid. Alternatively, one might say that we think of the blood as a part of the heart, where the stesses that characterize the muscular walls and the elastic heart-valve leaflets happen to be zero.

The computational consequence of this point of view is that we solve the fluid equations on a fixed cubic lattice, unencumbered by the complicated

Figure 14.9. Inflow structures of the model heart. At the left side of the figure one sees the superior vena cava (above), the inferior vena cava (below), and the right atrium between them. The right atrium empties through the tricuspid valve, which is seen just to the right of the inferior vena cava. The corresponding structures on the left side of the heart are seen on the right side of the figure. They are the four pulmonary veins, of which two are visible at the top, the left atrium, and the mitral valve. The left-atrial appendage (auricle) is seen at the far right side of the figure.

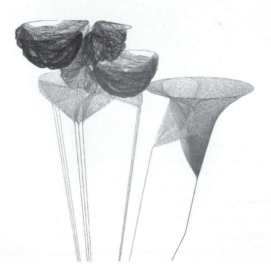

Figure 14.10. The four valves of the model heart. The outflow (aortic and pulmonic) valves are seen above and the inflow (mitral and tricuspid) valves are seen below. Note that the inflow valves are supported by fans of chordae tendineae, which insert into papillary muscles, two for the mitral and three for the tricuspid valve. The outflow valves, by contrast, are self-supporting.

Figure 14.11. Detail of an outflow (aortic or pulmonic) valve of the model heart.

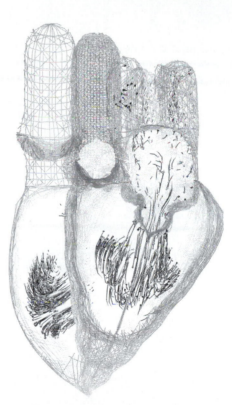

Figure 14.12. Cutaway view of the model heart during early ventricular diastole (re-laxation). In this and the figures that follow, the heart model is shown in action. Fluid flow is represented by streaklines, which were computed by tracing the trajectories of particles immersed in the flow. The current position of each such particle is shown as a small blob, and the recent past positions are shown by a fading tail. In this particular figure, only the markers that were initially in the atria are shown. Note the ring vortex visible in the left ventricle (right side of the figure) below the mitral valve.

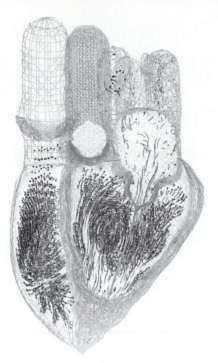

Figure 14.13. Cutaway view of the model heart during early ventricular diastole showing markers that were initially in the ventricles in addition to the atrial markers shown previously in Figure 14.12.

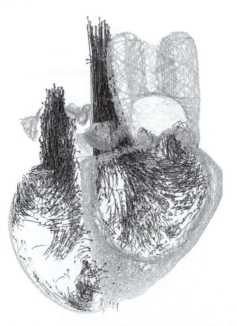

Figure 14.14. Cutaway view of the model heart during ventricular systole (contraction). Only markers that were initially in the ventricles are shown.

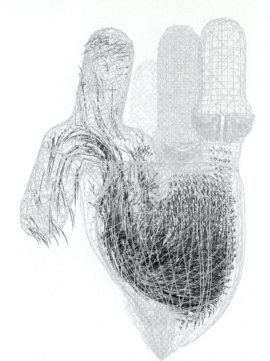

Figure 14.15. Flow pattern in the model right ventricle during early ventricular diastole.

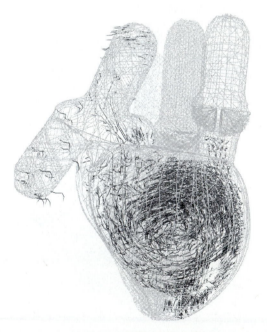

Figure 14.16. Flow pattern in the model right ventricle during late ventricular diastole. Note the prominent vortex that forms in the right ventricle during diastole.

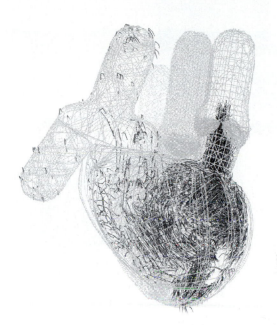

Figure 14.17. Flow pattern in the model right ventricle during ventricular systole.

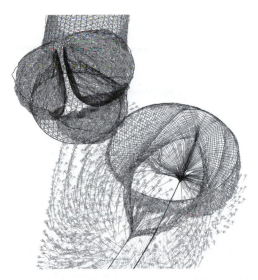

Figure 14.18. Detail of the mitral and aortic valves of the model during ventricular diastole.

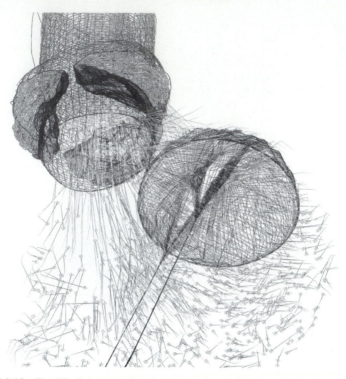

Figure 14.19. Detail of the mitral and aortic valves of the model during ventricular systole.

time-dependent geometry of the heart and its valves. The spatial configuration of those tissues is tracked separately, by means of a collection of tissue marker points that move at the local fluid velocity while simultaneously exerting force on the fluid in which they are immersed. The tissue-fluid interaction is mediated by a smoothed approximation to the Dirac δ-function, which is used both in spreading the fiber forces out onto the computational lattice of the fluid and also in the interpolation of the fluid-velocity field to the marker points of the cardiac tissue fibers.

The computer model that has been created in this way is now ready for use in applied studies concerning the normal and pathological physiology of the heart, and also as a test chamber in the design of prosthetic cardiac valves.

14.8 ACKNOWLEDGMENTS

The work described in this chapter was supported by the National Science Foundation under research grant BIR-9302545. Computation was performed on the Cray C-90 computer at the Pittsburgh Supercomputing Center under a grant (MCA93S004P) from the MetaCenter Allocation Committee. Visualization of results was done at the Academic Computing Facility, New York University. The authors are indebted to Eleen Collins for typesetting the chapter.

REFERENCES

Bell, John B., Colella, Phillip, and Glaz, Harland M. 1989. A second-order projection method for the incompressible Navier-Stokes equations. *Journal of Computational Physics*, **85**(2), 257–283. {*316, 337*}

Chorin, Alexandre J. and Marsden, Jerrold E. 1993. *A Mathematical Introduction to Fluid Mechanics*. New York NY: Springer-Verlag. ISBN 0-387-97918-2. Pages ix + 169. {*310, 337*}

Green, A. E. and Adkins, J. E. 1970. *Large elastic deformations*. Oxford, UK: Clarendon Press. ISBN 0-19-853334-9. Pages xiv + 324. {*310, 337*}

Guyton, Arthur C. 1991. *Textbook of Medical Physiology*. Philadelphia PA: W. B. Saunders Company. ISBN 0-7216-3087-1. Pages 97–109 (of xli + 1014). {*310, 337*}

McQueen, D. M. and Peskin, C. S. 1983. Computer-assisted design of pivoting-disc prosthetic mitral valves. *Journal of thoracic and cardiovascular surgery*, **86**, 126–135. {*310, 337*}

McQueen, D. M. and Peskin, C. S. 1985. Computer-assisted design of butterfly bileaflet valves for the mitral position. *Scandinavian Journal of Thoracic and Cardiovascular Surgery*, **19**, 139–148. {*310, 337*}

McQueen, D. M., Peskin, C. S., and Yellin, E. L. 1982. Fluid dynamics of the mitral valve: physiological aspects of a mathematical model. *American journal of physiology*, **242**, H1095–H1110. {*310, 337*}

Meisner, J. S., McQueen, D. M., Ishida, Y., Vetter, H. O., Bortolotti, U., Strom, J. A., Frater, R. W. M., Peskin, C. S., and Yellin, E. L. 1985. Effects of timing of atrial systole on LV filling and mitral valve closure: computer and dog studies. *American journal of physiology*, **249**, H604–H619. {*310, 337*}

Peskin, C. S. 1989. Fiber-architecture of the left ventricular wall: an asymptotic analysis. *Communications on Pure and Applied Mathematics (New York)*, **42**(1), 79–113. {*326, 337*}

Peskin, C. S. and McQueen, D. M. 1992. Cardiac fluid dynamics. *CRC critical reviews in biomedical engineering*, **20**(5/6), 451–459. {*310, 314, 337*}

Peskin, C. S. and McQueen, D. M. 1993. Computational biofluid dynamics. *Contemporary mathematics*, **141**, 161–186. {*310, 314, 337*}

Peskin, C. S. and McQueen, D. M. 1994. Mechanical equilibrium determines the fractal fiber architecture of the aortic heart valve leaflets. *American journal of physiology*, **266**(1), H319–H328. {*326, 337*}

Peskin, C. S. and McQueen, D. M. 1995. A general method for the computer simulation of biological systems interacting with fluids. *Pages 265–276 of:* Ellington, C. P. and Pedley, T. J. (eds), *Biological Fluid Dynamics: Proceedings of a meeting held at the University of Leeds, UK, 4–8 July 1994*. Symposia of the Society for Experimental Biology, vol. 49. Cambridge, UK: The Company of Biologists Limited. {*310, 314, 337*}

Press, William H., Flannery, Brian P., Teukolsky, Saul A., and Vetterling, William T. 1986. *Numerical Recipes: The Art of Scientific Computing*. Cambridge, UK: Cambridge University Press. ISBN 0-521-30811-9. Pages 390–396; 449–453; and 646–647 (of xx + 818). {*316, 337*}

Streeter, Jr., D. D., Spotnitz, H. M., Patel, D. P., Ross, Jr., J., and Sonnenblick, E. H. 1969. Fiber orientation in the canine left ventricle during diastole and systole. *Circulation research*, **24**, 339–347. {*323, 337*}

Streeter, Jr., D. D., Powers, W. E., Ross, M. A., and Torrent-Guasp, F. 1978. Three-dimensional fiber orientation in the mammalian left ventricular wall. *Pages 73–84 of:* Baan, Jan, Noordergraaf, Abraham, and Raines, Jeff (eds), *Cardiovascular System Dynamics*. Cambridge, MA, USA: MIT Press. {*323, 324, 337*}

Thomas, C. E. 1957. The muscular architecture of the ventricles of hog and dog hearts *American Journal of Anatomy*, **101**, 17–57. {*323, 337*}

15. BIOCONVECTION

N. A. Hill

15.1 INTRODUCTION

Bioconvection is the term used to describe the phenomenon of spontaneous pattern formation in suspensions of swimming microorganisms. It is commonly seen in the laboratory and produced by protozoa, bacteria, and single-celled algae. Typical patterns are illustrated in Figures 15.1, 15.2, and 15.3. In all cases, the microorganisms are 3% to 5% more dense than the fluid in which they swim (essentially water), and a variety of different directional responses to their environment cause them to aggregate in specific regions. These behaviors are known as taxes, if the cells respond by changing their orientation, or as kineses, if they change their speed or rate of turning. The resulting spatial variations in the density of the suspension drive a bulk flow with downwelling where the concentration of cells is high and upwelling where the concentration is low. It is important to realize that it is the swimming of the individual cells that provides the energy for bioconvection. Suspensions of nonmotile cells do *not* form patterns.

Bioconvection is an example of biological self-organization and the creation of complexity; the patterns can be highly regular and ordered when the overall cell concentration is high. The flows associated with bioconvection can beneficially modify the local environment, for example by stirring up nutrients or enhancing the transport of dissolved atmospheric gases.

Although bioconvection is usually observed in laboratory cultures, it is also seen in nature in small puddles and quiet ponds where the water is still. Even in open water such as the oceans, swimming microorganisms aggregate in highly visible, large-scale blooms. It is important to understand their tactic responses that play a causal role in aggregation, even though patterns, which depend on a delicate balance of fluid mechanics and directional responses, are unlikely to form in this case because of wave action and currents.

In this chapter I shall discuss the mathematical models used to analyze bioconvection in algal and bacterial suspensions and work that has been carried out to characterize the swimming responses of individual cells.

15.2 GYROTAXIS AND BIOCONVECTION

Single-celled algae tend to swim upward in the absence of a flow or external stimuli. This is known as gravitaxis (or geotaxis). At least one species,

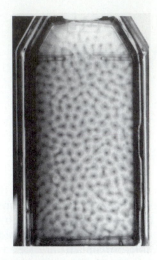

Figure 15.1. A well-developed bioconvection pattern in a suspension of the alga *Chlamydomonas nivalis*. The suspension is contained in a horizontal cuvette approximately 1 cm deep. The cuvette is seen from vertically above and is approximately 5 cm wide and 11 cm long. Dark regions correspond to high concentrations of algae associated with down-welling, and the dots indicate downwelling plumes. (Photograph courtesy of Professor J. O. Kessler, Physics Department, University of Arizona.)

Figure 15.2. Bioconvection patterns in a suspension of the bacterium *Bacillus subtilis*. The bacterial suspension lies in a horizontal dish about 5 cm in diameter and is approximately 3 mm deep. It is viewed from above under dark-field illumination, so that the brightest regions correspond to the highest concentrations of bacteria and are associated with downwelling of the suspension. (Photograph courtesy of Professor J. O. Kessler, Physics Department, University of Arizona.)

Figure 15.3. Vertical bioconvection plumes in a suspension of the alga *Chlamydomonas nivalis* grown in a 250 ml culture flask. The average concentration is approximately 5×10^5 cells cm^{-3}, and the dark lines are concentrated, downwelling plumes of algae.

Euglena gracilis (Lebert and Häder 1996), has an active gravity-sensing mechanism, but we shall concentrate on those such as *Chlamydomonas nivalis* (Figure 15.3) which orient by being bottom heavy. The evolutionary advantage of gravitaxis is that in dark, turbid water they need to swim toward the surface in order to receive enough light for photosynthesis. They also usually exhibit strong phototactic responses when the illumination is sufficiently bright. In this section we shall concentrate on their behavior in the absence of light; phototaxis is discussed below. A typical bottom-heavy cell is illustrated in Figure 15.4. Since algal cells are small, with typical body diameters of 10–20 μm, and swim at speeds of 100 μm s^{-1}, inertia can be neglected. Mathematically the neglect of inertia is justified when the dimensionless Reynolds number, Re $= \rho U L/\mu$, is much less than one, where U is a typical cell swimming speed, L is a typical cell diameter, ρ is the density of the fluid, and μ is the viscosity. In this case, Re $= O(10^{-3})$. Thus the cell swims in a direction $\boldsymbol{p}$ at an angle θ to the vertical determined by a balance between the gravitational torque, $\boldsymbol{T}_g$ due to its being bottom heavy, and a viscous torque, $\boldsymbol{T}_v$, due to fluid-velocity gradients, $\boldsymbol{\nabla u}$, across its body and rotation of the cell, i.e.,

$$\boldsymbol{T}_v + \boldsymbol{T}_g = 0. \tag{15.1}$$

This balance is known as gyrotaxis (Kessler 1985b). For example for a spherical cell of radius a,

$$\boldsymbol{T}_v = 4\pi \mu a^3 (\boldsymbol{\omega} - 2\boldsymbol{\Omega}), \tag{15.2}$$

where $\boldsymbol{\omega} = \text{curl}\, \boldsymbol{u}$ is the vorticity and equals twice the local angular velocity of the fluid (Acheson 1990), μ is the viscosity of the fluid, and $\boldsymbol{\Omega}$ is the cell's angular velocity. $\boldsymbol{\Omega}$ can be written as

$$\boldsymbol{\Omega} = \Omega_{\|}\boldsymbol{p} + \boldsymbol{p} \wedge \dot{\boldsymbol{p}}, \tag{15.3}$$

where $\dot{\boldsymbol{p}}$ is the component of the cell's angular velocity perpendicular to $\boldsymbol{p}$. Also

$$\boldsymbol{T}_g = hmg\boldsymbol{p} \wedge \hat{\boldsymbol{k}}, \tag{15.4}$$

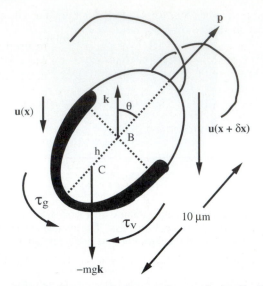

Figure 15.4. A schematic diagram of a bottom-heavy algal cell. The cell swims in a direction given by the unit vector p at an angle θ to the vertical. Buoyancy forces act through its geometric center B. Its center of mass C is displaced by a distance h from B toward its anterior because of the arrangement of dense organelles, such as the photosynthesizing chloroplast. Since the cells are very small, inertia can be neglected and the gravitational torque, T_g, tending to right the cell is balanced by the viscous torque, T_v which arises due both to differences in the fluid velocity u on either side of the cell and to rotation of the cell.

where h is the displacement of the cell's center of mass from its center of buoyancy and has been estimated by Kessler (1986) to be about 10% of the cell diameter; m is the cell's mass, $-g\hat{k}$ the acceleration due to gravity and $\hat{k}$ a unit vector vertically upward. Expressions for T_v for the more general case of a spheroidal cell are given in the review by Pedley and Kessler (1992).

In reality, the cell's direction of travel is not entirely deterministic and the gyrotactic balance (15.1) describes the average orientation of the cell. The randomness in the cell's motion is caused by internal stochastic processes which affect the way in which the flagella are used to propel the cell through the fluid. Both the bacteria and alga are too large and too vigorous swimmers for their swimming velocities to be significantly influenced by Brownian effects due to the thermal motion of the water molecules.

Gyrotaxis can be demonstrated in an experiment in a slow Poiseuille flow down a vertical tube of circular cross section with a diameter of 1 cm (Kessler 1985a; Kessler 1986). The fluid velocity in Poiseuille flow, in the usual cylindrical polar coordinates (r, θ, z) with z vertically downward, is $u = U_0(1 - r^2/R^2)\hat{z}$, where $U_0\hat{z}$ is the velocity down the axis of the tube and R is the tube's radius. The corresponding vorticity is $\omega = (2r/R^2)\hat{\theta}$, which can be substituted into (15.2) to calculate T_v. If $U_0 \approx 1 \, \text{mm s}^{-1}$, then $\Omega = 0$ and the viscous torque gives one stable equilibrium orientation with individual cells tipped away from the upward vertical toward the axis of the pipe. Thus the cells swim toward the axis as they are carried along in the flow, and focus into a narrow beam. Conversely, if the direction of the flow is reversed, the cells are oriented away from the axis toward the walls and no beam is seen, confirming the role of gravity in cell orientation.

Continuum models (Childress et al. 1975) are used to describe the macroscopic flow of the suspension that occurs in bioconvection because clearly it is not possible to follow the trajectories of tens of millions of individual cells. The concentration of cells is expressed as $n(x, t)$, which equals the number of cells per unit volume at position x and time t. Cell conservation implies that

$$\frac{\partial n}{\partial t} + \nabla \cdot J = 0, \tag{15.5}$$

where

$$J(x, t) = nu + n < V > -D \cdot \nabla n \tag{15.6}$$

is the flux of cells at (x, t); $u(x, t)$ is the fluid velocity, $< V > (x, t)$ is the ensemble average cell swimming velocity, and $D(x, t)$ is the cell-diffusion tensor, which is given by

$$D(t) = \int_0^\infty < V_r(t) V_r(t - t') > dt' \tag{15.7}$$

(Pedley and Kessler 1990). Here V_r is the swimming velocity of the cell relative to the mean velocity, $< V >$. Equation (15.6) for the flux $J(x, t)$ is based on the assumption that the cells' motion can be described by random-walk models (Berg 1983). The first term on the righthand side of (15.6) represents advection of cells by the flow of the fluid, the second is the flux of cells due to their mean swimming velocity, and the third is Fickian diffusion down cell-concentration gradients due to the random nature of the cells' trajectories. In the simplest model for bioconvection due to gyrotaxis,

$$< V >= V_s p^*, \tag{15.8}$$

where V_s is the drift speed due to swimming averaged over the population, $p^*(x, t)$ is the unit vector satisfying the gyrotactic balance (15.1), and

$$D = DI, \tag{15.9}$$

where D is a constant and I is the identity tensor.

The volume fraction of cells in the suspension, nv, where v is the volume of a cell, is normally less than 0.01, so that the suspension is dilute. The incompressibility of the fluid requires that

$$\nabla \cdot u = 0. \tag{15.10}$$

The momentum balance is expressed by the Navier-Stokes equations (Acheson 1990)

$$\rho\left(\frac{\partial}{\partial t} u + (u \cdot \nabla)u\right) = -\nabla p + \mu \nabla^2 u + nvg\Delta\rho, \tag{15.11}$$

where ρ is the constant density of the fluid, $p(x, t)$ is the excess pressure above hydrostatic balance, and $\rho + \Delta\rho$ is density of the cells. The last term in equation (15.11) is the buoyancy force due to the concentration of cells, which drives the flow of the suspension.

The walls of the container are rigid, and the upper surface of the suspension, even if it is open to the air, appears in experiments to behave as a rigid surface, because a small proportion of the cells form a film or scum at the interface and remain stuck there. Thus the boundary conditions are that

$$u = 0 \quad \text{and} \quad J \cdot \nu = 0, \tag{15.12}$$

where ν is the unit normal vector to the boundary.

This model for bioconvection is an extension of the first rational model of Childress, Levandowsky, and Spiegel (1975), who assumed the cells to be purely gravitactic and to swim vertically upwards on average, unaffected by any viscous torques due to fluid motion, so $p^* = \hat{k}$. Childress et al. considered the initial growth of bioconvection patterns in a layer of finite depth and infinite width; the upswimming of the micro-organisms leads to their accumulating near the upper surface, so that the upper region of the suspension is denser than the lower region, because the cells are more dense than the fluid. If the density gradient is sufficiently great, the suspension is gravitationally unstable and an overturning instability ensues, analogous to Rayleigh-Bénard convection in a layer of fluid heated from below (Drazin and Reid 1985). An unsatisfactory feature of their results is that when the cell-density gradient is just sufficiently large to generate an instability, the wavelength of the initial patterns is predicted to be infinite. In practice, this implies that the pattern spacing would be influenced by the width of the container, contrary to observations.

Pedley, Hill and Kessler (1988) examined the linear stability of a uniform suspension in a very deep layer and showed that gyrotaxis gives rise to a second, independent instability mechanism. If a small fluctuation causes the cell concentration in a particular blob of fluid to be greater than in its surroundings, that blob will sink, thereby generating a fluid-velocity distribution with horizontal vorticity, similar to downward Poiseuille flow in a pipe. This focuses other cells into the wake of the blob, and a concentrated, downwelling plume will be formed. Plumes that have been formed in this way are shown in Figure 15.3. Gyrotaxis also modifies the initial patterns formed by the overturning mechanism in a layer of moderate depth, and Hill, Pedley and Kessler (1989) found that it leads to quantitatively realistic predictions of finite wavelengths for the initial patterns.

15.3   ALGAL CELL TRAJECTORIES AND THE FOKKER-PLANCK EQUATION

The mathematical model for gyrotaxis given by (15.8) and (15.9) is unrealistic in that it prescribes a constant drift speed and a constant diffusivity, both of which should depend on the degree of randomness in the cell's motion, which is described by an intrinsic rotational diffusivity (see below). A more complete and rational model for the orientation probability density function $f(p)$ has been developed by Pedley and Kessler (1990) and Hill and Häder

(1996). Unlike bacterial cells such as *Escheria coli* (Berg and Brown 1972), which move in a series of almost straight lines punctuated by abrupt changes in direction (run-and-tumble), algal cells change direction gradually over the course of several flagellar beats. Hill and Häder recorded the trajectories of many individual algal cells swimming in a suspension and used a biased, correlated random-walk model in the limit in which the time step, τ tends to zero. The cells were constrained to swim in planes by the geometry of the cuvette, which could be positioned either horizontally or vertically in order that three-dimensional statistics could be reconstructed. Turning angles, δ, between successive straight-line segments linking the data points were analyzed as functions of the angle θ to a fixed axis, usually the preferred direction of motion. If the turning angles are treated as steps of a random walk on the unit circle, a Fokker-Planck equation can be derived for $f(p)$ of the form

$$\frac{\partial f}{\partial t} + \boldsymbol{\nabla} \cdot (\mu(\theta)f) = \frac{\sigma^2}{2}\nabla^2 f, \tag{15.13}$$

where $\mu(\theta)\tau$ is the mean of the turning angle, $< \delta(\theta) >$, and $\sigma^2\tau$ is its variance as $\tau \to 0$. $\mu(\theta)$ can be identified with $\dot{p}$ in (15.3) and $\boldsymbol{\nabla}$ is the gradient operator in p-space (i.e., on the unit sphere); $\mu(\theta)$ represents deterministic reorientation due to external torques and intrinsic behavioral responses, while $\sigma^2/2$ is a rotational diffusion coefficient due to random fluctuations in orientation caused by the internal workings of the cell. Thus an appropriate measure of the randomness in overall cell orientation is $\sigma/\max\{\mu(\theta)\}$. Hill and Häder (1996) found the swimming speed to be independent of direction, so that after solving (15.13), the mean velocity in the flux given by (15.6) is

$$< \boldsymbol{V} >= V_A < \boldsymbol{p} >, \tag{15.14}$$

where $V_A =< |\boldsymbol{V}| >$ is the average swimming speed and

$$< \boldsymbol{p} >\equiv \int \boldsymbol{p} f(\boldsymbol{p})\, d\boldsymbol{p} \tag{15.15}$$

is the mean orientation, the integral in (15.15) being over the whole unit sphere. D is computed from (15.7). Pedley and Kessler (1990) used the Fokker-Planck equation (15.13), assuming the forms of the coefficients μ and $\sigma^2/2$, to analyze the initial stages of bioconvection in an unbounded layer of infinite depth. The model gives refined estimates of the initial pattern wavelengths; the main qualitatively different result is that initially, in an unbounded fluid, purely horizontal instabilities are always more unstable than three-dimensional ones.

15.4 PHOTOTAXIS AND SHADING

Given the need to photosynthesize, many motile algae are strongly phototactic as well as being gyrotactic, and thus it is essential to include phototaxis in realistic models of their behavior. Some phototactic responses have been measured using the random-walk techniques of Hill and Häder discussed above, but further experimental work is needed to completely characterize phototaxis in the algae. Nevertheless, Vincent and Hill (1996) were able to construct the following generic model when phototaxis dominates any viscous torques due to fluid flow.

When the suspension is illuminated from above, the mean swimming velocity is again given by (15.14) with

$$< p >= T(I)\hat{k}, \tag{15.16}$$

$\hat{k}$ being the unit vector in the vertical direction and $T(I)$ the phototaxis function. $T(\cdot)$ depends on the intensity I of the light reaching the cell and has the generic form indicated in Figure 15.5. Most phototactic algae are positively phototactic; that is, they swim toward the light source when the light intensity is below a critical value, I_c, and are negatively phototactic when $I > I_c$, presumably because strong UV wavelengths in sunlight can bleach and thus kill them. Thus cells tend to aggregate at optimal places in their environment where $I \approx I_c$. Turbidity in the water column will cause I to decrease with depth in the natural environment, but cells also absorb and scatter light, which decreases I further. This latter effect is termed self-shading and is the dominant effect in laboratory cultures. It determines the positions in the suspension where I equals I_c. Provided that the volume fraction of cells in the suspension

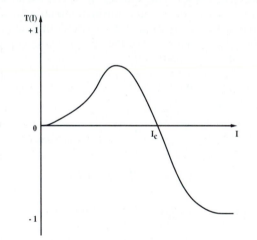

Figure 15.5. A sketch of the typical form of the phototaxis function, $T(I)$. I is the intensity of the light reaching the swimming cells. When $T = 0$, there is no preferred direction of motion. If $T = +1$ all the cells swim directly toward the light source, and if $T = -1$ they all swim away from the source. When $0 < I < I_c$, the cells exhibit positive phototaxis, moving on average toward the light in order to photosynthesize. The optimum light intensity is I_c, and when $I > I_c$, the cells are negatively phototactic, fleeing away from sunlight that is too intense and can kill them.

is sufficiently small, self-shading is given by the Lambert-Beer law:

$$I(x) = I_s \exp\left[-\alpha \int_0^r n(r')\,dr'\right], \tag{15.17}$$

where α is the extinction coefficient and r is a vector from x to the light source, which has an intensity I_s.

The most interesting and applicable case to study is that in which $I \approx I_c$ throughout a horizontal layer of the suspension of depth H, and then to simplify the analysis, the phototaxis function is linearized about I_c giving

$$T = -\lambda(I - I_c) + O[(I - I_c)^2]. \tag{15.18}$$

Further analytical progress has been made on the assumption that $\alpha n H \ll 1$. Good estimates for α and n are difficult to obtain, but based on an estimate of 4×10^{-9} mg of chlorophyll for a typical *Chlamydomonas* cell (see Harris (1989), p. 590) and an absorption cross section of $50 \, cm^2 \, mg^{-1}$ of chlorophyll (see Tett (1990), Table 4.1), we estimate that $\alpha = 2 \times 10^{-7} \, cm^2 \, / \, cell$. In bioconvection experiments in the laboratory, high cell concentrations of between 5×10^5 cells cm^{-3} in layers 1–2 cm deep and 5×10^6 cells cm^{-3} in layers 1–2 mm deep are common, so that $\alpha n H \approx 0.2$. So, assuming that $\alpha n H \ll 1$, the Lambert-Beer law (15.17) can be expanded to give at leading order

$$<p> = \lambda I_c \alpha \left[\int_z^0 n(x', t) \, dz' - CHN_0 \right] \hat{k}, \tag{15.19}$$

where $I = I_c$ at $z = -CH$, the upper surface being at $z = 0$. N_0 is the average cell concentration throughout the suspension.

In the model problem that Vincent and Hill (1996) used to illustrate how phototaxis can cause bioconvection, the suspension is initially well mixed, so $I = I_c$ at a constant depth, $z = -CH$, across the whole layer. Cells above $z = -CH$ swim downward toward this critical depth, and cells below swim upward setting up an equilibrium-state cell-concentration profile proportional to $sech^2(z + CH)$. Thus the suspension above the critical depth is stably stratified and fluid below unstably stratified, which once again leads to a Rayleigh-Bénard-like overturning instability when the cell-concentration gradient is sufficiently big. This is an example of penetrative convection, which was first studied in a different context by Veronis (1963). Work is in progress on a mathematical model in which gyrotaxis and phototaxis are combined.

15.5 BACTERIAL BIOCONVECTION

Bacterial suspensions, including those of the common soil bacterium *Bacillius subtilis*, also form bioconvection patterns, but the orientation mechanisms are very different. *B. subtilis* consume dissolved oxygen and are oxytactic (or aerotactic) *i. e.* they swim up gradients in the concentration of dissolved oxygen. Typically the patterns are seen to form from initially well-stirred suspensions of concentrations of about 10^9 cells mm^{-3} in fluid layers of depths between 1 and 10 mm, and the initial patterns take the form of rolls having width similar to the depth. Examples of such patterns are shown in (Hillesdon et al. 1995). They studied the development of the equilibrium patterns from a uniform state in a layer of finite depth where upper surface is open to the air. In this analysis there is no bulk fluid motion, and the equilibrium is a preliminary to bioconvection. There are few quantitative data on aerotaxis, so they made the following assumptions about the bacterial motion.

It is observed that when the concentration of dissolved oxygen, $C(x, t)$, falls below a minimum, C_{min} which may be greater than zero, the cells no longer swim. Thus it is convenient to define a dimensionless oxygen concentration

$$\theta = (C - C_{\min})/(C_0 - C_{\min}), \tag{15.20}$$

where C_0 is the concentration at the upper surface. The cells' swimming speed is

$$V_s = V_{so} W(\theta), \tag{15.21}$$

where V_{so} is the maximum swimming speed, approximately $20 \, \mu\mathrm{m \, s}^{-1}$, and W is a saturating function of θ lying between 0 and 1. $W(\theta)$ was taken to be either

$$W(\theta) = H(\theta) \quad \text{or} \quad W(\theta) = [1 - \exp(-\theta/\theta_1)]H(\theta), \tag{15.22}$$

where $H(\theta)$ is the Heaviside step function and θ_1 is a constant. By analogy with isotropic diffusion due to a random walk (Berg 1983), the bacterial swimming diffusivity is

$$D = V_s^2 \tau /3, \tag{15.23}$$

where τ is a constant representing the velocity fluctuation correlation time, which was taken to be 1 s. Aerotaxis was assumed to be proportional to $\nabla\theta$, viz.,

$$< V > = a V_s \nabla \theta, \tag{15.24}$$

where a is a constant-length scale, and the bacteria were assumed to consume dissolved oxygen at a rate K proportional to their swimming speed, so that

$$K = K_0 W(\theta), \tag{15.25}$$

where K_0 is the maximum consumption rate. K_0 has to be estimated indirectly (see below). Expressions (15.23) and (15.24) are then substituted into the cell conservation equation (15.6). The incompressibility condition (15.10) and the Navier-Stokes equations (15.11) again describe the flow of the suspension and, to complete the system, the conservation of dissolved oxygen is given by the advection-diffusion equation,

$$\frac{\partial C}{\partial t} = -\nabla \cdot (\boldsymbol{u} C - D^* \nabla C) - K(\theta) n, \tag{15.26}$$

where $D^* = 10^{-3} \, \mathrm{mm}^2 \, \mathrm{s}^{-1}$ is the molecular diffusivity of oxygen and the final term represents consumption by the bacteria. The boundary conditions on C are that $C = C_1$ at the upper surface, assuming that oxygen is absorbed sufficiently rapidly through the air-fluid interface to maintain the concentration of dissolved oxygen at its saturation level just below the interface, and $\partial C / \partial z = 0$ at the lower surface, the no-flux condition.

In this model, beginning with a well-mixed uniform state in which $C = C_1$ throughout the layer, the bacteria begin to consume oxygen, thus lowering C everywhere except at the upper surface, where $C = C_1$. Thus an oxygen gradient is established, and the cells begin to swim aerotactically upward.

Depending on the layer depth, either oxygen will diffuse sufficiently rapidly throughout the layer that C is always greater than C_{min} and the cells are always swimming, or C will fall to C_{min} in a quiescent sublayer at the bottom of the container in which the cells are no longer swimming. In experiments, this sublayer was seen to form at depths below about 5 mm, which was used to determine the maximum oxygen consumption rate, K_0, in (15.25).

In the first case, the cells in the "shallow" layer accumulate toward the upper surface, ultimately forming a cell-concentration profile that decreases in an exponential-like fashion with depth, so that chemotaxis balances diffusion. The oxygen-concentration profile develops in such a way that the concentration at any level first decreases to a minimum value and then recovers to a steady value (still less than C_1) as oxygen diffuses down the gradient.

In the second, "deep-layer," case, the quiescent sublayer develops from the bottom of the container and moves rapidly upward toward the mid-depths. Then, as oxygen diffuses down from the upper surface, the quiescent interface moves slowly downward, again resulting in a discontinuity in cell concentration at the interface, as cells above it begin to swim upward and decrease the concentration immediately above the interface. When $W = H(\theta)$, the system settles to a final equilibrium with a permanent quiescent sublayer. In the more realistic case, when $W = [1-\exp(-\theta/\theta_1,)]H(\theta)$, the quiescent interface propagates very slowly toward the bottom, so that, after a time scale of 200 times that required for the steady state to form in all the other cases, typically $O(10^3 \text{ s})$ compared to $O(10 \text{ s})$, the final profile is of the shallow layer. The difference between the two deep-layer cases arises because the consumption of oxygen by cells just above the interface when $W = [1 - \exp(-\theta/\theta_1)]H(\theta)$ is sufficiently small that oxygen can ultimately diffuse down through the quiescent interface.

Kessler (private communication) has also observed that at high dissolved-oxygen-concentration gradients, the molecular receptors on the bacteria saturate, so that the cells cannot detect any oxygen gradient until C falls below a critical value C_1 ($C_{min} < C_1 < C_0$). This results in an initial aggregation of cells somewhat below the upper surface, which gradually moves toward the top of the whole layer.

In all cases, the top-heavy concentration profiles create potentially unstable density profiles in the suspension, which can lead to the Rayleigh-Taylor-like overturning flow associated with initial formation of bioconvection. If the initial uniform concentration of cells is high enough, the overturning will begin before the equilibrium state is reached and so may develop as an example of penetrative convection, when as described by Kessler, there is a concentrated sublayer below the upper surface. As bioconvection develops, the bulk flow of the suspension will advect oxygen around the whole layer, enhancing the transport of oxygen, thus improving the environment for the bacteria and also creating horizontal as well as vertical oxygen-concentration gradients. Hillesdon and Pedley (1996) have examined the linear stability of the equilibrium state and they show that the stability depends on the parameter $\Gamma = v N_0 g h^3 \Delta\rho/\mu D$, where v is the average volume of a cell, $\Delta\rho$ is the difference in the density of the cells to that of the fluid, N_0 is the mean cell concentration, g is the acceleration due to gravity, h is the layer depth, and μ is the viscosity of the fluid. Γ is analogous to the Rayleigh number in thermal convection (Drazin and Reid 1985), and the equilibrium state is unstable if Γ

exceeds some critical value. In the deep-layer case, the instability is oscillatory and is a further example of penetrative convection, since the fluid motion disturbs the lower quiescent sublayer, which is locally gravitationally stable

15.6   CONCLUSIONS

Bioconvection is a system rich in complexities and compelling pattern-formation mechanisms, which are currently being studied using nonlinear mathematical analysis and numerical simulations. Reproducible quantitative experimental data on algal-pattern wavelengths have been obtained by Bees and Hill (1996), which will be used to test the theories. Better comparisons will be possible as the results of the numerical calculations become available. Studies of the motion of algal cells (Hill and Häder 1996) have led to well-characterized self-consistent modeling from the microscopic stochastic behavior of individual cells up to macroscopic, deterministic cooperative phenomena. This full modeling cycle has yet to be carried out for the more complicated bacterial system.

As a result of this work, much has been learned about how these microscopic creatures function as whole organisms and about biological self-organisation through fluid mechanics. Future applications include better quantitative models for the growth of and predation on oceanic blooms of plankton, which are vital to the marine ecology.

REFERENCES

Acheson, D. J. 1990. *Elementary Fluid Dynamics*. Oxford applied mathematics and computing science series. Oxford University Press. ISBN 0-19-859660-X (case), 0-19-859679-0 (paperback). Pages ix + 397. {*341, 343, 350*}

Bees, M. A. and Hill, N. A. 1996. *Wavelengths of biconvection pattern wavelengths in suspension of* Chlamydomonas. In preparation. {*350*}

Berg, H. C. and Brown, D. A. 1972. Chemotaxis in *Escheria coli* analyzed by three-dimensional tracking. *Nature*, **239**, 500–504. {*345, 350*}

Berg, Howard C. 1983. *Random Walks in Biology*. Princeton University Press. ISBN 0-691-00064-6. Page 152. {*343, 348, 350*}

Childress, S., Levandowsky, M., and Spiegel, E. A. 1975. Pattern formation in a suspension of swimming micro-organisms. *Journal of Fluid Mechanics*, **69**, 595–613. {*343, 344, 350*}

Drazin, P. G. and Reid, W. H. 1985. *Hydrodynamic Stability*. New York, NY: Cambridge University Press. {*344, 349, 350*}

Harris, Elizabeth H. 1989. *The* Chlamydomonas *Sourcebook: a comprehensive guide to biology and laboratory use*. San Diego: Academic Press. ISBN 0-12-326880-X. Pages xiv, 780. {*347, 350*}

Hill, N. A. and Häder, D.-P. 1996. A biased random walk model for the trajectories of swimming micro-organisms. *Journal of Theoretical Biology*. Submitted. {*345, 350*}

Hill, N. A., Pedley, T. J., and Kessler, J. O. 1989. Growth of bioconvection patterns in a suspension of gyrotactic micro-organisms in a layer of finite depth. *Journal of Fluid Mechanics*, **208**, 509–543. {*344, 350*}

Hillesdon, A. J. and Pedley, T. J. 1996. Pattern formation in suspensions of oxytactic bacteria: liner theory. *Journal of Fluid Mechanics*. To appear. {*349, 350*}

Hillesdon, A. J., Pedley, T. J., and Kessler, J. O. 1995. The development of concentration gradients in a suspension of chemotactic bacteria. *Bulletin of Mathematical Biology*, **57**(2), 299–344. {*347, 350*}

Kessler, J. O. 1985a. Co-operative and concentrative phenomena of swimming microorganisms. *Contemporary Physics*, **26**, 147–166. {*342, 351*}

Kessler, J. O. 1985b. Hydrodynamic focusing of motile algal cells. *Nature*, **313**, 218–220. {*341, 351*}

Kessler, J. O. 1986. Individual and collective fluid dynamics of swimming cells. *Journal of Fluid Mechanics*, **173**, 191–205. {*342, 351*}

Lebert, M. and Häder, D.-P. 1996. How *Euglena* tells up from down. *Nature*, **379** (6566), 590–590. {*341, 351*}

Pedley, T. J. and Kessler, J. O. 1990. A new continuum model for suspensions of gyrotactic micro-organisms. *Journal of Fluid Mechanics*, **212**, 155–182. {*343–345, 351*}

Pedley, T. J. and Kessler, J. O. 1992. Hydrodynamic phenomena in suspensions of swimming micro-organisms. *Annual Review of Fluid Mechanics*, **24**, 313–358. {*342, 351*}

Pedley, T. J., Hill, N. A., and Kessler, J. O. 1988. The growth of bioconvection patterns in a uniform suspension of gyrotactic micro-organisms. *Journal of Fluid Mechanics*, **195**, 223–237. {*344, 351*}

Tett, P. 1990. Chapter 4: The Photic Zone. *In:* Henning, P. J., Campbell, A. K., Whitfield, M., and Maddock, L. (eds), *Light and Life in the Sea*. New York, NY: Cambridge University Press. {*347, 351*}

Veronis, G. 1963. Penetrative convection. *Astrophysical Journal*, **137**, 641–663. {*347, 351*}

Vincent, R. V. and Hill, N. A. 1996. Bioconvection in a suspension of phototactic algae. *Journal of Fluid Mechanics*, **300**, 1–29. To appear. {*345, 347, 351*}

A. AGE-STRUCTURED MODELS

Frederick R. Adler

A.1 LESLIE MATRICES

Natural populations are generally **structured**, meaning that not all individuals are the same. Understanding the dynamics and evolution of populations must take this structure into account. We here outline some basic techniques for describing growth and computing fitness.

The simplest sort of structure to model is **age structure**, where the reproduction and survival of individuals depends on age. A **Leslie matrix** provides a convenient way to describe such a population (this material is described in great depth in Caswell (1989) and Charlesworth (1980)). As an example, consider a population with two age classes, juveniles and adults, each lasting one year. Juveniles produce f_0 offspring and adults produce f_1 offspring. Juveniles survive to adulthood with probability p_0 If J_t represents the numbers of juveniles and A_t the number of adults, then the numbers at time $t + 1$ can be expressed in matrix-vector notation as

$$\left(\begin{array}{c} J_{t+1} \\ A_{t+1} \end{array} \right) = \left(\begin{array}{cc} f_0 & f_1 \\ p_0 & 0 \end{array} \right) \left(\begin{array}{c} J_t \\ A_t \end{array} \right). \tag{A.1}$$

With more age classes, we need a value f_i for reproduction at age i and p_i for survival from age i to age $i + 1$.

A more general formulation is a model with **stage structure**. Individuals need not mature to the next stage, but can remain where they are. For the situation in (A.1), suppose adults remain alive with probability p_1. The matrix is then

$$\left(\begin{array}{c} J_{t+1} \\ A_{t+1} \end{array} \right) = \left(\begin{array}{cc} f_0 & f_1 \\ p_0 & p_1 \end{array} \right) \left(\begin{array}{c} J_t \\ A_t \end{array} \right). \tag{A.2}$$

In either case, if growth continues for a long time, the population approaches a **stable age distribution**(or a **stable stage distribution**)and asymptotic growth rate λ which solve the equation

$$A\vec{w} = \lambda\vec{w}. \tag{A.3}$$

The eigenvalue λ is the largest eigenvalue of the matrix and is the appropriate measure of fitness for a type with given values of f_i and p_i. The Perron-Frobenius theorem guarantees that the components of the eigenvector are positive and that the largest eigenvalue is a positive real number.

A.2   REPRODUCTIVE VALUE

Reproductive value provides a way to think about the "fitness" of individuals of different ages. This quantity measures how important individuals of different ages are to the growth of the population. Let v_0 and v_1 represent the reproductive value of juveniles and adults, respectively. The "value" of a juvenile must equal the value of what it produces (f_0 juveniles and p_0 adults), but *discounted by the growth of the population* (a juvenile next year is worth only $1/\lambda$ as much as a juvenile this year). Then

$$v_0 = \frac{1}{\lambda}(f_0 v_0 + p_0 v_1) \tag{A.4}$$

$$v_1 = \frac{1}{\lambda}(f_1 v_0 + p_1 v_1) \tag{A.5}$$

where we equated the reproductive value v_1 of an adult with the discounted value of what it can produce in equation A.5. In matrix form, these equations can be written

$$(v_0, v_1)\boldsymbol{A} = \lambda(v_0, v_1). \tag{A.6}$$

The vector of reproductive values is a **left** eigenvector of the matrix. Because these equations are linear, solutions are defined only up to a constant. The ratio of reproductive value satisfies the single equation

$$\frac{v_0}{v_1} = \frac{f_0 v_0 + p_0 v_1}{f_1 v_0 + p_1 v_1}. \tag{A.7}$$

A more concrete derivation compares the size of a population starting with a single juvenile with one starting with a single adult. After n generations, the size of the population starting with a single juvenile is

$$S_J(n) = \left\| \boldsymbol{A}^n \begin{pmatrix} 1 \\ 0 \end{pmatrix} \right\|, \tag{A.8}$$

where the double bars indicate the total population (adults plus juveniles). Starting from a single adult, the size is

$$S_A(n) = \left\| \boldsymbol{A}^n \begin{pmatrix} 0 \\ 1 \end{pmatrix} \right\|. \tag{A.9}$$

The ratio is the relative value of a juvenile. The **definition** of the relative reproductive value is

$$\lim_{n \to \infty} \frac{S_J(n)}{S_A(n)} = \frac{v_0}{v_1}. \tag{A.10}$$

But

$$S_J(n) = \left\| A^{n-1} A \begin{pmatrix} 1 \\ 0 \end{pmatrix} \right\| = \left\| A^{n-1} \begin{pmatrix} f_0 \\ p_0 \end{pmatrix} \right\| \qquad \text{(A.11)}$$

$$= \left\| f_0 A^{n-1} \begin{pmatrix} 1 \\ 0 \end{pmatrix} + p_0 A^{n-1} \begin{pmatrix} 0 \\ 1 \end{pmatrix} \right\| \qquad \text{(A.12)}$$

$$= f_0 S_J(n-1) + p_0 S_A(n-1) \qquad \text{(A.13)}$$

(we assumed that the population had converged to the stable age distribution to remove the sum from the double bars). Similarly,

$$S_A(n) = f_1 S_J(n-1) + p_1 S_A(n-1). \qquad \text{(A.14)}$$

Taking the ratio,

$$\frac{S_J(n)}{S_A(n)} = \frac{f_0 S_J(n-1) + p_0 S_A(n-1)}{f_1 S_J(n-1) + p_1 S_A(n-1)}. \qquad \text{(A.15)}$$

If the ratio of S_J to S_A converges as $n \to \infty$, these values satisfy equation A.7.

These ideas generalize to populations with multiple age classes or where the structure derives from size or position. The fitness of a genotype is the leading eigenvalue, and the reproductive value of a particular type is proportional to the size of a population starting with one member of that type and can be found conveniently as the left eigenvector of the matrix.

A.3 THE EULER-LOTKA EQUATION

Similar methods can be used if age is structured continuously. Let $m(x)$ be the rate of reproduction of an organism of age x, and $l(x)$ be the probability of survival to age x from age 0 (different from the p_i used in a Leslie matrix). Let $N(x, t)$ be the density function describing the population at time t, meaning that

$$\text{number between age } a_0 \text{ and } a_1 = \int_{a_0}^{a_1} N(x, t)\, dx. \qquad \text{(A.16)}$$

The rate at which individuals are born at time t is the integral of the number of surviving individuals born time x ago times their reproductive rate, or

$$N(0, t) = \int_0^\infty N(0, t-x) l(x) m(x)\, dx. \qquad \text{(A.17)}$$

We can use this equation to find the rate of growth, and thus the fitness, of a type with particular $m(x)$ and $l(x)$ functions. Suppose the population grows exponentially, or that $N(0, t) = N_0 e^{rt}$. Substituting into equation A.17,

$$N_0 e^{rt} = \int_0^\infty N_0 e^{r(t-x)} l(x) m(x)\, dx. \qquad \text{(A.18)}$$

Dividing by $N_0 e^{rt}$, we find

$$1 = \int_0^\infty e^{-rx} l(x) m(x) \, dx. \tag{A.19}$$

This is the **Euler-Lotka** equation for r, the rate of growth of this population.

REFERENCES

Caswell, Hal. 1989. *Matrix Population Models: Construction, Analysis, and Interpretation.* Sunderland, MA: Sinauer Associates, Inc. ISBN 0-87893-094-9, 0-87893-093-0 (paperback). Pages xiv + 328. {*3, 18, 22, 353, 356*}

Charlesworth, Brian. 1980. *Evolution in Age-Structured Populations.* Cambridge, UK: Cambridge University Press. ISBN 0-521-23045-4, 0-521-29786-9. Pages xiii + 300. {*3, 10, 22, 353, 356*}

B. QUALITATIVE THEORY OF ORDINARY DIFFERENTIAL EQUATIONS

Mark A. Lewis and Hans G. Othmer

B.1   INTRODUCTION

This appendix introduces techniques for the qualitative analysis of systems of nonlinear ordinary differential equations (ODEs). In many biological problems the state of the system evolves continuously on an appropriate time scale, and thus differential equations are appropriate for describing the evolution of the state. Furthermore, many phenomena of interest are nonlinear, and therefore the resulting equations are nonlinear.

To introduce some of the techniques described herein, we consider a simple example. Suppose that $c(t)$ is the concentration in moles/liter of a certain chemical X that is introduced into a beaker containing V liters of fluid at the rate $q(t)$ moles/minute. Suppose that the introduction of X does not change the fluid volume significantly, and further suppose that X disappears at a rate that is proportional to its concentration. Then the equation that describes how the total **amount** of X changes with time is as follows:

$$V\frac{dc}{dt} = q(t) - kVc, \tag{B.1}$$

where $k > 0$ is the first-order rate constant for the destruction of X. In order to complete the specification of the problem we also have to specify an initial condition $c(0)$.

This equation is in effect a **balance** equation, because it states that the rate of change of the total amount Vc of X in the beaker depends on the balance between competing processes: one, the influx, that leads to an increase in X, and another, the decay process, that leads to a decrease in X. If q is independent of time, then it may happen that these two processes balance precisely, and the total amount of X does not change. Clearly, this happens only if $c = q/kV$, and we call this the **steady-state** value of X. Under the assumptions above the **state** of the system is characterized by a single quantity, the amount of X. In general there may be many state variables whose evolution we are interested in describing. In this example, would the equation change if the volume V of the fluid is not constant? Would other state variables be required?

In addition to finding steady states of the system, one is also interested in determining their **stability**. Roughly speaking (this will be made more precise later), a steady state c^s is stable if solutions that begin nearby approach c^s as $t \rightarrow \infty$. This suggests that one might study the equation that governs the evolution of perturbations of the steady state, rather than the equation for c itself. If we set $\xi = c(t) - c^s$, then we find that

$$\frac{d\xi}{dt} = -k\xi \tag{B.2}$$

where we are still assuming that q is time independent. This equation shows that ξ increases or decreases monotonically, according to whether it is negative or positive initially, and that $\xi(t) \rightarrow 0$ as $t \rightarrow \infty$. Therefore the steady state is stable according to the informal definition. Since (B.1) is linear, the equation for ξ is automatically linear, but, as we shall see shortly, one can often determine stability by linearizing a nonlinear equation around a steady state. The foregoing shows that we can gain a fairly complete understanding of the behavior of the solutions of (B.1) without actually solving (B.1) explicitly; this is what we mean by a qualitative analysis of the equation.

This appendix is divided into two parts: first an informal discussion centered around systems of two equations that stresses the geometry of the phase plane, and second a part in which the ideas are made more precise via definitions and statements of theorems. This first part is meant for students who may have only had an introductory course in ODEs, but who can appreciate the important concepts using simple examples. The second part generalizes the context to systems of ODES, makes the notion of stability precise, and introduces some elementary ideas from bifurcation theory, which helps in understanding how the qualitative picture changes as the parameters are varied. This part is aimed at a mathematically more advanced audience.

By necessity the material given here is brief and focuses on the theory needed to analyze the ODE models in this book. These ODE models include metapopulations (Chapter 2), biomass dynamics of structured populations (Chapter 4), calcium dynamics (Chapter 6), cell division (Chapter 7), blood disease (Chapter 8), pupil light reflexes (Chapter 9), electrical bursting of cells (Chapter 10), muscle dynamics (Chapter 11), and forced oscillations (Chapter 12). A useful practical exposition on ODEs in mathematical biology is given by Odell (1980). More complete elementary treatments of the material in this appendix can be found in Arrowsmith and Place (1990), Boyce and DiPrima (1977), Jordan and Smith (1987), Hale and Kocak (1991), Hirsch and Smale (1974), and Perko (1991). More advanced treatments are given in Amann (1990), Andronov et al. (1973), Arnold (1973), Bhatia and Szegö (1967), Devaney (1986), Guckenheimer and Holmes (1983) and, Hartman (1982).

B.2   QUALITATIVE THEORY OF SECOND-ORDER SYSTEMS OF ODES

The description of a system with two state variables that evolve according to ordinary differential equations takes the form of the second-order system of equations

$$\frac{dx_1}{dt} = f_1(x_1(t), x_2(t), p_1, \ldots, p_m),$$

$$\frac{dx_2}{dt} = f_2(x_1(t), x_2(t), p_1, \ldots, p_m). \tag{B.3}$$

The functions f_1 and f_2 encode the dependence of the rates of change of the dependent variables x_1 and x_2 on the current values of x_1 and x_2. As they are written, there is no explicit time dependence in the rates of change, which means that the explicit origin of time is not important: such equations are called **autonomous** equations. Furthermore, they do not depend on the past history of the processes that make up the f_i, but if they did they would be called delay equations. In general the rate of change of x_1 depends on the value of x_2 and vice versa, and thus the equations are coupled.

In addition to the time-dependent variables, the rates also depend on parameters $p_1, \ldots, p_m$. Typically parameters are regarded as constants on the time scale of interest, although they may in fact vary slowly, and they may be under the explicit control of an experimenter.

We now know how the equations evolve in time (*i. e.*, according to (B.3)), and in order to complete the specification of the problem we have to prescribe their values at a given instant. Since the explicit origin of time is immaterial, we specify the particular value at $t = 0$, thus assigning the values

$$x_1(0) = x_1^0, \quad x_2(0) = x_2^0. \tag{B.4}$$

Together with equations (B.3) this defines an initial-value problem (IVP) describing the dynamical system.

Given a specific set of initial values, we would like to be able to calculate the solution to the problem as t evolves. However, first we need to know how many solutions to the IVP (B.3)–(B.4) actually exist: none, one, or several. The answer hinges upon the nature of the functions f_i in (B.3): when they are linear, the solution exists for all time and can be written as a sum of exponentials (see later); when they are nonlinear, there is a unique solution to the IVP at least for some interval of t, $[0, T]$ if each f_i is continuously differentiable with respect to each x_i. Weaker conditions on the f_is can actually be used to give the same result, and the value of T can be explicitly calculated; interested readers are referred to Jackson (1989).

A simple cautionary example to show how very reasonable-looking first-order IVPs can fail to have solutions is: $dx/dt = x^2$, $x(0) = 1$. The solution $x = 1/(1 - t)$ exists only for $t < 1$ (reader verify). An IVP that has more than one distinct solution is: $dx/dt = \sqrt{x}$, $x(0) = 0$. Solutions $x(t) = 0$ and $x(t) = t^2/4$ exist for all t. (Note that $\sqrt{x}$ is not differentiable at $x = 0$.) While the issues of existence and uniqueness clearly are important, we put them to one side for the remainder of the appendix and tacitly assume that the equations we describe have a unique solution.

Rather than choosing specific initial values, we may wish to analyze the qualitative behavior of the system for a whole spectrum of initial data, so as to understand the general dynamics governing the biology. In either case, it is very unusual to be able to write down an exact solution for systems of nonlinear ODEs. In most cases such formulae simply do not exist. Even in cases where approximate solutions can be calculated, they are typically so complicated as to defy simple translation back into conclusions about the underlying dynamical system.

One alternative is to use a computer to numerically calculate solutions and then to display the solutions graphically. This can be a relatively painless procedure if one uses a package such as xpp. (Web browsers can find out more about this package from http://www.pitt.edu/~phase.) However, the blind use of such a computational approach has a severe limitation: while most biological systems are defined by a range of values for each parameter $p_1, \ldots, p_m$ and for initial values $x_1^0, \ldots x_n^0$, numerical solutions can be calculated only for fixed parameters and fixed initial data. A blind numerical investigation of the way solutions depend upon parameters and initial data can quickly become very cumbersome. For example, even very simple systems typically have at least two parameters. If each parameter and the initial data varied over ten values, then there would be 1000 simulations to do, resulting in 1000 different outputs. Not only might these outputs be difficult to interpret, but they can miss the crucial features of the model. To quote Odell (1980), "Often the region in parameter space within which the parameter values must lie to make the model behave realistically is so small or curiously shaped that even a thundering numerical shotgun blast will miss it."

As we will show in this appendix, the very place where the numerics are least useful is where mathematical methods are the most useful — that is in understanding the qualitative dependence of solutions on parameter values and initial values. Thus the numerical and mathematical methods are complementary approaches: with numerics one can calculate specific solutions, and mathematical methods can be used to tease out the dependence of underlying dynamical structure on parameters and initial data.

B.2.1   The phase plane — a geometric interpretation

A system of two autonomous ODEs has a useful geometric interpretation in the x_1–x_2 plane, usually called the **phase plane**. We can think of a particular solution to (B.3) as a solution curve or **trajectory** parameterized by t and moving through the two-dimensional x_1–x_2 phase plane (see, for example, Figure B.1). The equation specifying the trajectory is

$$\frac{dx_1}{dx_2} = \frac{f_1(x_1, x_2)}{f_2(x_1, x_2)}. \tag{B.5}$$

Each trajectory has a direction; as we move forward along the curve, t increases, and as we move backward, t decreases. The starting point, given by the initial condition for (B.3,) is (x_1^0, x_2^0). Thus different initial conditions yield trajectories originating at different points in the phase plane. A direct consequence of uniqueness of solutions to the autonomous system (B.3) is that different trajectories can never cross in the phase plane. At any given point $(x_1(t), x_2(t))$ on the trajectory we can construct a tangent vector $(f_1(x_1, x_2), f_2(x_1, x_2))$. The length of this vector specifies the speed at which the solution passes along the trajectory at the point of attachment. These vectors are said to specify the **flow** in phase space. Thus, given the flow, trajectories are constructed by drawing

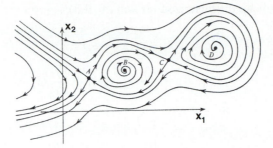

Figure B.1. A phase-plane diagram illustrating homoclinic and heteroclinic orbits. Points A and C are joined by two different heteroclinic orbits. A homoclinic orbit leaves C and then returns to it. The homoclinic orbit beginning and ending at C is also a separatrix, defining an invariant region for the flow in the phase plane. Refer to text for detailed discussion. (Based on Odell (1980).)

curves that are everywhere tangent to the flow vectors (see, for example, Figures 7.7 and B.1).

Steady-state solutions, or **singular points**, of (B.3) are defined as points (x_1^s, x_2^s) with zero flow vectors: these solutions do not change with t. In other words, $f_1(x_1^s, x_2^s) = f_2(x_1^s, x_2^s) = 0$. We refer to **nullclines** as curves along which one of nonlinear functions equals zero (see, for example, Figure 7.7). The curve $f_1(x_1, x_2) = 0$ defines the x_1 nullcline, and the curve $f_2(x_1, x_2) = 0$ defines the x_2 nullcline. Notice that the flow is vertical across the x_1 nullcline and the flow is horizontal across the x_2 nullcline. Steady-state solutions, if they exist, occur at intersections of the nullclines (see, for example, Figure 7.7). Note that once we draw the nullclines and know the signs of the f_is on either side of the nullclines, we have a fairly good qualitative understanding of the flow without having done any detailed mathematics!

B.2.2 Linear stability analysis — a method for investigating qualitative behavior

One biologically relevant question is whether or not a system near equilibrium will remain there. Mathematically, the question of whether solutions that start "near" steady-state solutions remain "near" them for all t is referred to as **Lyapunov stability**. A stronger version of stability, referred to as **local asymptotic stability**, requires that solutions starting "near" steady-state solutions approach them more and more closely as t progresses. As we will show, the stability of steady-state solutions plays an important role in understanding the flow of ODEs in phase space. Our approach to answering the question of stability is to analyze the behavior of a related linear system, and then to connect this behavior back to that of the nonlinear system.

If we are "near" the steady-state solution (x_1^s, x_2^s), then the quantities $\xi_1 = x_1(t) - x_1^s$ and $\xi_2 = x_2(t) - x_2^s$ are small. Specifically, when $(x_1(t), x_2(t))$ are displaced infinitesimally from (x_1^s, x_2^s) so that $0 \leq |\xi_1|, |\xi_2| \ll 1$, we can neglect relatively small quadratic terms in ξ_1 and ξ_2 and even smaller higher-order terms to obtain the linearized version of (B.3):

$$
\begin{aligned}
\frac{d\xi_1}{dt} &= \frac{d\xi_1}{dt} = f_1(x_1^s + \xi_1, x_2^s + \xi_2) \\
&= f_1(x_1^s, x_2^s) + \frac{\partial f_1}{\partial x_1}(x_1^s, x_2^s)\xi_1 + \frac{\partial f_1}{\partial x_2}(x_1^s, x_2^s)\xi_2 \\
\frac{d\xi_2}{dt} &= \frac{d\xi_2}{dt} = f_2(x_1^s + \xi_1, x_2^s + \xi_2) \\
&= f_2(x_1^s, x_2^s) + \frac{\partial f_2}{\partial x_1}(x_1^s, x_2^s)\xi_1 + \frac{\partial f_2}{\partial x_2}(x_1^s, x_2^s)\xi_2.
\end{aligned}
\tag{B.6}
$$

Since by definition of the steady state $f_1(x_1^s, x_2^s) = f_2(x_1^s, x_2^s) = 0$, the equations are written as a linear, constant-coefficient ODE system

$$
\frac{d\xi}{dt} = A\xi, \quad \xi = \begin{pmatrix} \xi_1 \\ \xi_2 \end{pmatrix}, \quad A = \begin{pmatrix} a & b \\ c & d \end{pmatrix},
\tag{B.7}
$$

and

$$
\begin{aligned}
a &= \frac{\partial f_1}{\partial x_1}(x_1^s, x_2^s), \quad b = \frac{\partial f_1}{\partial x_2}(x_1^s, x_2^s), \\
c &= \frac{\partial f_2}{\partial x_1}(x_1^s, x_2^s), \quad d = \frac{\partial f_2}{\partial x_2}(x_1^s, x_2^s).
\end{aligned}
\tag{B.8}
$$

The matrix A is referred to as the **Jacobian** or the **community matrix**.

Solutions to this linear system (B.7) can be calculated exactly (Boyce and DiPrima 1977). The eigenvalues of A, λ_1, and λ_2, are the solutions of the quadratic equation $\det(A - \lambda I) = 0$, *i. e.*, of

$$
\begin{vmatrix} a - \lambda & b \\ c & d - \lambda \end{vmatrix} = 0.
\tag{B.9}
$$

The eigenvalues can be expressed explicitly in terms of the trace $(a + d)$ and determinant $(ad - bc)$ of A as

$$
\lambda_{1,2} = \frac{1}{2}(\operatorname{tr} A \pm \sqrt{\operatorname{tr}^2 A - 4 \det A}).
\tag{B.10}
$$

The case $\operatorname{tr}^2 A - 4 \det A \neq 0$ yields distinct eigenvalues $\lambda_1 \neq \lambda_2$ and gives a general solution to (B.7) of the form

$$
\xi = c_1 \mathbf{v_1} \exp(\lambda_1 t) + c_2 \mathbf{v_2} \exp(\lambda_1 t),
\tag{B.11}
$$

with $\mathbf{v_1}$ and $\mathbf{v_2}$ the eigenvectors of A satisfying

$$
\begin{pmatrix} a - \lambda_i & b \\ c & d - \lambda_i \end{pmatrix} \xi = \mathbf{0},
\tag{B.12}
$$

and c_1 and c_2 arbitrary constants. The rather uncommon repeated-eigenvalue case is discussed, for example, in Boyce and DiPrima (1977) and Murray (1989).

The beauty of the explicit form of the solution to the linear problem is the information that it gives us about the flow of (B.7) in the $\xi_1-\xi_2$ phase plane. For example, if $\mathrm{Re}\lambda_1, \mathrm{Re}\lambda_2 < 0$, i.e., tr $A < 0$ and det $A > 0$, substitution into (B.11) shows that the solution trajectories to (B.7) approach zero for large t, that is, $\xi \to \mathbf{0}$ as $t \to \infty$. Expanding out the exponential in (B.11) shows that complex λ_is yield oscillatory motion in the phase plane. Similar logic can be used to catalog the qualitative behavior of (B.7) for all possible values of tr A and det A (Figure B.2). Note that while trajectories can intersect at the steady state $(\xi_1^s, \xi_2^s) = (0, 0)$, they do not cross each other. Returning to the idea of stability introduced above, we observe from Figure B.2 that the linear system (B.7) is locally asymptotically stable if and only if tr $A < 0$ and det $A > 0$.

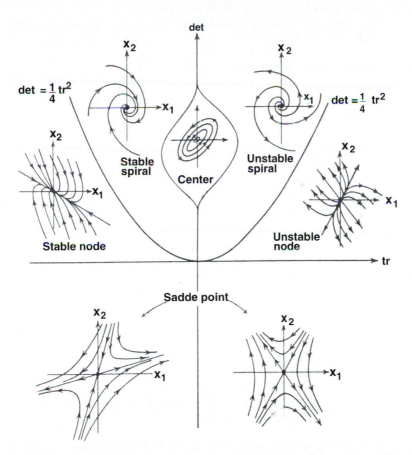

Figure B.2. Summary diagram showing how the trace (tr) and determinant (det) of the community matrix A determine the qualitative behavior of the flow of (B.7) near the steady state. Refer to text for detailed discussion. (Based on Odell (1980).)

The connection between behavior of the linear system (B.7) and the original nonlinear system of interest (B.3) is given elegantly by a mathematical result known as the **Hartman-Grobman theorem**. This states that close to (x_1^s, x_2^s) the local topology of the flow of the nonlinear system is inherited from the related linear system, providing (x_1^s, x_2^s) is **hyperbolic** — that is, providing the eigenvalues λ_i are not imaginary. Nonlinearities in (B.3) simply warp the flow of (B.7), as it is locally embedded with $(\xi_1, \xi_2) = (0, 0)$ centered at $(x_1, x_2) = (x_1^s, x_2^s)$. In other words, ignoring the quadratic and higher-order terms in the linearization process does not change the qualitative results. Thus Figure B.2 not only shows behavior of the linear system (B.7); it also shows qualitative flow for the nonlinear system (B.3) near the steady state.

The subspace of the phase plane along which trajectories leave a steady state solution is called the **unstable manifold**, and the subspace along which trajectories enter the steady-state solution is the **stable manifold**. Depending upon the number of eigenvalues with negative real part, the dimension of the stable manifold is 0, 1, or 2. The sum of the dimensions of the stable and unstable manifolds is two at a hyperbolic rest point.

The cases where one or more eigenvalues have real part which pass through zero yields a **bifurcation** or immediate qualitative change in the flow. Simple cases of these occur in Figure B.2 when the line det $= 0$ is crossed vertically (**saddle-node** bifurcation) and when the line tr $= 0$ is crossed horizontally above the det $= 0$ line (**Hopf** bifurcation). The Hartman-Grobman theorem and the theory of bifurcations is given in more explicit detail in the following section.

B.2.3   Nonlinear behavior in the phase plane — putting it all together

It is helpful to characterize structures in the phase plane, building up a "vocabulary" of possible solution types. Not surprisingly, the existence of such structures depends critically upon parameter values on the model; typically they will exist for some parameter ranges and not for others. Fortunately for the purposes of this appendix, the behavior of order-2 ODE systems is far more restrained than that of order-3 systems of ODEs, which can exhibit a plethora of exotic behaviors including chaos (Jackson 1989).

- A **limit cycle** is a t-periodic solution whose trajectory forms a closed loop in the phase plane (see, for example, Figure 10.4).

- A **homoclinic orbit** leaves a saddle-point steady state (Figure B.1) on the **unstable manifold**, tangent to the eigenvector which has a corresponding positive eigenvalue, and then returns back to the same steady-state solution along the **stable manifold**, tangent to the eigenvector which has a corresponding negative eigenvalue.

- A **heteroclinic orbit** connects an unstable manifold from one steady state to the stable manifold of another steady state (Figure B.1). The homoclinic and heteroclinic orbits described above are bounded (finite) for all t but do not correspond to t-periodic solutions, because it takes an infinite amount of t to leave or approach a steady state along the paths (Figure B.1).

- A **separatrix** divides the phase plane into invariant regions based upon the direction of flow. The flow in each basin is invariant — that is, it never leaves the basin (Figure B.1). As shown in Figure B.1, a steady state with local asymptotic stability need not be **globally** asymptotically stable — that is, trajectories starting at all points in the phase plane need not approach the steady state as $t \to \infty$. However, by definition the converse is true: global asymptotic stability implies local asymptotic stability.

Because trajectories cannot cross one another, the above structures govern the topology (or connectedness) of the flow in the phase plane. The significance of this becomes clear when we ask where trajectories "end off" after any initial transient behavior passes. A great deal of mathematical effort has gone into methods to prove when such structures exist. For example, the Poincaré-Bendixson theorem can be used to show existence of a limit cycle, while Dulac's criterion and Bendixson's negative criterion can be used to rule out the possibility of a limit cycle (see, for example, Jordan and Smith (1987)). These and related results are part of the mathematical "tool kit" that serious students of dynamical systems can use to understand the structure of the flow, but they are beyond the scope of this appendix. Without precise tools it is still possible to use "plausibility" arguments for the possible existence of structures, based upon understanding of the flow gleaned from analysis of the null clines coupled with linear analyses about steady states. While such arguments can only suggest the existence, they can be, in turn, tested with numerical simulations for sets of fixed parameters.

We are now in a position to put together all the information regarding null clines, steady states and their stability and flow to derive a **phase portrait** describing behavior of trajectories in the phase plane. We can break the procedure into several steps

- Draw nullclines, locate steady-state solutions, and deduce the qualitative form of the flow. Determine how the shapes of nullclines and their intersections depend upon model parameters.

- Calculate local stability of any steady-state solutions. Determine analytically how the stability depend upon model parameters.

- Try to locate possible heteroclinic and homoclinic orbits, limit cycles, basins of attraction, and separatrices.

- Put together results from both analyses to construct a **phase portrait** or flow diagram. Draw the flow and sample trajectories. How does the flow change as parameters are varied? Analyze the large-t behavior of trajectories: determine whether solutions can become unbounded, whether they approach steady states, or whether they approach limit cycles or other orbits.

- Confirm your predictions with numerical solutions for sets of fixed parameter values.

- Draw biological conclusions. What dynamical behaviors are possible? How are they translated into biological terms? Are the parameter values needed for these behaviors actually biologically possible?

B.3   QUALITATIVE ANALYSIS OF N-DIMENSIONAL SYSTEMS

Consider the autonomous, parameterized family of ordinary differential equations

$$\frac{dx}{dt} = f(x, p), \tag{B.13}$$

$$x(0) = x_0, \tag{B.14}$$

where $x \in M \subseteq R^n$ and $p \in I \subseteq R^m$. We shall assume that f is sufficiently smooth (at least C^2) so that the solution $\phi(x_0, t)$ of (B.13) through x_0 at $t = 0$ exists for small t and is unique. The differential equation defines a family of vector fields

$$X_f = \Sigma f_i(x, p) \frac{\partial}{\partial x_i}$$

and it will sometimes be more convenient to refer to X_f rather than the system of ordinary differential equations that defines it. Let $\mathcal{O}(x_0) = \{\phi(x_0, t)|t \in R\}$ denote the **oriented orbit** through x_0. Some orbits of particular interest are the steady states or rest points x^s, for which $\mathcal{O}(x^s) = x^s$, and the periodic orbits $\gamma(t)$. A solution $\gamma(t) = \phi(x_0)$ through x_0 is periodic of least period $T > 0$ if $\phi(x, t + T) = \phi(x, t)$ for $t \in R$ and no smaller T suffices. The **phase portrait** of (B.13) is the set of all orbits of (B.13), thought of as a family of curves in R^n.

The definition of $\phi(x_0, t)$ assumes that solutions of (B.13) are defined for all $t \in R$. This can be assured by redefining f in a smooth manner outside some large set if necessary, so as to preclude solutions that blow up in finite time. We define M so that solutions beginning in M remain in M for all $(t, p) \in (R, I)$, and thus $\phi : M \times R \rightarrow M$ defines a smooth map called the **flow** defined by (B.13). For instance, if $f(x, p) = Ax$ for a constant matrix A, then we can take $M = R^n$ and $\phi(x_0, t) = \exp(At)x_0$ for any $x_0 \in R^n$.

It is generally impossible to map out the complete phase portrait for a parameterized family of vector fields, but the long-term behavior is determined by the structure of sets in phase space that are attracting in some sense. For the present, let p be fixed and consider a single vector field. A subset $S \subseteq M$ is **positively invariant (negatively invariant), (invariant)**, respectively, if $\phi(S, t) \subset S$ for $t \in R^+$ $(t \in R^-), (t \in R)$, respectively, where $\phi(S, t)$ has the obvious meaning. The $\omega-$ **limit set** of $x \in M$ is the set $L_\omega(x) \equiv \{y \in M| \exists$ a sequence $t_n \rightarrow \infty \ni \phi(x, t) \rightarrow y\}$, i.e. $L_\omega(x)$ is the set of points that $\phi(x, t)$ approaches as $t \rightarrow +\infty$. By reversing time in the foregoing we obtain $L_\alpha(x)$, the $\alpha-$ **limit set** of $x \in M$. The limit sets $L_\alpha(\mathcal{O})$ and $L_\omega(\mathcal{O})$ of an orbit are just the α and ω limit sets of any $x \in \mathcal{O}$. Both $L_\omega(x)$ and $L_\alpha(x)$ are closed invariant sets.

A set N is a **neighborhood** of a set $\Lambda \subset M$ if there is an open set $G \subset N$ such that $\Lambda \subset G$. Let Λ be a nonempty closed invariant subset of M. The **domain of attraction** of Λ is the set $A(\Lambda) \equiv \{x \in M|L_\omega(x) \subseteq \Lambda\}$. Λ is an **attractor** if $A(\Lambda)$ is a neighborhood of Λ and Λ contains a dense orbit. Λ is stable if every neighborhood U of Λ contains a neighborhood V such that $\phi(V, t) \subseteq U$ for $t \in R^+$, and Λ is **asymptotically stable** if it is stable and is an attractor. For example, if x^s is a rest point of (B.13), then neighborhoods can be taken as $\{x \in R^n \mid \|x - x^s\| \le \delta\}$. Thus x^s is an attractor if there is a $\delta > 0$ such that for all $x(0)$ that satisfy $\|x(0) - x^s\| < \delta$, $\lim_{t \to \infty} x(t) = x^s$. Furthermore, x is stable if, given $\epsilon > 0$ there is a $\delta > 0$ such that $\|x(t) - x^s\| < \epsilon$ for $t \in R^+$ whenever $\|x(0) - x^s\| < \delta$, and asymptotically stable if it is stable and $\lim_{t \to \infty} x(t) = x^s$. A sufficient condition for asymptotic stability of x^s is that the real part of every eigenvalue of the Jacobian matrix $D_x f(x^s)$ is strictly negative (Hale 1969). Asymptotically stable rest points are also called **sinks**. The simplest types of attractors are rest points, periodic solutions, quasi-periodic solutions, and invariant tori. All other attractors are usually called **strange attractors**.

To understand when linearization gives us a complete local picture of the phase portrait, to within a smooth change of coordinates, we have to introduce the notion of the equivalence of two systems. Consider the systems on R^n given by

$$\frac{dx}{dt} = f(x), \tag{B.15}$$

$$\frac{dy}{dt} = g(y), \tag{B.16}$$

where f and g satisfy the general conditions given earlier. Let ϕ and Ψ denote the flows associated with (B.15) and (B.16), respectively. We say that ϕ is **topologically equivalent** to Ψ, and write $\phi \sim_T \Psi$, if there is a homeomorphism h (a coordinate change) and an increasing homeomorphism $\alpha : R \to R$ (a scaling of time) such that

$$(h \circ \phi)(x, t) = \Psi(h(x), \alpha(t)).$$

If α is the identity the flows are said to be **topologically conjugate**. The underlying vector fields X_f and X_g are said to be equivalent whenever the corresponding flows are topologically equivalent, and this defines an equivalence relation on the set $\chi(M)$ of vector fields on M. Given a topology on $\chi(M)$, $X \in \chi(M)$ is said to be **structurally stable** if there is an open neighborhood of X such that every Y in this neighborhood is equivalent to X.

In general it is hard to find an h that works for a given pair of vector fields, but in the linear case a necessary and sufficient condition for equivalence of two flows is known if neither matrix has eigenvalues with a zero real part. Consider

$$\frac{dx}{dt} = Ax$$

where A is an $n \times n$ constant matrix. Let $\sigma(A)$ denote the spectrum of A (i.e., the set of eigenvalues of A) and decompose $\sigma(A)$ as follows.

$$\sigma(A) = \sigma^+(A) \cup \sigma^0(A) \cup \sigma^-(A),$$

where

$$\sigma^+(A) \equiv \{\lambda \in \sigma(A) \mid \mathrm{Re}\lambda > 0\}$$
$$\sigma^0(A) \equiv \{\lambda \in \sigma(A) \mid \mathrm{Re}\lambda = 0\}$$
$$\sigma^-(A) \equiv \{\lambda \in \sigma(A) \mid \mathrm{Re}\lambda < 0\}.$$

The invariant subspaces with the properties

$$W^s(0) \equiv \{x_0 \in R^n \mid \phi(x_0, t) \to 0 \text{ exponentially as } t \to +\infty\},$$

$$W^u(0) \equiv \{x_0 \in R^n \mid \phi(x_0, t) \to 0 \text{ exponentially as } t \to -\infty\},$$

$$W^c(0) \equiv \{x^0 \in R^n \mid \phi(x_0, t) \uparrow \text{ or } \downarrow \text{ at most algebraically for } |t| \to \infty\},$$

are called the **stable manifold**, the **unstable manifold**, and the **center manifold** of the origin in R^n, respectively. The dimensions of these subspaces are the sums of the algebraic multiplicities of all eigenvalues in the corresponding spectral set, and we denote these n_A^-, n_A^+ and n_A^0 (or n^+, n^0 and n^- if no confusion is possible). The origin is said to be a *hyperbolic rest point* if $\sigma^0(A) = \emptyset$, and a linear flow is hyperbolic if the origin is a hyperbolic rest point.

This leads us to the following result, which we state in terms of the phase portraits rather than the flows.

Theorem 1 *A necessary and sufficient condition for topological equivalence of the phase portraits of two linear systems on R^n, all of whose eigenvalues have nonzero real parts, is that the number of eigenvalues with negative real parts be the same in both systems.*

The proof of this may be found in Irwin (1980) and Arnold (1973).

Since an eigenvalue on the imaginary axis can be moved into the right or left half complex plane by an arbitrarily small perturbation of the elements of A, the only stable linear systems are those without eigenvalues on the imaginary axis. Thus the preceding theorem leads to a partition of the subset of the linear vectorfields in $\chi(M)$ into structurally stable types. For instance, the planar linear systems discussed earlier, for which

$$A = \begin{bmatrix} a & b \\ c & d \end{bmatrix}, \tag{B.17}$$

can be partitioned into three stable types. The eigenvalues of A are completely determined by trace $A = a + d$ and det $A = ad - bc$, and the three types are as shown in Figure B.3. Notice that the classification shown in Figure B.3 is coarser than that shown in Figure B.2, in that stable spirals and nodes are lumped together into sinks (they are topologically equivalent), and unstable spirals and nodes are lumped together into sources. The transitions between these types will be discussed later.

Since nonlinear systems can have multiple rest points and other more complicated attractors, one cannot hope for a global result like Theorem 1 for such systems. However, if

$$\frac{dx}{dt} = Ax + F(x),$$

where $F(x) \sim o(\|x\|)$ for $x \to 0$, the nonlinear term is a small perturbation near zero. Consequently, if the associated linear system is structurally stable the phase portrait of the nonlinear system should be essentially the same as that for the linear system, and this equivalence is made precise by the Hartman-Grobman theorem.

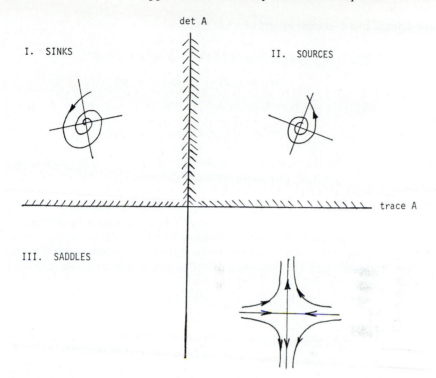

Figure B.3. The partition of the *trace-determinant* plane into regions in which the topological type of the corresponding linear two-dimensional system is constant.

Theorem 2 Let x and ξ satisfy

$$\frac{dx}{dt} = Ax + F(x)$$
$$\frac{d\xi}{dt} = A\xi$$

where F is C^1 on $B_\delta(0)$ (the ball of radius zero around the origin), and $F(0) = D_x F(0) = 0$. Suppose that A has no eigenvalues with zero real part. Let $\phi(x_0, t)$ and $\Psi(\xi_0, t) = e^{At}\xi_0$ be the associated flows, and let $N_0 = B_\delta(0)$ and M_0 denote neighborhoods of $x = 0$ and $\xi = 0$, respectively. Then ϕ restricted to $N_0 \sim_T \Psi$ restricted to M_0.

For a proof see Hartman (1982).

As in linear systems, one can define stable and unstable manifolds near a rest point of a nonlinear system. If x^s is a rest point of $\dot{x} = f(x)$, x^s is said to be hyperbolic if the linear flow generated by the Jacobian matrix $D_x f(x^s)$ is hyperbolic. The local stable and unstable manifolds for a hyperbolic rest point are defined in a δ-neighborhood of x^s as follows.

$$
\begin{aligned}
W_{loc}^s(x^s) &= \{x \in B_\delta(x^s) \mid \phi(x,t) \in B_\delta(x^s), t \in R^+ \\
&\quad \text{and } \phi(x,t) \to x^s \text{ as } t \to \infty\}, \\
W_{loc}^u(x^s) &= \{x \in B_\delta(x^s) \mid \phi(x,t) \in B_\delta(x^s), t \in R^- \quad\quad \text{(B.18)} \\
&\quad \text{and } \phi(x,t) \to x^s \text{ as } t \to -\infty\}.
\end{aligned}
$$

It follows from the Hartman-Grobman theorem that if x^s is a hyperbolic rest point of a C^2 vector field, then there are local stable and unstable manifolds at x^s of dimension n^- and n^+, respectively. The theorem only shows that these manifolds are C^0, but in fact they are as smooth as the underlying flow.

In linear systems the invariant manifolds associated with the spectral decomposition are disjoint invariant subspaces and as such are defined globally. For a nonlinear system a globally defined stable manifold can be obtained by taking the union of backward images of points in a local stable manifold. Thus

$$
W^s(x^s) \equiv \bigcup_{t \leq 0} \phi\left(W_{loc}^s(x^s), t\right)
$$

for any local stable manifold at x^s. Similarly, the global unstable manifold is defined as the union of forward images of points in a local unstable manifold:

$$
W^u(x^s) \equiv \bigcup_{t \geq 0} \phi\left(W_{loc}^u(x^s), t\right).
$$

Our standard assumption is that solutions of $\dot{x} = f(x)$ exist and are unique, and therefore the stable manifolds of distinct rest points are disjoint, and similarly for the unstable manifolds. However, as the system studied by Sherman in Chapter 10 shows, the stable and unstable manifolds may intersect. In that example, the homoclinic orbit is at once the stable and unstable manifold of the origin.

B.4   BIFURCATION ANALYSIS

Now suppose that we have a parameterized family of vector fields on M. This means that we have a smooth map $\tau : P \subseteq R^m \to \chi(M)$. Let $\sum$ denote the set of structurally stable elements in $\chi(M)$; then $\mathbf{B} \equiv \chi(M) \backslash \sum$ is the set of vector fields that are structurally unstable under the relation of topological equivalence. By looking at the set $B \equiv \tau^{-1}(\mathbf{B})$ we locate parameter values for which the corresponding vector fields are unstable. We call B the **bifurcation set** and any $p \in B$ a **bifurcation value** of p. Any $p \in P \backslash B$ is called a **regular value** of p, and these have the property that $X_p \in \sum$ whenever $p \in P \backslash B$. If we allow p to vary along some curve C in P, then, as long as C does not intersect B, the phase portraits remain topologically equivalent, but they change whenever C crosses B. Of course, a given parameterization may not lead to all possible types of changes in the phase portrait, but B can still be very complicated, particularly if the number of parameters is large.

Example 1 Consider the planar linear systems with A given by (B.17). There are four distinct kinds of transitions between the three structurally stable classes.

(i) **I or II → III**   with trace $A \neq 0$,

(ii) **I → II**   with det $A \neq 0$,

(iii) **I → II**   with det $A = 0$,

(iv) **I or II → III**   with trace $A = 0$.

The first two bifurcations are called codimension-1 bifurcations because the subsets of B on which these transitions occur are submanifolds of R^4 of codimension 1, *i.e.*, locally they look like three-dimensional linear manifolds. The third and fourth transitions are codimension-2 transitions. It is clear that the first two are more likely to occur than the latter two in a vague sense, because it is easier to choose α, β, γ, and δ so as to make either trace A or det A vanish than it is to make both vanish simultaneously.

B.4.1   Static bifurcations

Some bifurcations in parameterized vector fields are local in the sense that the qualitative change in the phase portrait is localized in a neighborhood of a point x^* whenever p is close to the corresponding bifurcation value p^*. In these cases we call (x^*, p^*) a **bifurcation point** in $R^n \times R^m$. The simplest of these correspond to the appearance or disappearance of rest points or periodic solutions near (x^*, p^*). The former case, which occurs when a real eigenvalue passes through zero, is dealt with here; the latter case, which occurs when a pair of complex conjugate eigenvalues cross the imaginary axis (Re$\lambda = 0$), is the subject of Section B.4.2. In both cases we suppose that $m = 1$, since one parameter is sufficient for a codimension-1 bifurcation, and in both cases we suppose that there is a known rest point or steady state $x^s(p)$ for all values of p in some open interval I. That is, we suppose that there is a function $x^s(p)$ (not necessarily unique) such that $f(x^s(p), p) = 0$ for $p \in I$.

Without loss of generality we assume that $0 \in I$ and we let $\mathcal{N}(\cdot)$ denote the null space of a linear transformation and let '*' denote the adjoint. Then we can state the following result.

Theorem 3 Suppose that (B.13) can be written

$$\frac{dx}{dt} = L_0 x + p L_1(p)x + Q(u, p) + C(u, p) + \cdots, \qquad \text{(B.19)}$$

where $L_0 = Df_x(x^s(0), 0)$ is the Jacobian of f at $(x^s(0), 0)$, $L_1(p)$ is an $n \times n$ matrix, and Q and C are homogeneous polynomials in x of degree two and three respectively for fixed p. Let zero be a geometrically simple eigenvalue of L_0, and let $\eta(\eta^*)$ be the associated eigenvector of L_0 (resp. L_0^*). If

$$\langle \eta^*, L_1 \eta \rangle \neq 0 \qquad \text{(B.20)}$$

then $(0,0)$ is a bifurcation point of X_p, and in any sufficiently small neighborhood of $(0,0)$ there is a unique curve of nontrivial solutions $(x(s), p(s))$ parameterized by s with the property that $(x^s(s), p(s)) \to (0, 0)$ as $s \to 0$. This condition is called a **nondegeneracy condition**, and it implies that the eigenvalue of $L_0 + pL_1(p)$ that vanishes at $p = 0$ crosses $\mathrm{Re}\lambda = 0$ with a nonzero slope there. The foregoing gives no information on the solution structure near $p = 0$; for this one must compute $\langle \eta^*, \mathcal{Q}(\eta, 0)\rangle$, and if this vanishes, $\langle \eta^*, \mathcal{C}(\eta, 0)\rangle$, and so on. If $\langle \eta^*, \mathcal{Q}(\eta, 0)\rangle \neq 0$ then, solutions exist in a full neighborhood of $p = 0$, while if this functional vanishes but $\langle \eta^*, \mathcal{C}(\eta, 0)\rangle \neq 0$, then solutions exist (locally) either for $P > 0$ or for $P < 0$ but not for both. In this case one says that the bifurcation is **one-sided**. This pattern is typical: a nondegeneracy condition on the linear part of the operator determines whether or not bifurcation occurs, and a functional of the nonlinear terms determines the details of the soution structure near the bifurcation point.

An analysis of the spectrum of the Jacobian along the nontrivial branch of solutions shows that when the steady state changes stability at $p = 0$, there is an **exchange of stability** between the trivial solution and the bifurcating solution at $p = 0$. Locally the typical types of phase portraits for a planar system are as shown in Figure B.4. One sees there that all the changes in the flow occur along the x-axis. In fact this is true, within a coordinate change, even in an infinite-dimensional system, as long as the eigenvalue that passes through zero is simple in the appropriate sense. This is a consequence of the center manifold theorem, one particular version of which is quoted in the following subsection. At the level of the linearized equation the two cases shown in Figure B.4 are identical, and both correspond to a change in sign of the determinant of the Jacobian (Type I or II in Example 1 for a planar system ; see Figures B.2 and B.3). The canonical example on R^1 for (a) is $\dot{x} = x(p-x)$ and

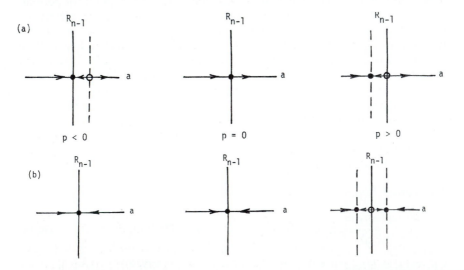

Figure B.4. Changes in the local phase portrait at a bifurcation from a known solution. (From Othmer (1985).) Here is a measure of the amplitude of the unstable mode.

for (b) is $\dot{x} = x(p - x^2)$. The corresponding bifurcation diagrams, which are simply plots of some functional of the solution versus the parameter, is shown in Figure B.5. In many problems $f(x, p) = 0$ at a point (x_0, p_0) but not in

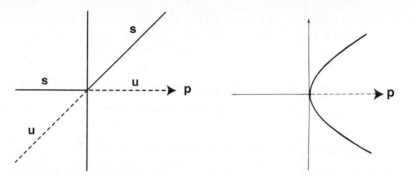

Figure B.5. The bifurcation diagrams for the bifurcations shown in Figure B.4.

every neighborhood of this point. In particular, there may be no reference solution like zero in the foregoing, and if (x_0, p_0) is a bifurcation point, there is an abrupt appearance or disappearance of a pair of solutions. In that case one can use the following result.

Theorem 4 Suppose that $f(x_0, p_0) = 0$ and that the Jacobian $Df_x(x_0, p_0)$ has rank $n - 1$. Further, suppose that the matrices

$$M_0 \equiv \left[Df_x^{(1)}, \dots, Df_x^{(n-1)}, f_p \right] \tag{B.21}$$

and

$$M_1 \equiv \left[f_x^{(0)}, \dots, Df_x^{(n-1)}, Df_{xx}\eta\eta \right],$$

in which $Df_x^{(i)}$ is the ith column of Df_x and all derivatives are evaluated at (x_0, p_0), have rank n. Then there is a unique curve of solutions passing through (x_0, p_0), and these solutions exist only for $p > p_0$ or $p < p_0$, according as $\det M_1 / \det M - 0$ is less than or greater than zero.

This type of bifurcation leads to the sudden appearance of two rest points and to the local bifurcation diagram shown in Figure B.6. It is often called a finite-amplitude bifurcation, to distinguish it from the zero-amplitude bifurcations shown in Figure B.5. The local changes in the phase portrait that occur at this bifurcation are shown in Figure B.7, wherein it is assumed that $p_0 = 0$.

B.4.2 Bifurcation of time-dependent solutions

In this section we discuss the other common codimension-1 bifurcation, that in which a pair of complex conjugate eigenvalues crosses from the left-half plane to the right-half plane. This leads to the appearance of solutions that are periodic in time, *i. e.*, to temporal oscillations of the sort discussed in numerous chapters in the book. In order to motivate the analytical statement of the result that covers this situation, we first give a heuristic discussion of a problem that can be solved analytically.

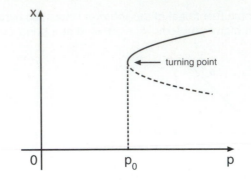

Figure B.6. The bifurcation diagram near a turning point.

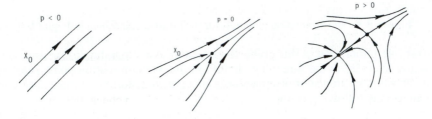

Figure B.7. The changes in the phase portrait at a turning point in the bifurcation diagram. X_0 is not a rest point for $p < 0$.

B.4.2.1   Heuristics on bifurcation of periodic solutions

Consider the planar system

$$\frac{dx}{dt} = \alpha x + \omega y - (x + \gamma y)(x^2 + y^2),$$

$$\frac{dy}{dt} = -\omega x + \alpha y + (\gamma x - y)(x^2 + y^2),$$

(B.22)

where α, ω and γ, are constants. In polar coordinates the system becomes

$$\frac{dr}{dt} = r(\alpha - r^2),$$

$$\frac{d\theta}{dt} = -\omega + \gamma r^2.$$

(B.23)

The reader should note the similarity (and differences) between these equations and the Poincaré oscillator discussed in Glass's chapter (Chapter 12). In the Poincaré oscillator, the righthand side of the r equation is $r(\alpha - r)$, but as we shall see later, in a general f the form given above is the normal form. In addition, the θ equation in (B.23) has an r-dependent twist, and as the solution below shows, the isochrons for this system are not radial lines, as in the Poincaré oscillator.

The solutions of these equations are

$$
r^2(t) =
\begin{cases}
\dfrac{\alpha r_0^2}{r_0^2 + (\alpha - r_0^2)e^{-2\alpha t}} & \alpha \neq 0, \\[4mm]
\dfrac{r_0^2}{1 + 2r_0^2 t} & \alpha = 0,
\end{cases}
$$

$$
\theta(t) =
\begin{cases}
\theta_0 + (\gamma\alpha - \omega)t + \dfrac{\gamma}{2}\ln \dfrac{r_0^2 + (\alpha - r_0^2)e^{-2\alpha t}}{\alpha} & \alpha \neq 0, \\[4mm]
\theta_0 - \omega t + \dfrac{\gamma}{2}\ln(1 + 2r_0^2 t) & \alpha = 0,
\end{cases}
$$

where $r_0 = r(0)$ and $\theta_0 = \theta(0)$. The point $(x, y, \alpha) = (0, 0, 0)$ is a bifurcation point, and an attracting invariant circle bifurcates from the origin as α passes through zero. In fact, the orbit $\Gamma = \{\gamma(t) \equiv (\sqrt{\alpha}, -\omega t)|t \in R\}$ has a stronger stability property known as **asymptotic orbital stability with asymptotic phase**. This means that Γ is asymptotically stable in the sense used earlier, and that there is a $c \geq 0$ such that $\lim_{t \to \infty} \|\phi_t(x_0, y_0) - \gamma(t + c)\| = 0$. That is, the solution through any point in a neighborhood of Γ approaches some translate of γ by the asymptotic phase c. The reader can check that c is given by $-[\theta_0 + \gamma \ln(r_0^2/\alpha)/2]/\omega$ in the previous example, and this reduces to $-\theta_0$ when the amplitude-dependent term in the phase equation vanishes. In that case the isochrons reduce to the radial lines $\theta = \theta_0$. Note that this equation provides an example of a one-sided bifurcation, in that the periodic solution exists only for $\alpha > 0$.

The question arises as to how typical this example is when the loss of stability is via a conjugate pair of eigenvalues that crosses the imaginary axis. This can be answered for planar systems by converting a general equation to normal form. Higher-dimensional systems are first reduced to a center manifold and then converted to a normal form. We sketch the analysis for ordinary differential equations, but the same approach works for certain classes of partial differential equations as well.

Suppose that a smooth planar system with a rest point at u^s is written[1]

$$
\frac{du}{dt} = K(p)u + Q(u, p) + C(u, p) + \cdots , \tag{B.24}
$$

where $K(p) \equiv Df_u(u^s(p), p)$ is the Jacobian of f at $(u^s(p), p)$. Suppose that for p near $p-0$, $K(p)$ has complex-conjugate eigenvalues $\lambda_{\pm}(p) = \alpha(p) \pm i\omega(p)$, where $\alpha(p_0) = 0$ and $\omega(p_0) \neq 0$. Let $\eta = \eta_1 + i\eta_2$ be the eigenvector of $K(p)$ associated with λ_+. Then the coordinate change $u = P\eta$, where P is a matrix whose columns are η_2 and η_1, converts the linear part of (B.24) into the canonical form

$$
P^{-1}KP =
\begin{bmatrix}
\alpha & -\omega \\
\omega & \alpha
\end{bmatrix}. \tag{B.25}
$$

---

[1] We first write the equation in terms of u so that after the coordinate transformation (B.26) it is written in terms of x.

To put the nonlinear terms into a "normal form" that depends continuously on p for p near p_0, define the change of variables

$$\eta = x + \sum_{k=2}^{2m+1} \Gamma_k(x), \tag{B.26}$$

where Γ_k is a homogeneous polynomial of degree k in the x. Since this transformation is the identity at lowest order, it preserves the canonical form of the linear terms. It can be shown (e.g., Shafranov, (1974)) that the Γ_k's can be chosen so that (i) the resulting equations in x have no terms of even order, and (ii) the odd terms are of the form $g_{2l+1}x_1(x_1^2 + x_2^2)^l - G_{2l+1}x_2(x_1^2 + x_2^2)^l$ in the first equation and $g_{2l+1}x_2(x_1^2 + x_2^2)^l + G_{2l+1}x_1(x_1^2 + x_2^2)^l$ in the second equation. The coefficients g_k and G_k are called the Lyapunov numbers of the system. Thus (B.24) can be reduced to the form

$$\frac{dx_1}{dt} = \alpha x_1 + \omega x_2 + (g_3 x_1 - G_3 x_2)\left(x_1^2 + x_2^2\right) + \mathcal{O}\left(\|x\|^4\right),$$

$$\tag{B.27}$$

$$\frac{dx_2}{dt} = -\omega x_1 + \alpha x_2 + (G_3 x_1 + g_3 x_2)\left(x_1^2 + x_2^2\right) + \mathcal{O}\left(\|x\|^4\right),$$

or in polar form

$$\dot{r} = r\left[\alpha + g_3 r^2 + \cdots\right] + \mathcal{O}(r^4, \theta),$$

$$\tag{B.28}$$

$$\dot{\theta} = -\omega + G_3 r^2 + \mathcal{O}(r^4, \theta).$$

Equations (B.27) and (B.28) are normal forms to fourth order for a planar system with complex-conjugate eigenvalues. Clearly (B.22) and (B.23) are special cases of (B.27) and (B.28). If g_3 is nonzero, then complete information on the existence and stability of periodic solutions that bifurcate at $(x, y, p) = (0, 0, p_0)$ is contained in the truncated normal form that one obtains by dropping the higher-order terms. In particular, if we write $\alpha(p) = \alpha'(p_0)p + \cdots$ where $\alpha'(p_0) \neq 0$, then it follows from (B.28) that there are four possible cases, depending on the signs of $\alpha'(p_0)$ and g_3. For instance, if $\alpha'(p_0) > 0$ and $g_3 > 0$ the solutions exist for $p < p_0$ and are unstable, while if $\alpha'(p_0) > 0$ and $g_3 < 0$ they exist for $p > p_0$ and are stable. In all cases there is an exchange of stability at the bifurcation point.

In view of what happens in linear systems, it can be expected that the essential behavior in an n-dimensional nonlinear system that has one complex-conjugate pair of eigenvalues whose real part changes sign at $p = 0$, and $n - 2$ eigenvalues with negative real part, is governed by a two-dimensional subsystem of the full system. The sense in which this is true can be made precise by first splitting the equations into 2- and $(n-2)$-dimensional systems that agree at lowest order with the splitting associated with the linear terms. This is accomplished via the center-manifold theorem. The version given here (and a proof of it) can be found in Othmer (1993) and in references given there.

We begin with a definition of a center manifold. Suppose that we write (B.24) in the form

$$\frac{du}{dt} = Au + F(u) \tag{B.29}$$

where $A = K(0)$, $F \in C^k$, $k \geq 1$, and $F(0) = D_u F(0) = 0$. We suppress the parameter dependence for the present but will reintroduce it later. Suppose that $\sigma(A) = \sigma^-(A) \cup \sigma^0(A)$ (*i.e.*, there are no eigenvalues with positive real parts), and that there are n^0 eigenvalues with zero real parts.

Definition 1 A center manifold for (B.29) at $u = 0$ is an n^0–dimensional manifold $W^c(0)$ with the properties that:

(i) (Invariance) If $u(0) \in W^c(0)$ then $u(t) \in W^c(0)$ $\forall t \in R$.

(ii) The tangent space to $W^c(0)$ at $u = 0$ is the range of P^0, where P^0 is the projection onto the generalized eigenspace of A corresponding to $\sigma^0(A)$.

The **local center manifold theorem** is the following.

Theorem 5 Consider the system

$$\frac{du}{dt} = Au + F(u), \tag{B.30}$$

where $\sigma^+(A) = \emptyset$, $\sigma^0(A) \neq \emptyset$, and $\sigma^-(A) \neq \emptyset$. Let n^0 be the dimension of the generalized eigenspace associated with $\sigma^0(A)$. Further, suppose that $F \in C^k$, $k \geq 1$, and that $F(0) = D_u F(0) = 0$. For a sufficiently small $\delta > 0$ there exists a Lipschitz continuous function $h : R^{n^0} \to R^{n^-}$ such that the graph of h in a ball of radius δ centered at the origin is an n^0– dimensional manifold that is locally invariant for (B.30) and tangent to the center subspace of A at the origin.

Hereafter "center manifold" will refer to any local center manifold unless we specifically note otherwise. Note that this result applies equally well if the eigenvalues with zero real parts are real or complex, and as a result, can be used for the static bifurcation discussed earlier, where $n^0 = 1$.

To apply this to the n-dimensional version of (B.24), consider the augmented system

$$\frac{du}{dt} = K(0)u + F(u, p) \tag{B.31}$$

$$\frac{dp}{dt} = 0, \tag{B.32}$$

where $K_0 \equiv K(0)$ and $F(u, p)$ contains all higher-order terms. Suppose that the spectrum of $K(0)$, $\sigma(K(0))$ can be written $\sigma = \sigma_1 \cup \sigma_2$, where $\sigma_1 = \{\lambda \in \sigma(K(0)) | \text{Re}\lambda = 0\}$ and $\sigma_2 = \{\lambda \in \sigma(K(0)) | \text{Re}\lambda \leq -\varepsilon, \varepsilon > 0\}$. The spectrum of the linear part of (B.31) and (B.32) is $\sigma \cup \{0\}$ and R^{n+1} can be split as the direct sum $R^0 \oplus R^-$, where the eigenvalues of $K | R^0$ have zero real part. Then the center manifold theorem guarantees the existence of an invariant manifold in R^{n+1} tangent to the center subspace R^0 at the origin. A constant p slice of this is not a center manifold except at $p = 0$, but it serves equally well. Equation (B.32) guarantees that the p=constant slice is invariant.

If (B.31) is split in accordance with the spectral splitting, we can write

$$\frac{dx}{dt} = K_0 x + F_1(x, y, p),$$

$$\frac{dy}{dt} = K_1 y + F_2(x, y, p),$$

(B.33)

where $\sigma(K_0) = \sigma_1$ and $\sigma(K_1) = \sigma_2$. The invariant manifold can be written in the form $y = \Psi(x, p)$, where Ψ is gotten from the second equation. The first equation then reduces to

$$\frac{dx}{dt} = K_0 x + F_1(x, \Psi(x, p), p).$$

(B.34)

In the case where σ_1 consists of a pair of pure imaginary eigenvalues this manifold is two dimensional, and (B.34) can be put into a normal form similar to (B.27). It can also be shown that for p sufficiently small the stability of a bifurcating periodic solution on this manifold implies its stability as a solution of the full system, and similarly for instability. Thus the essential information about bifurcation and stability can be gotten from a two-dimensional system that describes the flow on the p = constant slice of the center manifold. This is illustrated in Figure B.8, where we show the changes in the flow near the origin as the real part α of the critical eigenvalues changes sign as p passes through zero.

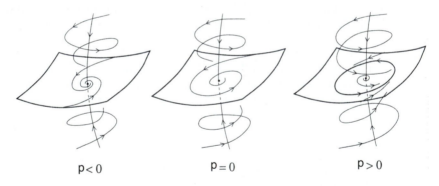

p < 0   p = 0   p > 0

Figure B.8. The local flow in phase space in a neighborhood of zero.

B.4.3   Codimension-2 bifurcations

We now have a fairly complete picture as to how the phase portrait changes when a single real eigenvalue crosses through zero, or a pair of complex-conjugate eigenvalues cross the imaginary axis. If the zero eigenvalue or the real part of the complex pair crosses $\text{Re}\lambda = 0$ with nonzero speed ($\partial\lambda/\partial p \neq 0$ or $\partial\alpha/\partial p \neq 0$), then bifurcation occurs, and whether the bifurcation is one- or two-sided is determined by certain functionals of the nonlinear terms. Thus cases (i) and (ii) in Example 1 in a planar can be understood from this analysis. As we remarked earlier, these are called codimension-1 bifurcations, essentially because one parameter suffices to describe them.

The remaining two cases in Example 1 are more complicated because they are more degenerate. By this we mean that in the space R^4 of all 2×2 matrices, the set on which both $\operatorname{tr} A$ and $\det A$ vanish simultaneously is obviously smaller than that on which either alone vanishes (how are these sets defined?). To appreciate that the solution structure may also be more complicated, one only has to realize that the preceding analysis shows that either periodic solutions or steady states different from $x^s(p)$ can exist in a neighborhood of the origin of the $\operatorname{tr} A - \det A$ plane. One can then imagine that more complicated behavior may result if these solutions interact in some way. A discussion of this is beyond the scope of this appendix, and we refer the reader to Kuznetsov (1995).

REFERENCES

Amann, Herbert. 1990. *Ordinary Differential Equations: An introduction to nonlinear analysis*. Berlin: de Gruyter. ISBN 0-89925-552-3, 3-11-011515-8. Pages xii + 458. Translated from the German by Gerhard Metzen. {*358, 379*}

Andronov, A. A., Leontovich, E. A., Gordon, I. I., and Maier, A. G. 1973. *Theory of Bifurcations of Dynamic Systems on a Plane*. New York: Wiley. ISBN 0-470-03194-8. Pages xiv + 482. {*358, 379*}

Arnold, V. I. (Vladimir Igorevich). 1973. *Ordinary Differential Equations*. Cambridge, MA: MIT Press. ISBN 0-262-01037-2. Pages ix + 280. Translated and edited by Richard A. Silverman. {*358, 368, 379*}

Arrowsmith, D. K. and Place, C. M. 1990. *An Introduction to Dynamical Systems*. Cambridge University Press. ISBN 0-521-31650-2 (paperback), 0-521-30362-1 (hardcover). Page 423. {*358, 379*}

Bhatia, N. P. (Nam Parshad) and Szegö, G. P. 1967. *Dynamical Systems: Stability Theory and Applications*. Lecture notes in mathematics, vol. 35. Berlin; New York: Springer-Verlag. Pages 6 + 416. {*358, 379*}

Boyce, W. E. and DiPrima, R. C. 1977. *Elementary Differential Equations and Boundary Value Problems*. New York: John Wiley and Sons. ISBN 0-471-09334-3. Page 652. {*358, 362, 363, 379*}

Devaney, Robert L. 1986. *An Introduction to Chaotic Dynamical Systems*. New York: W. Benjamin. ISBN 0-8053-1601-9. Pages xii + 320. {*358, 379*}

Guckenheimer, J. and Holmes, P. J. 1983. *Nonlinear Oscillations, Dynamical Systems and Bifurcations of Vector Fields*. Applied mathematical sciences, vol. 42. New York: Springer. ISBN 0-387-90819-6. Pages xii + 459. {*358, 379*}

Hale, Jack K. and Kocak, Huseyin. 1991. *Dynamics and Bifurcations*. New York: Springer-Verlag. ISBN 0-387-97141-6, 3-540-97141-6. Pages xiv + 568. {*358, 379*}

Hartman, Philip. 1982. *Ordinary Differential Equations*. Second edn. Boston: Birkhäuser. ISBN 3-7643-3068-6. Pages xiv + 612. {*358, 369, 379*}

Hirsch, M. W. and Smale, S. 1974. *Differential Equations, Dynamical Systems and Linear Algebra*. Pure and applied mathematics, vol. 60. Academic Press. ISBN 0-12-349550-4. Pages xi + 358. {*358, 379*}

Irwin, M. C. 1980. *Smooth Dynamical Systems*. Pure and applied mathematics, vol. 94. Academic Press. ISBN 0-12-374450-4. Pages x + 259. {*368, 379*}

Jackson, E. A. 1989. *Perspectives of Nonlinear Dynamics*. Cambridge: Cambridge University Press. ISBN 0-521-42632-4 (vol 1, paperback), 0-521-34504-9 (vol. 1), 0-521-42633-2 (vol. 2, paperback), 0-521-35458-7 (vol. 2). Two volumes. {*359, 364, 379*}

Jordan, D. W. and Smith, P. 1987. *Nonlinear Ordinary Differential Equations*. Second edn. Oxford: Clarendon Press. ISBN 0-19-859657-X, 0-19-859656-1 (paperback). Pages ix + 381. {*358, 365, 380*}

Kuznetsov, Y. A. 1995. *Elements of Applied Bifurcation Theory*. Springer-Verlag. {*379, 380*}

Murray, J. D. 1989. *Mathematical Biology*. Biomathematics, vol. 19. New York: Springer. ISBN 0-387-19460-6 (New York), 3-540-19460-6 (Berlin). Pages xiv + 767. {*363, 380*}

Odell, G. M. 1980. Qualitative theory of systems of ordinary differential equations, including phase plane analysis and the use of the hopf bifurcation theorem. *Pages 649–727 of:* Segel, Lee A. (ed), *Mathematical Models in Molecular and Cellular Biology*. Cambridge: Cambridge University Press. {*358, 360, 361, 363, 380*}

Othmer, H. G. 1985. The mathematical aspect of temporal oscillations in reacting systems. *Pages 7–54 of:* Field, Richard J. and Burger, Maria (eds), *Oscillations and Traveling Waves in Chemical Systems*. Wiley. {*372, 380*}

Othmer, H. G. 1993. *Techniques and Applications of Bifurcation Theory*. Lecture Notes, University of Utah. {*376, 380*}

Perko, L. 1991. *Differential Equations and Dynamical Systems*. Texts in applied mathematics, vol. 7. New York: Springer. ISBN 0-387-97443-1 (New York), 3-540-97443-1 (Berlin). Pages xii + 403. {*358, 380*}

C. AN INTRODUCTION TO PARTIAL DIFFERENTIAL EQUATIONS

Hans G. Othmer

C.1 INTRODUCTION

For a system of ordinary differential equations such as (B.1) the unknowns are n functions x_i that depend on time and the initial condition. However, in many problems the unknowns depend on one or more space variables as well as on time, and in this case we write $u_i = u_i(x, y, z, t) = u_i(\mathbf{x}, t)$ in case they are defined in three-space. Examples include the calcium distribution throughout a cell (Chapter 6), the states along a one-dimensional muscle fiber (Chapter 11), and the velocity and density of blood in a ventricle of the heart (Chapter 14). It is shown in these chapters how one derives an evolution equation for the application at hand, but in this appendix we outline a more general approach applicable to a wide variety of problems and give a very brief introduction to the solution of the resulting equations. Good introductory texts on partial differential equations are Zachmanoglou and Thoe (1986) and Zauderer (1989). Somewhat more advanced treatments are given in DiBenedetto (1995) and Gustafson (1980).

C.2 DERIVATION OF THE EQUATIONS OF CHANGE OR BALANCE EQUATIONS

For the ordinary differential equation

$$\frac{du}{dt} = f(u), \tag{C.1}$$

the righthand side is derived by considering all processes that contribute to the rate of change du/dt of u. The same principle holds when u is defined throughout a region in space, but the expression of this principle is more complex. To derive this we first recall the definitions of some operations on vectors and we state a mathematical result called the divergence theorem.

Let V be a volume in three-space with smooth boundary S, and let $\mathbf{v}$ be a vector field defined in V. By this we mean that at each point $\mathbf{x}$ in V we have the R^3-valued function $\mathbf{v}(\mathbf{x}) \equiv (v_1(\mathbf{x}), v_2\mathbf{x}, v_3(\mathbf{x}))$. Furthermore, let ∇ denote the vector differential operator whose Cartesian components are

$$\nabla = \left(\mathbf{i} \frac{\partial}{\partial x}, \mathbf{j} \frac{\partial}{\partial y}, \mathbf{k} \frac{\partial}{\partial z} \right) \tag{C.2}$$

where $\mathbf{i}, \mathbf{j}, \mathbf{k}$ are unit vectors in the x, y, z directions respectively. This operator can be used in a variety of ways. When it acts on a scalar field $u(\mathbf{x})$ it produces a vector called the **gradient of** u, which is given by

$$\nabla u = \left(\mathbf{i} \frac{\partial u}{\partial x}, \mathbf{j} \frac{\partial u}{\partial y}, \mathbf{k} \frac{\partial u}{\partial z} \right). \tag{C.3}$$

From the components of ∇u one sees that ∇u provides a measure of the rate of change of u in the three coordinate directions. What happens if ∇ operates on a vector? There are several possible outcomes, depending on how it operates. We can form a product, called the dot or scalar product, which is defined by

$$\nabla \cdot \mathbf{v} = \frac{\partial v_1}{\partial x} + \frac{\partial v_2}{\partial y} + \frac{\partial v_3}{\partial z}. \tag{C.4}$$

This is a scalar quantity called the **divergence of v**. We can also form a product defined as

$$\nabla \mathbf{v} = \mathbf{ii} \frac{\partial v_1}{\partial x} + \mathbf{ij} \frac{\partial v_2}{\partial x} + \cdots + \mathbf{kk} \frac{\partial v_3}{\partial z}. \tag{C.5}$$

This quantity is called a **second-rank tensor** and in general it has nine components in three space dimensions. This suggests that a second-rank tensor can be represented by a 3×3 matrix whose components are $\partial v_1/\partial x, \ldots, \partial v_3/\partial z$, and this can always be done in a Cartesian frame. The fluid stress defined in Chapter 14 and the conductivity tensors used in Chapter 13 provide examples of tensors.

The divergence theorem, which can be found in msot undergraduate calculus books, states that

$$\int_V \nabla \cdot \mathbf{v} \, dV = \int_S \mathbf{n} \cdot \mathbf{v} \, dS \tag{C.6}$$

or, in words, the volume integral of the divergence of $\mathbf{v}$ is equal to the surface integral of the scalar product of $\mathbf{v}$ with the unit outward normal $\mathbf{n}$. If $\mathbf{n} \cdot \mathbf{v}$ is positive at a point, then $\mathbf{v}$ points outward, and if this is true everywhere, then the lefthand side must also be positive. More generally, if the vector field $\mathbf{v}$ represents the flow of a quantity (examples are given later), then the surface integral is positive, so the volume integral must also be positive. The surface integral represents the net flow or flux (positive or negative) out of V, and thus the integral of the divergence represents the net change of $\mathbf{v}$ in the volume V. With this at hand we can now state a general form of a balance equation.

Let $\rho(\mathbf{x}, t)$ be the density of a quantity X, so defined that $\rho(\mathbf{x}, t) \Delta V$ is the amount of X in the volume ΔV. For example, if ρ is the mass, density then it can be measured as mass per unit volume. However, the quantity could also be the fluid momentum, as in Chapter 14. In any case, the total amount of the quantity in the volume V is $\int_V \rho(\mathbf{x}, t) \, dv$. Suppose that the quantity is created at the net rate $R(\rho)$ per unit volume. Then the equation which describes how the total amount changes is

$$\frac{d}{dt} \int_V \rho(\mathbf{x}, t) \, dv = -\int_S \mathbf{n} \cdot \mathbf{j} \, dS + \int_V R(\rho(\mathbf{x}, t)) \, dV. \qquad (C.7)$$

Here $\mathbf{j}$ is the flux of X, defined as the density of X per unit area per unit time. (Note that in all cases where ρ is the density of a physical quantity the units of each term in (C.7) must be the same. If $\rho(\mathbf{x}, t)$ is mass per unit volume, then the integral of ρ over the volume is the total mass, and the lefthand side of (C.6) has units of mass per unit time.) The first term has a negative sign, because if $\mathbf{n} \cdot \mathbf{j}$ is positive it represents an outflow from V, and therefore the total amount of X in V must decrease.

If we apply the divergence theorem (C.6), we obtain

$$\frac{d}{dt} \int_V \rho(\mathbf{x}, t) \, dv = -\int_V \nabla \cdot \mathbf{j} \, dV + \int_V R(\rho(\mathbf{x}, t)) \, dV \qquad (C.8)$$

If V is a fixed volume, we can rewrite this as

$$\int_V \left[\frac{\partial}{\partial t} \rho(\mathbf{x}, t) + \nabla \cdot \mathbf{j} - R(\rho) \right] dV = 0. \qquad (C.9)$$

This is the integral form of the balance equation for X.

If the spatial distribution of X is smooth and the statement of (C.9) holds for all smaller volumes contained in V, then the integrand must vanish and we obtain the differential form of the balance equation

$$\frac{\partial \rho}{\partial t} = -\nabla \cdot \mathbf{j} + R(\rho(\mathbf{x}, t)). \qquad (C.10)$$

If there is no creation or destruction of X, and if $\mathbf{j} = 0$ on S, then (C.8) reduces to

$$\frac{d}{dt} \int_V \rho(\mathbf{x}, t) \, dV = 0, \qquad (C.11)$$

which states that the total amount of X in V does not change, i.e., X is conserved, and in this case we call

$$\frac{\partial \rho}{\partial t} = -\nabla \cdot \mathbf{j} \qquad (C.12)$$

a conservation equation.

In order to obtain an equation whose only unknown is ρ we have to postulate a relationship between the density ρ and the flux $\mathbf{j}$. This is the point at which the physical and biological details of the process become important. Suppose, for example, that ρ is the density of a dye in a flowing stream of water. Let $\mathbf{u}$ be the local velocity of the water; then as a first approximation we could use the relationship

$$\mathbf{j} = \rho \mathbf{u} \qquad (C.13)$$

to represent the flux of dye. This relation is called a **constitutive relation** since it relates quantities such as the flux to the corresponding density of the quantity. Equation (C.13) simply states that the flux of dye is proportional to the local velocity; the dye is convected along at the local fluid speed. This relationship can't be completely accurate, because we know from experience that other processes contribute to the movement of dye. For example, (C.13) would state that if $\mathbf{u} \equiv 0$, then a small blob of dye injected into a quiescent fluid would always remain a small blob, but we know this isn't true (put a drop of ink into a beaker of water: what happens?). Thus (C.13) should really be interpreted to mean that the **dominant** mechanism of dye transport when $\mathbf{u} \neq 0$ is convection by the fluid.

If $R(\rho) \equiv 0$ and we use (C.13) in (C.12), we obtain

$$\frac{\partial \rho}{\partial t} + \nabla \cdot \rho \mathbf{u} = 0, \tag{C.14}$$

which is called the continuity equation. If we expand the second term we obtain

$$\frac{\partial \rho}{\partial t} + \mathbf{u} \cdot \nabla \rho + \rho \nabla \cdot \mathbf{u} = 0 \tag{C.15}$$

and if $\nabla \cdot \mathbf{u} = 0$ throughout a volume V, as in Chapter 14 this reduces to

$$\frac{\partial \rho}{\partial t} + \mathbf{u} \cdot \nabla \rho = 0. \tag{C.16}$$

Finally, if the density changes significantly only in the x-direction, then this further reduces to

$$\frac{\partial \rho}{\partial t} + u \frac{\partial \rho}{\partial x} = 0. \tag{C.17}$$

What does the condition on $\nabla \cdot \mathbf{u}$ mean physically? Can you think of an experimental configuration for which the dye density changes in only one space dimension? What sort of approximations are needed to justify this description?

How do we solve (C.17) when u is identically constant? Recall that (C.17) is the balance equation and states that the local rate of change $\partial \rho / \partial t$, is balanced by the net influx or outflux represented by $u \partial \rho / \partial x$. This suggests that if we move at the fluid velocity we'll see no change. To accomplish this define $\xi = x - ut$; then $\rho(x, t) \to P(\xi)$ and the equation satisfied by P is gotten by noting that

$$\frac{\partial \rho}{\partial x} = \frac{\partial P}{\partial \xi} \frac{\partial \xi}{\partial x} = \frac{\partial P}{\partial \xi} \tag{C.18}$$

and

$$\frac{\partial \rho}{\partial t} = \frac{\partial P}{\partial \xi} \frac{\partial \xi}{\partial t} = -u \frac{\partial P}{\partial \xi}.$$

Therefore (C.17) is satisfied whenever $\rho(x, t)$ is a function only of the combination $x - ut$. In other words, the general solution of (C.17) is

$$\rho(x, t) = F(x - ut) \tag{C.19}$$

where F is an arbitrary function. How do we determine a particular solution? If we specify the distribution of X in space at $t = 0$, we specify a function $\rho(x, 0)$, and the solution of (C.17) which satisfies this initial condition is

$$\rho(x, t) = \rho_o(x - ut). \tag{C.20}$$

The reader should plot the distribution of ρ in the $x - t$-plane for a given initial distribution, say $\rho_0(x) = e^{-x^2}$.

Equation (C.17) is called a first-order partial differential equation because it confirms only first derivatives of the unknown function ρ. If the physical process being modeled is diffusion, then the flux can be described by the constitutive relation

$$\mathbf{j} = -D\,\nabla\rho, \tag{C.21}$$

where D is the **diffusion constant**. (What are the units of D?) This relation states that X moves down the gradient of its density: does this make sense in the context of the ink-drop experiment described above? If we use (C.21) in (C.10), we obtain

$$\frac{\partial\rho}{\partial t} = \nabla \cdot D\nabla\rho + R(\rho) \tag{C.22}$$

and if D is a constant, this reduces to

$$\frac{\partial\rho}{\partial t} = D\nabla^2\rho + R(\rho). \tag{C.23}$$

Here ∇^2 is the Laplace operator, which is given by

$$\nabla^2\rho = \frac{\partial^2\rho}{\partial x^2} + \frac{\partial^2\rho}{\partial y^2} + \frac{\partial^2\rho}{\partial z^2}.$$

in Cartesian coordinates. In one space dimension (C.23) reduces to

$$\frac{\partial\rho}{\partial t} = D\frac{\partial^2\rho}{\partial x^2} + R(\rho). \tag{C.24}$$

This is called a reaction diffusion equation when X is the density of a chemical substance. In this case $R(\rho)$ represents the rate of change of X due to reaction at the point x in space. When ρ is the energy density, one can often write $\rho = C_p T$ where C_p is a heat capacity and T is the temperature. In this case (C.24) represents the balance of energy in V, and it shows how the temperature evolves in time. What does $R(\rho)$ represent in this case?

To complete the specification of the solution of (C.24) we have to specify an initial distribution of ρ, which is given by

$$\rho(\mathbf{x}, 0) = \rho_0(\mathbf{x})$$

and we have to specify conditions on the boundary of the region V. For example, we could stipulate the there is no outflow of X across the boundary. This is expressed mathematically by the condition that

$$\mathbf{n} \cdot D\nabla\rho = 0$$

on the surface S. Can you see why this is the appropriate relation?

When the function $R(\rho)$ is a nonlinear function, it is usually impossible to solve (C.24) analytically. However, if $R(\rho)$ is a linear function, an analytical solution for simple geometries of V is possible via a technique called separation of variables. An example of how this technique is used is given in Chapter 6.

REFERENCES

DiBenedetto, Emmanuele. 1995. *Partial Differential Equations*. Boston: Birkhäuser-Verlag. ISBN 0-8176-3708-7 (Boston), 3-7643-3708-7 (Basel). Pages xiv + 416. {*381, 386*}

Gustafson, Karl E. 1980. *Partial differential equations and Hilbert space methods*. New York: Wiley. ISBN 0-471-04089-4. Pages xv + 270. {*381, 386*}

Zachmanoglou, E. C. and Thoe, Dale W. 1986. *Introduction to Partial Differential Equations with Applications*. Dover. ISBN 0-486-65251-3. Pages x + 405. {*381, 386*}

Zauderer, Erich. 1989. *Partial differential equations of applied mathematics*. Second edn. New York: John Wiley and Sons. ISBN 0-471-61298-7. Pages xiii + 891. {*381, 386*}

Author/Editor Index

Subject Index

List of Contributors

Frederick R. Adler
Department of Mathematics and
Department of Biology
University of Utah
Salt Lake City, UT 84112
USA
Email: adler@math.utah.edu
WWW URL: http://www.math.utah.edu/~adler

Kathy Chen
Dept. of Biology
Virginia Polytechnic Institute and State University
Blacksburg, VA 24061
USA
Email: kchen@mail.vt.edu

Odo Diekmann
Vakgroep Wiskunde
Postbus 80.010
3508 TA Utrecht
The Netherlands
Email: O.Diekmann@math.ruu.nl

Stephen P. Ellner
Biomathematics Graduate Program
Department of Statistics
North Carolina State University
Raleigh, NC 27695-8203
USA
Email: ellner@stat.ncsu.edu
WWW URL: http://www.stat.ncsu.edu/ellner.html

Jennifer Foss
Department of Neurology, MC 2030
University of Chicago Hospitals
5841 South Maryland Ave.
Chicago, IL 60637
USA
Email: foss@ace.bsd.uchicago.edu

Leon Glass
Center for Nonlinear Dynamics in Physiology and Medicine
McGill University
3655 Drummond Room 1124
Montreal, Quebec H3G 1Y6
Canada
Email: glass@cnd.mcgill.ca
WWW URL: http://www.cnd.mcgill.ca/bios/glass/glass.html

Richard Gomulkiewicz
Department of Pure and Applied Mathematics and
Department of Genetics and Cell Biology
Washington State University
Pullman, WA 99164
USA
Email: gomulki@kuhub.cc.ukans.edu

W. S. C. Gurney
Department of Statistics and Modelling Science
University of Strathclyde
Glasgow, G11XH
Scotland
Email: bill@stams.strathcly

Nick A. Hill
Department of Applied Mathematics
University of Leeds
Leeds LS2 9JT
England
Email: nah@amsta.leeds.ac.uk

Robert D. Holt
Department of Systematics and Ecology and
Museum of Natural History
Dyche Hall
University of Kansas
Lawrence, KS 66045
USA

Wanda Krassowska
Department of Biomedical Engineering and
Duke-North Carolina NSF/ERC
Duke University
Durham, NC 27708-0329
USA
Email: wanda@eel-mail.mc.duke.edu

Mark A. Lewis
Department of Mathematics and
Department of Biology
University of Utah
Salt Lake City, UT 84112
USA
Email: mlewis@math.utah.edu
WWW URL: http://www.math.utah.edu/~mlewis

Michael C. Mackey
Centre for Nonlinear Dynamics in Physiology and Medicine
Departments of Physiology, Physics and Mathematics
McGill University
Montreal, Quebec
Canada
Email: mackey@cnd.mcgill.ca
WWW URL: http://www.cnd.mcgill.ca/bios/mackey/mackey.html

E. McCauley
Department of Biology
University of Calgary
Calgary, AB T2N 1N4
Canada
Email: mccauley@acs.ucalgary.ca
WWW URL: http://www.ucalgary.ca/~biology/ecology/mccauley.html

David M. McQueen
Courant Institute of Mathematical Sciences
New York University
251 Mercer Street
New York, NY 10012-1110
USA
Email: mcqueen@cims.nyu.edu

John Milton
Department of Neurology, MC 2030
University of Chicago Hospitals
5841 South Maryland Ave.
Chicago, IL 60637
USA
Email: sp1ace@ace.bsd.uchicago.edu

W. W. Murdoch
Department of Ecology, Evolution and Marine Biology
University of California
Santa Barbara, CA 93106
USA
Email: murdoch@lifesci.lscf.ucsb.edu
WWW URL: http://lifesci.ucsb.edu/EEMB/faculty/murdoch/index.html

Roger M. Nisbet
Department of Ecology, Evolution and Marine Biology
University of California
Santa Barbara, CA 93106
USA
Email: nisbet@lifesci.ucsb.edu, nisbet@islay.lscf.ucsb.edu

Bela Novak
Dept. of Agricultural Chemical Technology
Technical University of Budapest
1521 Budapest Gellert ter 4
Hungary

Hans G. Othmer
Department of Mathematics
University of Utah
Salt Lake City, UT 84112
USA
Email: othmer@math.utah.edu
WWW URL: http://www.math.utah.edu/~othmer

Edward Pate
Department of Pure and Applied Mathematics
Washington State University
Pullman, WA 99164-0001
USA
Email: pate@beta.math.wsu.edu

Charles S. Peskin
Courant Institute of Mathematical Sciences
New York University
251 Mercer Street
New York, NY 10012-1110
USA
Email: peskin@cims.nyu.edu
WWW URL: http://www.math.nyu.edu/faculty/peskin/index.html

A. M. de Roos
University of Amsterdam
Department of Pure and Applied Ecology
Kruislaan 320
NL-1098 SM Amsterdam
The Netherlands
Email: aroos@bio.uva.nl

Arthur Sherman
National Institutes of Health
BSA Building, Suite 350
9190 Rockville Pike MSC 2690
Bethesda, MD 20892-2690
USA
Email: asherman@nih.gov
WWW URL: http://mrb.niddk.nih.gov/sherman/

Simon Tavaré
Professor of Mathematics and Biological Sciences
Department of Mathematics
University of Southern California
1042 W. 36th Place, DRB 155
Los Angeles, CA 90089-1113
USA
Email: stavare@gnome.usc.edu

John J. Tyson
Dept. of Biology
Virginia Polytechnic Institute and State University
Blacksburg, VA 24061
USA
Email: tyson@mail.vt.edu

Colophon

This book was typeset using LaTeX 2_ε and a LaTeX book style file developed for the publisher's requirements, together with the amstex package, and an extended version of Tom Rokicki's epsf package for including POSTSCRIPT files as figures. An extended version of the mathtime package was used to provide body fonts of 10.5pt Times-Roman, 10pt Helvetica, and 10pt Courier. A small number of characters are used as 1270dpi bitmap representations of Computer Modern mathematics fonts, and LaTeX line and circle fonts. There are 56 font face and size combinations used in the book, from 11 different scalable fonts and 11 different bitmap fonts.

Authors submitted chapter drafts in LaTeX 2.09 or LaTeX 2_ε format, mostly via Internet electronic mail, although one chapter was in a commercial word processor format which was then converted to LaTeX 2_ε. Because of wide variation in author styles and TeXpertise, many hundreds of hours of work were required at the editors' site to bring the book to fruition.

Bibliographies were prepared using Oren Patashnik's BIBTeX bibliographic database system, with LaTeX styles modified to support chapter bibliographies. The citation and bibliography styles are extensions of David Rhead's authordate package. The Internet resources of the U.S. Library of Congress, the OCLC databases, the Compendex database, the University of California Melvyl catalog, and the American Mathematical Society's MathSciNet database were invaluable for checking and extending the bibliographic data.

A complete BIBTeX file for the bibliographies in this book, and for references to the chapters themselves, is available at the editors' World-Wide Web site, http://www.math.utah.edu/books/.

A project of this complexity would have been much more difficult, were it not for Stuart Feldman's make, Richard Stallman's GNU emacs, Alfred Aho, Peter Weinberger, and Brian Kernighan's awk language, Arnold Robbins' GNU gawk implementation of awk, Daniel Trinkle's detex, Geoffrey Tobin's dv2dt utility, L. Peter Deutsch's ghostscript, Kresten Krab Thorup and Per Abrahamsen's lacheck LaTeX syntax checker, Pehong Chen, Michael Harrison, and Leslie Lamport's makeindex indexing system, Frank Mittelbach's multicol package used for the book indexes, Piet Tutelaers' ps2pk utility, which makes POSTSCRIPT Type 1 fonts available for the xdvi screen previewer, Digital Equipment Corporation's pstotext utility, UNIX spell and GNU ispell spelling checkers, Nelson H. F. Beebe's authidx author/editor indexing package, bibcheck, bibclean, biblabel, biblex, biborder, bibparse, bibsort, and bibunlex bibliography tools, chkdelim delimiter balance checker, dw doubled word finder, epsutil POSTSCRIPT utility, and extended BIBTeX and LaTeX support for GNU Emacs, and many other smaller tools in the UNIX operating system. Donald E. Knuth wrote TeX and METAFONT. Leslie Lamport wrote LaTeX 2.09, and he collaborated with the international LaTeX Development Team in the production of LaTeX 2_ε. Nelson H. F. Beebe's dvialw TeX DVI driver for POSTSCRIPT output was used for most of the project, and for technical reasons at the printer, Tom Rokicki's dvips was used for production of the final POSTSCRIPT files. It is a tribute to the generosity of these many authors that their software tools are freely available to the world.